METHODS IN MOLECULAR BIOLOGY

For further volumes:
http://www.springer.com/series/7651

CpG Islands

Methods and Protocols

Edited by

Tanya Vavouri

Josep Carreras Leukaemia Research Institute, Badalona, Barcelona, Spain

Miguel A. Peinado

IGTP, IMPPC, Badalona, Barcelona, Spain

Humana Press

Editors
Tanya Vavouri
Josep Carreras Leukaemia
Research Institute
Badalona, Barcelona, Spain

Miguel A. Peinado
IGTP, IMPPC
Badalona, Barcelona, Spain

ISSN 1064-3745 ISSN 1940-6029 (electronic)
Methods in Molecular Biology
ISBN 978-1-4939-7767-3 ISBN 978-1-4939-7768-0 (eBook)
https://doi.org/10.1007/978-1-4939-7768-0

Library of Congress Control Number: 2018935913

Printed on acid-free paper

This Humana Press imprint is published by the registered company Springer Science+Business Media, LLC part of Springer Nature.
The registered company address is: 233 Spring Street, New York, NY 10013, U.S.A.

Preface

The discovery of the extreme simplicity of DNA structure by Watson, Crick, Franklin, Gosling, and Wilkins unlocked the keys to understanding how life is encoded. As predicted by Rosalyn Franklin, the infinite variety of nucleotide sequences would explain the biological specificity of DNA. More than 60 years later, the obtention of accurate genome maps including structural and functional features is still a prime target of biomedical research and absolutely necessary to understanding life's complexity.

In this context, the discovery by Adrian Bird of the wisely named CpG islands probably represents the turning point of genetics and epigenetics towards an integrated view of both fields. CpG islands consist of short stretches of DNA with high density of CpG dinucleotides, overlap with the promoter of more than half of mouse and human genes, and remain largely free of DNA methylation. The generalized concept of DNA methylation as a sign of repressed chromatin and gene silencing emerges directly from the analysis of CpG islands. Nowadays, CpG islands have become true navigation points to understand gene regulation in fundamental processes such as development and cell differentiation as well as in diseases like cancer.

The aim of this book, addressed to a broad range of scientists, is to summarize bioinformatic and molecular biology methods useful to identify and to explore the functions of CpG islands. The book is structured in three parts. The first one offers a historical perspective by two chief pioneers in the field, Adrian Bird and Francisco Antequera, and reviews important properties of CpG islands in three more chapters. The second part includes six chapters describing computational and wet lab methods related to the study of DNA methylation. Finally, the third part presents in-depth protocols for the analysis of CpG island functional features including epigenetic profiling and chromatin interactions.

We hope the readers find in this book information and method(s) that provide a key clue to decipher how a genome's structure and organization contribute to regulate biological processes.

We would like to thank all the contributors, colleagues, and past and present members of our groups, as well as funding agencies, for making this book possible.

Barcelona, Spain

Tanya Vavouri
Miguel A. Peinado

Contents

Part III Chromatin and Related Genomics Methods Used for Understanding the Functional Role of CpG Islands and Other Regulatory Regions

Contributors

SERGIO ALONSO • *Program of Predictive and Personalized Medicine of Cancer (PMPPC), Germans Trias i Pujol Research Institute (IGTP), Badalona, Barcelona, Spain*

CHRISTINA AMBROSI • *Department of Molecular Mechanisms of Disease, University of Zurich, Zurich, Switzerland; Molecular Life Science PhD Program of the Life Science Zurich Graduate School, University of Zurich, Zurich, Switzerland*

MASAMI ANDO-KURI • *Department of Biology, Emory University, Atlanta, GA, USA*

FRANCISCO ANTEQUERA • *Instituto de Biología Funcional y Genômica, Consejo Superior de Investigaciones Científicas (CSIC)/Universidad de Salamanca, Salamanca, Spain*

TUNCAY BAUBEC • *Department of Molecular Mechanisms of Disease, University of Zurich, Zurich, Switzerland*

MAGDA BIENKO • *Science for Life Laboratory, Department of Medical Biochemistry and Biophysics, Karolinska Institutet, Stockholm, Sweden*

ADRIAN BIRD • *Wellcome Trust Centre for Cell Biology, University of Edinburgh, Edinburgh, UK*

NEIL P. BLACKLEDGE • *Department of Biochemistry, University of Oxford, Oxford, UK*

VICTOR G. CORCES • *Department of Biology, Emory University, Atlanta, GA, USA*

NICOLA CROSETTO • *Science for Life Laboratory, Department of Medical Biochemistry and Biophysics, Karolinska Institutet, Stockholm, Sweden*

JOAQUIN CUSTODIO • *Science for Life Laboratory, Department of Medical Biochemistry and Biophysics, Karolinska Institutet, Stockholm, Sweden*

MANEL ESTELLER • *Cancer Epigenetics and Biology Program (PEBC), Bellvitge Biomedical Research Institute (IDIBELL), Barcelona, Catalonia, Spain; Department of Physiological Sciences II, School of Medicine, University of Barcelona, Barcelona, Catalonia, Spain; Institució Catalana de Recerca i Estudis Avançats (ICREA), Barcelona, Catalonia, Spain*

PEGGY J. FARNHAM • *Department of Biochemistry and Molecular Medicine, Norris Comprehensive Cancer Center, Keck School of Medicine, University of Southern California, Los Angeles, CA, USA*

HUMBERTO J. FERREIRA • *Cancer Epigenetics and Biology Program (PEBC), Bellvitge Biomedical Research Institute (IDIBELL), Barcelona, Catalonia, Spain*

INTZA GARIN • *Molecular (Epi)Genetics Laboratory, BioAraba National Health Institute, OSI Araba-Txagorritxu, Vitoria-Gasteiz, Spain*

ELENI GELALI • *Science for Life Laboratory, Department of Medical Biochemistry and Biophysics, Karolinska Institutet, Stockholm, Sweden*

GABRIELE GIRELLI • *Science for Life Laboratory, Department of Medical Biochemistry and Biophysics, Karolinska Institutet, Stockholm, Sweden*

CRISTINA GÓMEZ-MARTÍN • *Department of Genetics, Faculty of Science, University of Granada, Granada, Spain; Lab. de Bioinformática, Centro de Investigación Biomédica, PTS, Instituto de Biotecnología, Granada, Spain*

ALICE GROB • *Department of Life Sciences, Imperial College London, London, UK*

DOMINIC GRÜN • *Max-Planck Institute of Immunobiology and Epigenetics, Freiburg, Germany*

MICHAEL HACKENBERG • *Department of Genetics, Faculty of Science, University of Granada, Granada, Spain; Lab. de Bioinformática, Centro de Investigación Biomédica, PTS, Instituto de Biotecnología, Granada, Spain*

JOSIP STEFAN HERMAN • *Max-Planck Institute of Immunobiology and Epigenetics, Freiburg, Germany*

MARK ISALAN • *Department of Life Sciences, Imperial College London, London, UK*

ROBERT J. KLOSE • *Department of Biochemistry, University of Oxford, Oxford, UK*

RICARDO LEBRÓN • *Department of Genetics, Faculty of Science, University of Granada, Granada, Spain; Lab. de Bioinformática, Centro de Investigación Biomédica, PTS, Instituto de Biotecnología, Granada, Spain*

HANNAH K. LONG • *Department of Chemical and Systems Biology and Institute for Stem Cell Biology and Regenerative Medicine, Stanford University School of Medicine, Stanford University, Stanford, CA, USA; Department of Biochemistry, University of Oxford, Oxford, UK; Molecular Haematology Unit, Weatherall Institute of Molecular Medicine, John Radcliffe Hospital, University of Oxford, Oxford, UK*

IZASKUN MALLONA • *Predictive and Personalized Medicine of Cancer Program, Health Research Institute Germans Trias i Pujol (IGTP), Can Ruti Campus, Badalona, Spain*

MASSIMILIANO MANZO • *Department of Molecular Mechanisms of Disease, University of Zurich, Zurich, Switzerland; Molecular Life Science PhD Program of the Life Science Zurich Graduate School, University of Zurich, Zurich, Switzerland*

MASUE MARBIAH • *Department of Life Sciences, Imperial College London, London, UK*

YULIA MEDVEDEVA • *Institute of Bioengineering, Research Center of Biotechnology, Russian Academy of Science, Moscow, Russia; Department of System Biology and Bioinformatics, Vavilov Institute of General Genetics, Russian Academy of Science, Moscow, Russia; Department of Biological and Medical Physics, Moscow Institute of Physics and Technology, Dolgoprudny, Russia*

DAVID MONK • *Imprinting and Cancer Group, Cancer Epigenetic and Biology Program (PEBC), Bellvitge Institute for Biomedical Research (IDIBELL), Hospital Duran and Reynals, Barcelona, Spain*

ANA MONTEAGUDO-SÁNCHEZ • *Imprinting and Cancer Group, Cancer Epigenetic and Biology Program (PEBC), Bellvitge Institute for Biomedical Research (IDIBELL), Hospital Duran and Reynals, Barcelona, Spain*

SEBASTIAN MORAN • *Cancer Epigenetics and Biology Program (PEBC), Bellvitge Biomedical Research Institute (IDIBELL), Barcelona, Catalonia, Spain*

GUIOMAR PEREZ DE NANCLARES • *Molecular (Epi)Genetics Laboratory, BioAraba National Health Institute, OSI Araba-Txagorritxu, Vitoria-Gasteiz, Spain*

JOSÉ L. OLIVER • *Department of Genetics, Faculty of Science, University of Granada, Granada, Spain; Lab. de Bioinformática, Centro de Investigación Biomédica, PTS, Instituto de Biotecnología, Granada, Spain*

LORENZO PASQUALI • *Program of Predictive and Personalized Medicine of Cancer (PMPPC), Endocrine Regulatory Genomics Laboratory, Department of Endocrinology and Nutrition, Germans Trias i Pujol University Hospital and Research Institute, Badalona, Spain; Josep Carreras Leukaemia Research Institute, Badalona, Spain; CIBER de Diabetes y Enfermedades Metabólicas Asociadas (CIBERDEM), Barcelona, Spain*

MIGUEL A. PEINADO • *Predictive and Personalized Medicine of Cancer Program, Health Research Institute Germans Trias i Pujol (IGTP), Can Ruti Campus, Badalona, Spain*

MANUEL PERUCHO • *Program of Predictive and Personalized Medicine of Cancer (PMPPC), Germans Trias i Pujol Research Institute (IGTP), Badalona, Barcelona, Spain*

JOHN ANDREW POSPISILIK • *Max-Planck Institute of Immunobiology and Epigenetics, Freiburg, Germany*
MIREIA RAMOS-RODRÍGUEZ • *Program of Predictive and Personalized Medicine of Cancer (PMPPC), Endocrine Regulatory Genomics Laboratory, Department of Endocrinology and Nutrition, Germans Trias i Pujol University Hospital and Research Institute, Badalona, Spain*
HELENA RAURELL-VILA • *Program of Predictive and Personalized Medicine of Cancer (PMPPC), Endocrine Regulatory Genomics Laboratory, Department of Endocrinology and Nutrition, Germans Trias i Pujol University Hospital and Research Institute, Badalona, Spain*
SUHN KYONG RHIE • *Department of Biochemistry and Molecular Medicine, Norris Comprehensive Cancer Center, Keck School of Medicine, University of Southern California, Los Angeles, CA, USA*
I. SARAHI M. RIVERA • *Department of Biology, Emory University, Atlanta, GA, USA*
NATHAN R. ROSE • *Department of Biochemistry, University of Oxford, Oxford, UK*
M. JORDAN ROWLEY • *Department of Biology, Emory University, Atlanta, GA, USA*
SAGAR • *Max-Planck Institute of Immunobiology and Epigenetics, Freiburg, Germany*
SHANNON SCHREINER • *Department of Biochemistry and Molecular Medicine, Norris Comprehensive Cancer Center, Keck School of Medicine, University of Southern California, Los Angeles, CA, USA*
ALEXANDER SHERSHEBNEV • *Department of Biological and Medical Physics, Moscow Institute of Physics and Technology, Dolgoprudny, Russia; School of Public Health and Health Sciences, University of Massachusetts, Amherst, MA, USA*
ALBERTO SIERCO • *Predictive and Personalized Medicine of Cancer Program, Health Research Institute Germans Trias i Pujol (IGTP), Can Ruti Campus, Badalona, Spain*
KOICHI SUZUKI • *Department of Surgery, Saitama Medical Center, Jichi Medical University, Saitama, Japan*
ERIK WERNERSSON • *Science for Life Laboratory, Department of Medical Biochemistry and Biophysics, Karolinska Institutet, Stockholm, Sweden*
CHENHUAN XU • *Department of Biology, Emory University, Atlanta, GA, USA*
FUMIICHIRO YAMAMOTO • *Program of Predictive and Personalized Medicine of Cancer (PMPPC), Josep Carreras Leukaemia Research Institute (IJC), Badalona, Barcelona, Spain*

Part I

Sequence and Other Properties of CpG Islands

Chapter 1

CpG Islands: A Historical Perspective

Francisco Antequera and Adrian Bird

Abstract

The discovery of CpG islands (CGIs) and the study of their structure and properties run parallel to the development of molecular biology in the last two decades of the twentieth century and to the development of high-throughput genomic technologies at the turn of the millennium. First identified as discrete G + C-rich regions of unmethylated DNA in several vertebrates, CGIs were soon found to display additional distinctive chromatin features from the rest of the genome in terms of accessibility and of the epigenetic modifications of their histones. These features, together with their colocalization with promoters and with origins of DNA replication in mammals, highlighted their relevance in the regulation of genomic processes. Recent approaches have shown with unprecedented detail the dynamics and diversity of the epigenetic landscape of CGIs during normal development and under pathological conditions. Also, comparative analyses across species have started revealing how CGIs evolve and contribute to the evolution of the vertebrate genome.

Key words CpG islands, DNA mehylation, Chromatin, Transcription, Evolution

1 DNA Methylation Patterns

5-methyl-cytosine (5mC) was first detected as a natural component of DNA by Hotchkiss [1] when trying to determine the purity of DNA preparations in calf thymus. Soon afterward, it was also found in wheat [2] and subsequently in other types of eukaryotes. Technology at that time only permitted 5mC to be recognized as a minor fraction of all cytosine in DNA, leaving many of the questions concerning its genomic distribution and possible biological significance unanswered. Despite these limitations, chemical and enzymatic hydrolysis of DNA and biochemical methods such as column or paper chromatography, mass spectrometry, or thin-layer chromatography, established that more than 90% of all 5mC was found in CpG dinucleotides [3–5], and that a large fraction of them were methylated in the mammalian genome [6, 7]. Early studies also showed that nuclease-resistant regions and repetitive sequences were enriched in 5mC [8, 9], while actively transcribed regions were hypomethylated relative to the genome average [7],

Tanya Vavouri and Miguel A. Peinado (eds.), *CpG Islands: Methods and Protocols*, Methods in Molecular Biology, vol. 1766, https://doi.org/10.1007/978-1-4939-7768-0_1, © Springer Science+Business Media, LLC, part of Springer Nature 2018

providing the first indications that 5mCpG dinucleotides were not uniformly distributed within the genome.

The advent of molecular biology techniques around the middle of 1970s provided the unprecedented opportunity to dissect the structure, and hence elucidate the function of specific genomic regions, including individual genes. In the case of DNA methylation, the use of restriction enzymes to analyze CpG in their recognition sites allowed methylated and nonmethylated versions of the same sites to be distinguished and were used to generate the first maps of DNA methylation [10–14]. In particular, the isoschizomers Hpa II and Msp I were widely used, since both recognize the 5′-CCGG sequence, but Hpa II does not cleave if the internal CpG is methylated. Molecular hybridization, coupled with the newly developed Southern blot analysis [15], detected enrichment of nonmethylated CpGs in the promoter region of some genes in expressing tissues, which gave rise to the first hints that DNA methylation might repress transcription and suggested a possible role in the regulation of gene expression [16–21]. This notion was reinforced by the finding that gene promoters in the inactive X chromosome were methylated and could be reactivated by inducing demethylation with 5-azacytidine [22, 23].

A second and more straightforward application of the Hpa II/Msp I isochizomer pair was the analysis of the restriction patterns they generated using genomic DNA and agarose gels. A pioneer analysis in sea urchin revealed that approximately 40% of its genome was resistant to Hpa II and other 5mCpG-sensitive enzymes, but not to Msp I. This fraction was present in different tissues and remained stable during development, although small changes would have gone unnoticed at the level of resolution available at the time [24]. This pattern of interspersed methylated and nonmethylated domains was initially called "echinoderm-type" and was later found in many other invertebrates and some fungi [25–27]. Despite its widespread phylogenetic distribution, this organization was not universal, however, as two additional global genomic methylation patterns were detected. One was the "insect-type pattern," as exemplified by *Drosophila*, which showed undetectable differences between the Msp I and Hpa II digestions, suggesting a widespread lack of genomic methylation. The second pattern was found in mice and humans, as well as birds, reptiles, amphibia, and fish. This "vertebrate-type" pattern was opposite to that of *Drosophila*, as the genome was poorly digested by 5mCpG-sensitive enzymes, indicating widespread genomic methylation. In particular, there was no evidence of the methylated and nonmethylated compartments seen in organisms with the echinoderm-type pattern.

2 Discovery of the CpG Islands

To enhance the resolution of the Msp I/Hpa II comparison, Adrian Bird and his collaborators in Edinburgh used the simple strategy of end-labeling the Hpa II fragments with ^{32}P before separating them in an agarose gel. The aim was to separate the fragments of vertebrate DNA according their molecular weight, visualized by ethidium bromide staining, but at the same time to assess fragment number through direct autoradiography of the gel [28]. The difference between the two resulting patterns was astonishing: the ethidium bromide pattern consisted of very high molecular weight fragments, indicative of poorly digested vertebrate DNA (Fig. 1c, d), whereas autoradiography of the same gel revealed a prominent set of small fragments, which greatly exceeded the number of Hpa II fragments at the top of the gel (Fig. 1g, j–o). Most were less than 500 base pairs (bp) in length, the average being approximately 120 bp. Thus the experiment revealed two asymmetric compartments: a small fraction, amounting to less than 2% of the genome and rich in closely spaced and nonmethylated sites for CpG enzymes, and the rest of the genome, which was highly methylated. Given its large size, the methylated fraction was likely to include satellite DNA, repeated elements, nontranscribed regions, and genes.

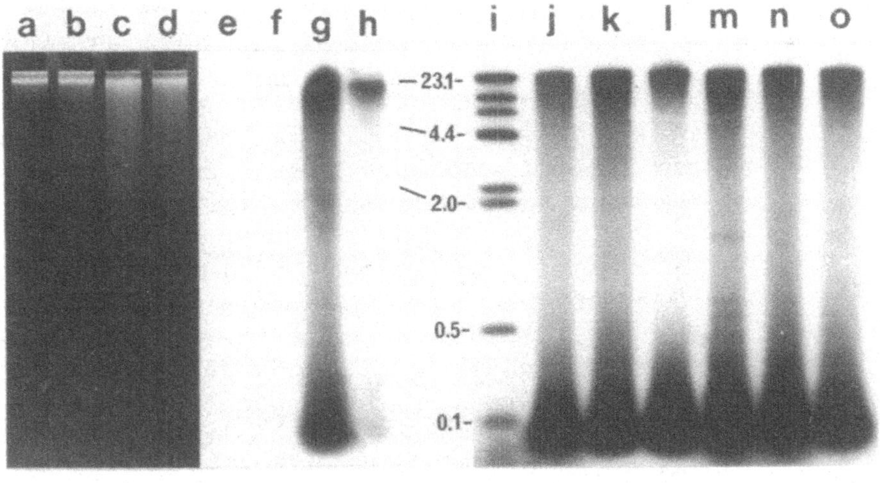

Fig. 1 End-labeling of Hpa II fragments of chicken DNA. Lanes (a–d), agarose gel of end-labeled chicken kidney DNA stained with ethidium bromide. Lanes (e–h), autoradiograph of the same gel. Samples (a) and (e), undigested DNA labeled with (α-^{32}P) dCTP; (b) and (f), undigested DNA labeled with (α-^{32}P) dTTP; (c) and (g), Hpa II-digested DNA labeled with (α-^{32}P) dCTP. (d) and (h), Hpa II-digested DNA labeled with (α-^{32}P) dTTP. Lane (i), bacteriophage lambda DNA digested with Hind III and end labeled with (α-^{32}P) dCTP. Lanes (j–o), autoradiograph of DNA from various chicken tissues after Hpa II digestion and end-labeling with (α-^{32}P) dCTP. Sample (j), blood cell nuclei; (k), whole blood; (l), sperm, (m), liver, (n), kidney; (o), brain. Gels were 1.2% agarose. Fragment lengths are given in kilobase pairs (Reproduced from [28], with permission)

A similar nonmethylated fraction was found in many other vertebrates and was initially called the "HTF fraction" (for Hpa II Tiny Fragments) [29], although it was later renamed as the CpG island fraction (CGI) [30]. Its discovery immediately provided a framework for the explanation of several other reports concerning asymmetric base composition along some gene sequences. For example, McClelland and Ivarie [31] found that the average CpG frequency along 15 mammalian genes was higher at the 5′ end than at the 3′ end. Regions enriched in CpGs were also found in the polymorphic exons of some MHC class I and II genes, and in the 5′ end of some genes from chicken and mice [32]. Since a few nonmethylated CpG sites had been mapped by Southern hybridization to these regions [21, 33], it was likely that they contributed to the CpG island fraction.

To identify the genomic origin of small nonmethylated fragments, several mouse genomic clones were isolated through the hybridization of individual HTF fragments to genomic libraries. Analysis of their sequences revealed that they were derived from regions approximately 1000 bp long with a base composition of 65% G + C, which was significantly higher than the genomic average of 40% G + C [29]. Based on this size and on the percentage of the genome represented by the HTF fraction, it was estimated that there were approximately 30,000 CpG islands in the mouse genome [29], a figure that is very close to more recent estimates based of genomic methylation analyses [34]. A striking feature of these regions was the absence of CpG suppression found in bulk DNA. In the methylated majority of the genome, CpG dinucleotides are present at 20% of the expected frequency due to the frequent conversion of 5mCpG to TpG caused by deamination of 5mC [35, 36]. This fact, together with their cleavage by 5mC-sensitive endonucleases, raised the possibility that CpG islands regions could be nonmethylated along their entire length in all of the tissues tested, including sperm. This expectation was proven correct by the development of methods capable of detecting methylation of every CpG by ligation-mediated PCR [37, 38], and later by the bisulphite treatment of DNA. The latter technique, which chemically converts C, but not mC, to T, has become the gold-standard method for methylation analyses at single nucleotide resolution [39], and is now widely used coupled with next generation sequencing [40].

3 Functional Properties of CpG Islands

An essential question that arose after their discovery concerned the connection between CGIs and genes. Northern hybridization of several CGI clones isolated at random from the mouse genome detected transcripts in several tissues, suggesting that many CGis could be associated with the promoters of widely expressed genes [29]. This association, together with the high frequency of sites for "rare cutter" enzymes (those with a high G + C content in their recognition sites) at CGIs, opened the unprecedented possibility of directly identifying the 5′ region of human genes in cosmid libraries and in genomic DNA [41, 42]. The first CGI promoter characterized in detail sustained bidirectional transcription [43], which was later found to be a general property of most CGI promoters [44–46]. Many studies of specific genes over the years and more recent genome-wide analyses have generalized the association between CGIs and genes to the extent that even those CGIs initially considered "orphans" have been found to colocalize with sites of transcription initiation [34]. That lack of methylation of CGIs was essential for the activity of the associated genes was initially shown by the transcriptional silencing caused by the artificial methylation of the 5′ CGI region of the hamster APRT gene in mouse cells, but not by methylation of the body of the gene [20, 47].

Methylation of the CGIs at the 5′ end of genes on the inactive X chromosomes was also associated with transcriptional silencing [48–51] and later studies showed that the repressive effect of methylation depended on the density of CpGs within the promoter regions, such that transcriptional silencing from fully methylated CGIs could not be reactivated even by strong enhancers [52, 53]. These findings were more broadly relevant, since methylated CGIs were also found in autosomal genes in tumours, for example within a region of human chromosome 11 [54], and also at tissue-specific genes in cultured cell lines [55, 56]. The affected CGIs were nonmethylated in normal tissues. Since those early studies, hundreds of examples have been documented in a variety of tumours, making unscheduled CGI methylation one of the hallmarks of cancer [57]. The finding of aberrant CGI methylation in cultured cells and in cancer highlighted the possibility that DNA methylation could be a natural regulator of CGI promoter activity during development. Indeed it emerged that, in addition to its role in X chromosome inactivation, DNA methylation is also functionally relevant for the monoallelic expression of imprinted genes (reviewed in [58]) and in the silencing of endogenous retrotransposons [59]. Despite these examples, it is unlikely that DNA methylation of CGI promoters is widely used to dynamically regulate gene expression during development, as the majority of CGIs remain in a nonmethylated state throughout [60]. There are also many examples where demethylation of non CGi promoters is not accompanied by their activation in tissues where these genes are normally silenced [59, 61–63].

Another important property of CGIs was the finding that replication origins colocalized with them at the CGI promoter region of the human *c-MYC* [64], *TIMM13* [65, 66], and *HSP70* genes [67]. As in the case of their promoter function, the colocalization of CGIs with ORIs was later generalized to include many CGIs, which colocalize with active ORIs in different human and mouse cell types [68–71].

A question raised by the initial characterization of CGIs was whether their distinctive properties would be paralleled by an equally distinctive chromatin organization. Histone fractionation of CGI chromatin in acid-urea gels showed that histone H1 was undetectable at CGIs and histones H3 and H4 were hyperacetylated [72]. DNAse I and micrococcal nuclease digestion revealed that CGIs included sites devoid of nucleosomes close to the site of transcription initiation. Although the extensive catalogue of histone modifications that are known today was missing in those days, these features potentially endowed CGIs with properties of "accessible" or open chromatin. How, then, did they manage to avoid the layer of methylation that covers the rest of the genome? The current view is that several mechanisms contribute to preventing methylation of CGIs. These include the binding of proteins to nonmethylated CpGs through the zinc finger CxxC domain such as the H3K36 histone demethylases KDM2A/KDM2B [73] and the histone H3K4 methyltransferase CFP1 [74]. Importantly, the histone modification H3K4me3, which frequently coincides with CGIs, prevents the DNA methyltransferase subunit Dnmt3L from accessing and methylating the associated DNA [75]. In addition, unscheduled methylation of CGIs could be actively removed by the TET proteins, which are also members of the CxxC protein family [76].

Paradoxically, it was the search for proteins that could shield CGIs from DNA methyltransferases that led to the identification of two protein activities, MeCP1 and MeCP2, which behaved in exactly the opposite way, as they were able to bind to methylated CpGs without sequence specificity for their flanking sequences. Those proteins included a methylated DNA binding domain (MBD) [77–79], which was later found in more proteins that bind to methylated DNA in vivo [80, 81]. The MBD family of proteins interprets the information encoded in the DNA methylation patterns through the interaction with transcriptional corepressors, histone epigenetic modifiers, and chromatin remodelers, and has overlapping but not redundant functions as shown by the different phenotypes generated by deletion of their coding genes (review of [82]).

4 Closing Remarks: From Maniatis to Methylomes

The basic structure and functional properties of CGIs were established following mostly the instructions and protocols described in the legendary first edition of the "Maniatis" manual [83]. This single volume was devoutly read by a whole generation of molecular biologists who laid the foundations for our current understanding of many genetic and developmental processes. In the field of DNA methylation, the one-step-at-a-time approach typical of classic molecular biology meant that several days of work involving DNA digestion, Southern blotting and hybridization were needed before the exciting moment arrived of developing the autoradiography to spot the differences—often corresponding to just one or two CpGs—between the Msp I and Hpa II banding patterns. At that time, the possibility of mapping the methylation status of every CpG in the genome in a single experiment or even in single cells [40] was beyond the reach of anyone's imagination.

The advent of high-throughput genomic technologies has provided a more nuanced picture of the methylation landscape than the vertebrate nonmethylated and methylated compartments initially described. For example, we now know that CGI-like regions can occasionally be methylated in different tissues or developmental stages [34, 84–87], and also that small nonmethylated regions, often encompassing one or a few CpGs, result from the binding of proteins to DNA [88–90]. The increasing availability of whole genome methylomes allows direct comparison between genomes within the same species at different developmental stages or physiological conditions, and makes the computational prediction of CGIs—that vary widely depending on small differences in base composition and CpG frequency [34, 84, 91]—unnecessary. These genome-wide analyses have confirmed previous findings that orthologous CGIs differ greatly in terms of base composition and CpG frequency among vertebrate species, although they maintain their nonmethylated condition [84, 92–94]. In fact CGI promoters evolve faster than CpG-poor promoters [95], which could account for the fact that CGIs are single-copy sequences in the genome, although they can share some binding sites for transcription factors. Several evolutionary scenarios for the emergence of CGIs have been proposed which, despite their colocalization with promoters and replication origins, suggest that CGIs have evolved under very little selective pressure [96].

The work of hundreds of laboratories, after almost 35 years since CGIs were first time visualized ([28], Fig. 1), has established CGIs as distinctive regulatory regions within the vast excess of genomic DNA sequence. Taking advantage of the breathtaking pace with which molecular techniques have improved our ability to scrutinize the dynamics of the genome, CGIs will doubtless continue to contribute to our understanding of the regulation, pathology, and evolution of the vertebrate genome for many years to come.

References

1. Hotchkiss RD (1948) The quantitative separation of purines, pyrimidines, and nucleosides by paper chromatography. J Biol Chem 175:315–332
2. Wyatt GR (1951) Recognition and estimation of 5-methylcytosine in nucleic acids. Biochem J 48:581–584
3. Doskocil J, Sorm F (1962) Distribution of 5-methylcytosine in pyrimidine sequences of deoxyribonucleic acids. Biochim Biophys Acta 55:953–959
4. Grippo P, Iaccarino M, Parisi E, Scarano E (1968) Methylation of DNA in developing sea urchin embryos. J Mol Biol 36:195–208
5. Sinsheimer RL (1955) The action of pancreatic deoxyribonuclease. II Isomeric dinucleotides. J Biol Chem 215:579–583
6. Gruenbaum Y, Stein R, Cedar H, Razin A (1981) Methylation of CpG sequences in eukaryotic DNA. FEBS Lett 124:67–71
7. Naveh-Many T, Cedar H (1981) Active gene sequences are undermethylated. Proc Natl Acad Sci U S A 78:4246–4250
8. Razin A, Cedar H (1977) Distribution of 5-methylcytosine in chromatin. Proc Natl Acad Sci U S A 74:2725–2728
9. Solage A, Cedar H (1978) Organization of 5-methylcytosine in chromosomal DNA. Biochemistry 17:2934–2938
10. Bird AP (1978) Use of restriction enzymes to study eukaryotic DNA methylation: II The symmetry of methylated sites supports semiconservative copying of the methylation pattern. J Mol Biol 118:49–60
11. Bird AP, Southern EM (1978) Use of restriction enzymes to study eukaryotic DNA methylation: I The methylation pattern in ribosomal DNA from Xenopus laevis. J Mol Biol 118:27–47
12. Gautier F, Bunemann H, Grotjahn L (1977) Analysis of calf-thymus satellite DNA: evidence for specific methylation of cytosine in C-G sequences. Eur J Biochem 80:175–183
13. Mandel JL, Chambon P (1979) DNA methylation: organ specific variations in the methylation pattern within and around ovalbumin and other chicken genes. Nucleic Acids Res 7:2081–2103
14. Waalwijk C, Flavell RA (1978) DNA methylation at a CCGG sequence in the large intron of the rabbit beta-globin gene: tissue-specific variations. Nucleic Acids Res 5:4631–4634
15. Southern EM (1975) Detection of specific sequences among DNA fragments separated by gel electrophoresis. J Mol Biol 98:503–517
16. Busslinger M, Hurst J, Flavell RA (1983) DNA methylation and the regulation of globin gene expression. Cell 34:197–206
17. Kruczek I, Doerfler W (1982) The unmethylated state of the promoter/leader and 5-′-regions of integrated adenovirus genes correlates with gene expression. EMBO J 1:409–414
18. Ott MO, Sperling L, Cassio D, Levilliers J, Sala-Trepat J, Weiss MC (1982) Undermethylation at the 5′ end of the albumin gene is necessary but not sufficient for albumin production by rat hepatoma cells in culture. Cell 30:825–833
19. Shen CK, Maniatis T (1980) Tissue-specific DNA methylation in a cluster of rabbit beta-like globin genes. Proc Natl Acad Sci U S A 77:6634–6638
20. Stein R, Razin A, Cedar H (1982) In vitro methylation of the hamster adenine phosphoribosyltransferase gene inhibits its expression in mouse L cells. Proc Natl Acad Sci U S A 79:3418–3422
21. Stein R, Sciaky-Gallili N, Razin A, Cedar H (1983) Pattern of methylation of two genes coding for housekeeping functions. Proc Natl Acad Sci U S A 80:2422–2426
22. Mohandas T, Sparkes RS, Shapiro LJ (1981) Reactivation of an inactive human X chromosome: evidence for X inactivation by DNA methylation. Science 211:393–396
23. Venolia L, Gartler SM, Wassman ER, Yen P, Mohandas T, Shapiro LJ (1982) Transformation with DNA from 5-azacytidine-reactivated X chromosomes. Proc Natl Acad Sci U S A 79:2352–2354
24. Bird AP, Taggart MH, Smith BA (1979) Methylated and unmethylated DNA compartments in the sea urchin genome. Cell 17:889–901
25. Antequera F, Tamame M, Villanueva JR, Santos T (1984) DNA methylation in the fungi. J Biol Chem 259:8033–8036
26. Bird AP, Taggart MH (1980) Variable patterns of total DNA and rDNA methylation in animals. Nucleic Acids Res 8:1485–1497
27. Whittaker PA, Hardman N (1980) Methylation of nuclear DNA in Physarum polycephalum. Biochem J 191:859–862

28. Cooper DN, Taggart MH, Bird AP (1983) Unmethylated domains in vertebrate DNA. Nucleic Acids Res 11:647–658
29. Bird A, Taggart M, Frommer M, Miller OJ, Macleod D (1985) A fraction of the mouse genome that is derived from islands of non-methylated, CpG-rich DNA. Cell 40:91–99
30. Bird A (1987) CpG islands as gene markers in the vertebrate nucleus. Trends Genet 3:342–347
31. McClelland M, Ivarie R (1982) Asymmetrical distribution of CpG in an "average" mammalian gene. Nucleic Acids Res 10:7865–7877
32. Tykocinski ML, Max EE (1984) CG dinucleotide clusters in MHC genes and in 5′ demethylated genes. Nucleic Acids Res 12:4385–4396
33. McKeon C, Ohkubo H, Pastan I, de Crombrugghe B (1982) Unusual methylation pattern of the alpha 2 (l) collagen gene. Cell 29:203–210
34. Illingworth RS, Gruenewald-Schneider U, Webb S, Kerr AR, James KD, Turner DJ, Smith C, Harrison DJ, Andrews R, Bird AP (2010) Orphan CpG islands identify numerous conserved promoters in the mammalian genome. PLoS Genet 6:e1001134
35. Coulondre C, Miller JH, Farabough PJ, Gilbert W (1978) Molecular basis of base substitution hotspots in *Escherichia coli*. Nature 274:775–780
36. Bird A (1980) DNA methylation and the frequency of CpG in animal DNA. Nucleic Acids Res 8:1499–1504
37. Nick H, Bowen B, Ferl RJ, Gilbert W (1986) Detection of cytosine methylation in the maize alcohol dehydrogenase gene by genomic sequencing. Nature 319:243–246
38. Pfeifer GP, Steigerwald SD, Mueller PR, Wold B, Riggs AD (1989) Genomic sequencing and methylation analysis by ligation mediated PCR. Science 246:810–813
39. Frommer M, McDonald LE, Millar DS, Collis CM, Watt F, Grigg GW, Molloy PL, Paul CL (1992) A genomic sequencing protocol that yields a positive display of 5-methylcytosine residues in individual DNA strands. Proc Natl Acad Sci U S A 89:1827–1831
40. Yong WS, Hsu FM, Chen PY (2016) Profiling genome-wide DNA methylation. Epigenetics Chromatin 9:26
41. Brown WR, Bird AP (1986) Long-range restriction site mapping of mammalian genomic DNA. Nature 322:477–481
42. Lindsay S, Bird AP (1987) Use of restriction enzymes to detect potential gene sequences in mammalian DNA. Nature 327:336–338
43. Lavia P, Macleod D, Bird A (1987) Coincident start sites for divergent transcripts at a randomly selected CpG islands as gene markers in the vertebrate nucleus. Trends Genet 3: 342–347
44. Adachi N, Lieber MR (2002) Bidirectional gene organization: a common architectural feature of the human genome. Cell 109:807–809
45. Core LJ, Waterfall JJ, Lis JT (2008) Nascent RNA sequencing reveals widespread pausing and divergent initiation at human promoters. Science 322:1845–1848
46. Seila AC, Calabrese JM, Levine SS, Yeo GW, Rahl PB, Flynn RA, Young RA, Sharp PA (2008) Divergent transcription from active promoters. Science 322:1849–1851
47. Keshet I, Yisraeli J, Cedar H (1985) Effect of regional DNA methylation on gene expression. Proc Natl Acad Sci U S A 82:2560–2564
48. Pfeifer GP, Tanguay RL, Steigerwald SD, Riggs AD (1990) In vivo footprint and methylation analysis by PCR-aided genomic sequencing: comparison of active and inactive X chromosomal DNA at the CpG island and promoter of human PGK-1. Genes Dev 4:1277–1287
49. Toniolo D, Martini G, Migeon BR, Dono R (1988) Expression of the G6PD locus on the human X chromosome is associated with demethylation of three CpG islands within 100 kb of DNA. EMBO J 7:401–406
50. Wolf SF, Jolly DJ, Lunnen KD, Friedmann T, Migeon BR (1984) Methylation of the hypoxanthine phosphoribosyltransferase locus on the human X chromosome: implications for X-chromosome inactivation. Proc Natl Acad Sci U S A 81:2806–2810
51. Yen PH, Patel P, Chinault AC, Mohandas T, Shapiro LJ (1984) Differential methylation of hypoxanthine phosphoribosyltransferase genes on active and inactive human X chromosomes. Proc Natl Acad Sci U S A 81:1759–1763
52. Boyes J, Bird A (1992) Repression of genes by DNA methylation depends on CpG density and promoter strength: evidence for involvement of a methyl-CpG binding protein. EMBO J 11:327–333
53. Hsieh CL (1994) Dependence of transcriptional repression on CpG methylation density. Mol Cell Biol 14:5487–5494
54. de Bustros A, Nelkin BD, Silverman A, Ehrlich G, Poiesz B, Baylin SB (1988) The short arm of chromosome 11 is a "hot spot" for hypermethylation in human neoplasia. Proc Natl Acad Sci U S A 85:5693–5697
55. Antequera F, Boyes J, Bird A (1990) High levels of de novo methylation and altered

chromatin structure at CpG islands in cell lines. Cell 62:503–514

56. Jones PA, Wolkowicz MJ, Rideout WM III, Gonzales FA, Marziasz CM, Coetzee GA, Tapscott SJ (1990) De novo methylation of the MyoD1 CpG island during the establishment of immortal cell lines. Proc Natl Acad Sci U S A 87:6117–6121
57. Stirzaker C, Taberlay PC, Statham AL, Clark SJ (2014) Mining cancer methylomes: prospects and challenges. Trends Genet 30:75–84
58. Bartolomei MS, Ferguson-Smith AC (2011) Mammalian genomic imprinting. Cold Spring Harb Perspect Biol 3:a002592
59. Walsh CP, Chaillet JR, Bestor TH (1998) Transcription of IAP endogenous retroviruses is constrained by cytosine methylation. Nat Genet 20:116–117
60. Borgel J, Guibert S, Li Y, Chiba H, Schübeler D, Sasaki H, Forné T, Weber M (2010) Targets and dynamics of promoter DNA methylation during early mouse development. Nat Genet 42:1093–1100
61. Bestor TH, Edwards JR, Boulard M (2015) Notes on the role of dynamic DNA methylation in mammalian development. Proc Natl Acad Sci U S A 112:6796–6799
62. Bird AP (1986) CpG-rich islands and the function of DNA methylation. Nature 321:209–213
63. Walsh CP, Bestor TH (1999) Cytosine methylation and mammalian development. Genes Dev 13:26–34
64. Vassilev L, Johnson EM (1990) An initiation zone of chromosomal DNA replication located upstream of the c-myc gene in proliferating HeLa cells. Mol Cell Biol 10:4899–4904
65. Biamonti G, Giacca M, Perini G, Contreas G, Zentilin L, Weighardt F, Guerra M, Della Valle G, Saccone S, Riva S et al (1992) The gene for a novel human lamin maps at a highly transcribed locus of chromosome 19 which replicates at the onset of S-phase. Mol Cell Biol 12:3499–3506
66. Giacca M, Zentilin L, Norio P, Diviacco S, Dimitrova D, Contreas G, Biamonti G, Perini G, Weighardt F, Riva S et al (1994) Fine mapping of a replication origin of human DNA. Proc Natl Acad Sci U S A 91:7119–7123
67. Taira T, Iguchi-Ariga SM, Ariga H (1994) A novel DNA replication origin identified in the human heat shock protein 70 gene promoter. Mol Cell Biol 14:6386–6397
68. Besnard E, Babled A, Lapasset L, Milhavet O, Parrinello H, Dantec C, Marin JM, Lemaitre JM (2012) Unraveling cell type-specific and reprogrammable human replication origin signatures associated with G-quadruplex consensus motifs. Nat Struct Mol Biol 19:837–844
69. Cayrou C, Coulombe P, Vigneron A, Stanojcic S, Ganier O, Peiffer I, Rivals E, Puy A, Laurent-Chabalier S, Desprat R, Mechali M (2011) Genome-scale analysis of metazoan replication origins reveals their organization in specific but flexible sites defined by conserved features. Genome Res 21:1438–1449
70. Delgado S, Gomez M, Bird A, Antequera F (1998) Initiation of DNA replication at CpG islands in mammalian chromosomes. EMBO J 17:2426–2435
71. Sequeira-Mendes J, Diaz-Uriarte R, Apedaile A, Huntley D, Brockdorff N, Gomez M (2009) Transcription initiation activity sets replication origin efficiency in mammalian cells. PLoS Genet 5:e1000446
72. Tazi J, Bird A (1990) Alternative chromatin structure at CpG islands. Cell 60:909–920
73. Blackledge NP, Zhou JC, Tolstorukov MY, Farcas AM, Park PJ, Klose RJ (2010) CpG islands recruit a histone H3 lysine 36 demethylase. Mol Cell 38:179–190
74. Thomson JP, Skene PJ, Selfridge J, Clouaire T, Guy J, Webb S, Kerr AR, Deaton A, Andrews R, James KD, Turner DJ, Illingworth R, Bird A (2010) CpG islands influence chromatin structure via the CpG-binding protein Cfp1. Nature 464:1082–1086
75. Ooi SKT, Qiu C, Bernstein E, Li K, Jia D, Yang Z, Erdjument-Bromage H, Tempst P, Lin SP, Allis CD, Cheng X, Bestor TH (2007) DNMT3L connects unmethylated lysine 4 of histone H3 to de novo methylation of DNA. Nature 448:714–717
76. Rasmussen KD, Helin K (2016) Role of TET enzymes in DNA methylation, development, and cancer. Genes Dev 30:733–750
77. Lewis JD, Meehan RR, Henzel WJ, Maurer-Fogy I, Jeppesen P, Klein F, Bird A (1992) Purification, sequence, and cellular localization of a novel chromosomal protein that binds to methylated DNA. Cell 69:905–914
78. Meehan RR, Lewis JD, McKay S, Kleiner EL, Bird AP (1989) Identification of a mammalian protein that binds specifically to DNA containing methylated CpGs. Cell 58:499–507
79. Nan X, Meehan RR, Bird AP (1993) Dissection of the methyl-CpG binding domain from the chromosomal protein MeCP2. Nucleic Acids Res 21:4886–4892
80. Baubec T, Schubeler D (2014) Genomic patterns and context specific interpretation of

DNA methylation. Curr Opin Genet Dev 25:85–92

81. Hendrich B, Bird A (1998) Identification and characterization of a family of mammalian methyl-CpG binding proteins. Mol Cell Biol 18:6538–6547
82. Du Q, Luu PL, Stirzaker C, Clark SJ (2015) Methyl-CpG-binding domain proteins: readers of the epigenome. Epigenomics 7:1051–1073
83. Maniatis T, Fritsch EF, Sambrook J (1982) Molecular cloning. A laboratory manual. Cold Spring Harbor Laboratory, Cold Spring Harbor, NY
84. Long HK, Sims D, Heger A, Blackledge NP, Kutter C, Wright ML, Grutzner F, Odom DT, Patient R, Ponting CP, Klose RJ (2013) Epigenetic conservation at gene regulatory elements revealed by non-methylated DNA profiling in seven vertebrates. elife 2:e00348
85. Maunakea AK, Nagarajan RP, Bilenky M, Ballinger TJ, D'Souza C, Fouse SD, Johnson BE, Hong C, Nielsen C, Zhao Y, Turecki G, Delaney A, Varhol R, Thiessen N, Shchors K, Heine VM, Rowitch DH, Xing X, Fiore C, Schillebeeckx M, Jones SJ, Haussler D, Marra MA, Hirst M, Wang T, Costello JF (2010) Conserved role of intragenic DNA methylation in regulating alternative promoters. Nature 466:253–257
86. Shen L, Kondo Y, Guo Y, Zhang J, Zhang L, Ahmed S, Shu J, Chen X, Waterland RA, Issa JP (2007) Genome-wide profiling of DNA methylation reveals a class of normally methylated CpG island promoters. PLoS Genet 3:2023–2036
87. Smallwood SA, Tomizawa S, Krueger F, Ruf N, Carli N, Segonds-Pichon A, Sato S, Hata K, Andrews SR, Kelsey G (2011) Dynamic CpG island methylation landscape in oocytes and preimplantation embryos. Nat Genet 43:811–814
88. Han L, Lin IG, Hsieh CL (2001) Protein binding protects sites on stable episomes and in the chromosome from de novo methylation. Mol Cell Biol 21:3416–3424
89. Lienert F, Mohn F, Tiwari VK, Baubec T, Roloff TC, Gaidatzis D, Stadler MB, Schubeler D (2011) Genomic prevalence of heterochromatic H3K9me2 and transcription do not discriminate pluripotent from terminally differentiated cells. PLoS Genet 7:e1002090
90. Stadler MB, Murr R, Burger L, Ivanek R, Lienert F, Scholer A, van Nimwegen E, Wirbelauer C, Oakeley EJ, Gaidatzis D, Tiwari VK, Schubeler D (2011) DNA-binding factors shape the mouse methylome at distal regulatory regions. Nature 480:490–495
91. Takai D, Jones PA (2002) Comprehensive analysis of CpG islands in human chromosomes 21 and 22. Proc Natl Acad Sci U S A 99:3740–3745
92. Antequera F, Bird A (1993) Number of CpG islands and genes in human and mouse. Proc Natl Acad Sci U S A 90:11995–11999
93. Cross S, Kovarik P, Schmidtke J, Bird A (1991) Non-methylated islands in fish genomes are GC-poor. Nucleic Acids Res 19:1469–1474
94. Cuadrado M, Sacristan M, Antequera F (2001) Species-specific organization of CpG island promoters at mammalian homologous genes. EMBO Rep 2:586–592
95. Carninci P, Sandelin A, Lenhard B, Katayama S, Shimokawa K, Ponjavic J, Semple CA, Taylor MS, Engstrom PG, Frith MC, Forrest AR, Alkema WB, Tan SL, Plessy C, Kodzius R, Ravasi T, Kasukawa T, Fukuda S, Kanamori-Katayama M, Kitazume Y, Kawaji H, Kai C, Nakamura M, Konno H, Nakano K, Mottagui-Tabar S, Arner P, Chesi A, Gustincich S, Persichetti F, Suzuki H, Grimmond SM, Wells CA, Orlando V, Wahlestedt C, Liu ET, Harbers M, Kawai J, Bajic VB, Hume DA, Hayashizaki Y (2006) Genome-wide analysis of mammalian promoter architecture and evolution. Nat Genet 38:626–635
96. Cohen NM, Kenigsberg E, Tanay A (2011) Primate CpG islands are maintained by heterogeneous evolutionary regimes involving minimal selection. Cell 145:773–786

Check for updates

Chapter 2

Biochemical Identification of Nonmethylated DNA by BioCAP-Seq

Hannah K. Long, Nathan R. Rose, Neil P. Blackledge, and Robert J. Klose

Abstract

CpG islands are regions of vertebrate genomes that often function as gene regulatory elements and are associated with most gene promoters. CpG island elements usually contain nonmethylated CpG dinucleotides, while the remainder of the genome is pervasively methylated. We developed a biochemical approach called biotinylated CxxC affinity purification (BioCAP) to unbiasedly isolate regions of the genome that contain nonmethylated CpG dinucleotides. The resulting highly pure nonmethylated DNA is easily analyzed by quantitative PCR to interrogate specific loci or via massively parallel sequencing to yield genome-wide profiles.

Key words DNA methylation, Nonmethylated DNA, CpG island, Nonmethylated island, Biotinylated CxxC affinity purification

1 Introduction

The methylation of cytosine at the C-5 position is one of the most widely studied and best understood of the epigenetic modifications [1–6]. DNA methylation occurs mostly within the context of CpG dinucleotides in vertebrates and is found pervasively throughout the genome [7–10]. In contrast, short contiguous regions of DNA are found interspersed across vertebrate genomes that lack methylation on CpG dinucleotides [11]. These nonmethylated regions, called CpG islands (CGIs), exhibit an increased density of CpG dinucleotides compared to the remainder of the genome [11–13] and are associated with up to two thirds of vertebrate gene promoters [14, 15]. CGIs specifically recruit a group of proteins that contain a ZF-CxxC domain that binds to nonmethylated DNA (Fig. 1a) [16–19]. ZF-CxxC proteins tend to associate with chromatin modifying activities suggesting nonmethylated DNA may function to alter chromatin structure at gene regulatory elements [20–27]. CGIs generally remain free of DNA methylation in most

Tanya Vavouri and Miguel A. Peinado (eds.), *CpG Islands: Methods and Protocols*, Methods in Molecular Biology, vol. 1766,
https://doi.org/10.1007/978-1-4939-7768-0_2,

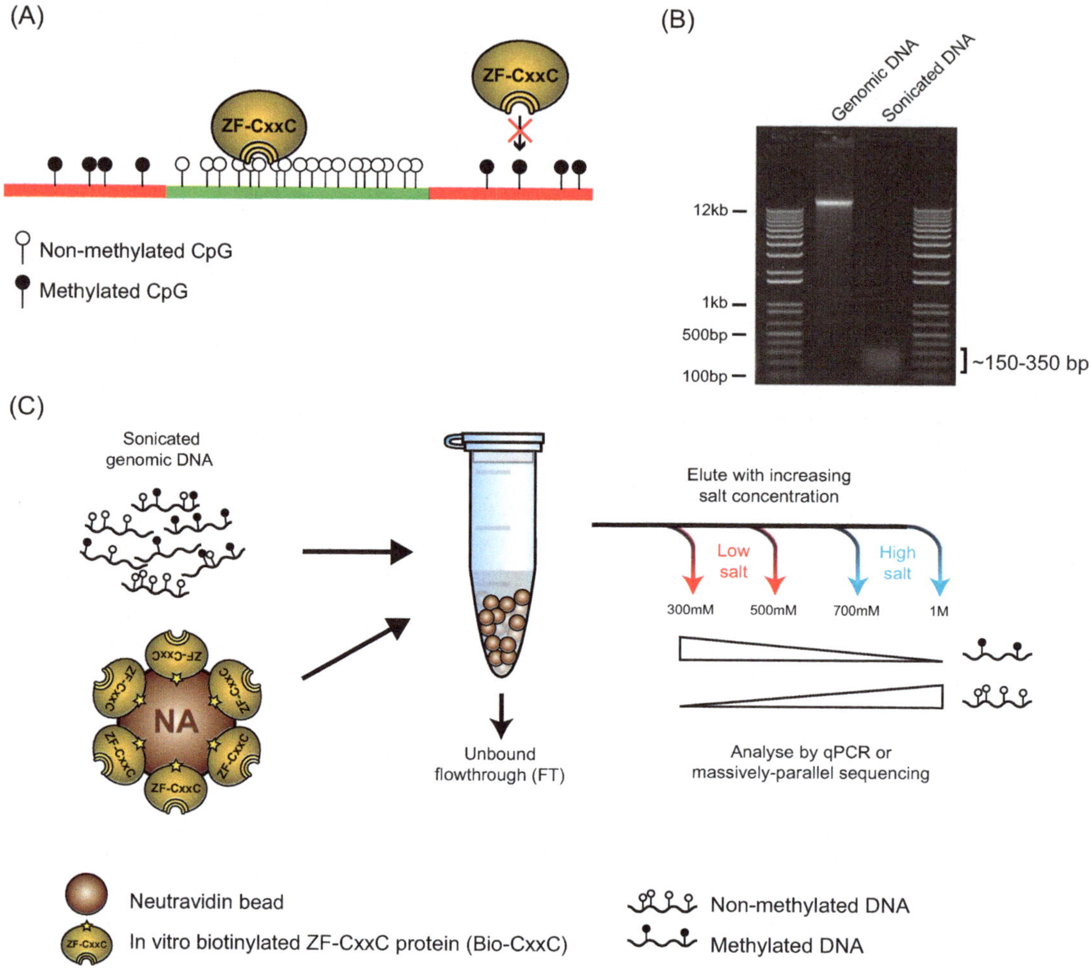

Fig. 1 The BioCAP technique. (**a**) The ZF-CxxC domain is able to bind to nonmethylated CpG dinucleotides. ZF-CxxC DNA binding is abrogated by DNA methylation of cytosine at CpG dinucleotides. Nonmethylated CpGs are depicted as open circles, and methylated CpGs as filled circles, a region corresponding to a CpG island is shown in green. (**b**) For BioCAP experiments, genomic DNA was sonicated for two hours to yield DNA fragments of 150–350 bp typically. Genomic DNA shown was extracted from mouse testes. (**c**) A schematic of the BioCAP procedure. In vitro biotinylated recombinant ZF-CxxC protein (Bio-CxxC) was immobilized onto neutravidin beads and incubated with sheared genomic DNA (**b**), allowing the ZF-CxxC domain to bind to DNA. Unbound DNA was removed in the FT, a series of elutions then followed using increasing salt concentrations. Highly methylated DNA is removed in the early low salt elutions (300 and 500 mM) and nonmethylated DNA was enriched in the late high salt elutions (700 and 1000 mM). DNA from each of these fractions was interrogated by qPCR or the 700 and 1000 mM elutions were combined for massively parallel sequencing library preparation

tissues and are thought to be protected from DNA methylation by DNA demethylases [28, 29], nucleotide composition and transcription factor binding [8, 30–33]. However acquisition of DNA methylation at distal CGIs, and in rare cases at gene promoters during development, can lead to gene silencing. Therefore CGI methylation can play a regulatory function in specific developmental contexts, and in human diseases such as cancer [3, 34].

A number of techniques are available for the analysis of methylated DNA at the genome scale. While many of these approaches also provide information about where nonmethylated DNA is located, this often requires extensive sequencing depth and is costly to implement. Given that only a very small fraction of most vertebrate genomes contains nonmethylated DNA, techniques have been developed to directly detect nonmethylated DNA. Importantly these approaches require significantly less sequencing depth. These methods generally utilize a selective purification approach to specifically isolate the nonmethylated fraction of the genome prior to massively parallel sequencing. For example, the MRE-seq approach utilizes methylation-sensitive restriction enzymes (MSREs) to digest DNA only when their recognition site is nonmethylated [35]. Small, double-cut DNA fragments released from genomic DNA can then be analyzed by massively parallel sequencing to generate nonmethylated DNA profiles [36, 37]. However this technique is inherently biased due to the use of restriction sites which are not uniformly distributed through the genome. Another method utilizes a chemical-labeling approach called methyltransferase-directed transfer of activated groups (mTAG) to covalently label unmodified CpG dinucleotides, which can then be used to isolate nonmethhyated regions of the genome [38]. While this method uses unbiased shearing prior to isolation, it is dependent on an enzymatic reaction which must proceed to completion to ensure sensitivity and accuracy. A third approach known as CxxC Affinity Purification (CAP) was developed to affinity purify nonmethylated DNA via the DNA binding selectivity of the ZF-CxxC domain for nonmethylated CpG dinucleotides [11, 39]. Briefly, genomic DNA is fragmented by sonication and then applied to an affinity matrix containing a recombinant ZF-CxxC protein. The nonmethylated DNA binds under low salt conditions and methylated DNA is removed by extensive washing. The nonmethylated DNA is then eluted with high salt to disrupt the ZF-CxxC domain interaction with nonmethylated DNA. The purified nonmethylated DNA is then amenable to downstream analysis including massively parallel sequencing. Importantly, this approach is not subject to the same biases as the other methods described above, because the genome is not fractionated with restriction enzymes prior to affinity purification and is not dependent on an enzymatic reaction to modify nonmethylated CpGs prior to affinity purification.

We modified the CAP technique by engineering a high affinity biotinylated ZF-CxxC domain which is immobilized to a neutravidin matrix, chosen for its low nonspecific binding to DNA (Fig. 1). Known as BioCAP, this approach has several advantages over traditional CAP in that it is carried out in batch in microcentrifuge tubes, is amenable to the use of small amounts of genomic input DNA, and is streamlined for automation allowing parallel sample processing [40]. Extensive BioCAP analysis by massively parallel sequencing (BioCAP-seq) has demonstrated that this approach is highly effective at generating genome-wide profiles of nonmethylated DNA in diverse vertebrate genomes and can function to compare the location of nonmethylated DNA between different tissues in the same organism [41].

Here we provide a detailed protocol for the analysis of nonmethylated DNA using BioCAP.

2 Materials

Prepare all solutions described below using ultrapure water (prepared by purifying deionized water to attain a sensitivity of 18 MΩ cm at 25 °C) and analytical grade reagents. Prepare and store all reagents at 4 °C unless indicated otherwise.

2.1 Expression and Purification of BioCAP ZF-CxxC Affinity Capture Protein

1. 1000× antibiotic stock solutions: Dissolve 50 mg kanamycin sulfate in 1 mL water, and filter using a 0.22 μm syringe filter to make a 50 mg/mL stock. Dissolve 34 mg chloramphenicol in 1 mL ethanol to make a 34 mg/mL stock. Store both at −20 °C.
2. 2×TY medium: dissolve 16 g tryptone, 10 g yeast extract, and 5 g NaCl in 1 L distilled water. Adjust pH to 7.0 and autoclave to sterilize. Store at room temperature.
3. Lysis buffer: 20 mM Tris pH 8.0, 500 mM NaCl, 0.1% NP40, 1× Complete EDTA-free Protease Inhibitor Cocktail added fresh prior to lysis.
4. 1 M $ZnCl_2$ solution: dissolve 272.6 mg $ZnCl_2$ in 2 mL distilled water and filter using 0.22 μm syringe filter. Store at room temperature.
5. IPTG stock: dissolve 476.6 mg IPTG in 2 mL distilled water and filter with 0.22 μm syringe filter. Store at −20 °C.
6. PBS: dissolve 8 g NaCl, 0.2 g KCl, 1.44 g Na_2HPO_4, and 0.24 g KH_2PO_4 in 1 L distilled water. Store at room temperature.
7. IMAC sepharose beads: six Fast Flow, GE Healthcare, 17-0921-07.
8. Bio-Rad column: Polyprep chromatography column, 731-1550.
9. Wash buffer: 50 mM NaH_2PO_4, 300 mM NaCl, 20 mM imidazole; adjust pH to 8.0, and add 1× Complete EDTA-free Protease Inhibitor Cocktail fresh prior to use.

10. Elution buffer: 50 mM NaH_2PO_4, 300 mM NaCl, 250 mM imidazole; adjust pH to 8.0, and add 1× Complete EDTA-free Protease Inhibitor Cocktail fresh prior to use.
11. TEV protease: Sigma, T4455-1MG.
12. HiPrep 26/10 desalting column: GE Healthcare, 17-5087-01.
13. Biotinylation buffer: 20 mM Tris, pH 8.0, 250 mM potassium glutamate. Store at room temperature.
14. BioMix: 10 mM ATP, 10 mM $Mg(OAc)_2$, 50 μM D-biotin. Store at −80 °C.
15. 1 M imidazole: dissolve 68 mg imidazole in 1 mL distilled water. Store at room temperature.
16. BC150 buffer: 20 mM HEPES pH 7.9, 150 mM KCl, 0.5 mM dithiothreitol (DTT), 10% v/v glycerol.

2.2 Preparation of Genomic DNA, Performing BioCAP and Picogreen Quantification of DNA

1. Extraction buffer: 20 mM Tris–HCl (pH 8.0), 10 mM EDTA, 100 mM NaCl, 0.5% SDS (add after homogenization). Add fresh: 200 μg/mL Proteinase K. If purifying genomic DNA from sperm, add 40 mM DTT fresh. Store extraction buffer without Proteinase K and DTT at room temperature.
2. RNaseA/T1: Thermo Scientific, EN0551.
3. 70% ethanol: 70% EtOH and 30% distilled water by volume. Store at room temperature.
4. Genomic-tip 100/G: QIAGEN, 10243.
5. NeutrAvidin Agarose Resin: Thermo Scientific, 29200.
6. NeutrAvidin-coated magnetic beads: Thermo Scientific, 7815-2104-011150.
7. Protein LoBind tube: Eppendorf, 022431081.
8. CAP buffers: *X* mM NaCl, 0.1% Triton X-100, 20 mM HEPES pH 7.9, 12.5% v/v glycerol. CAP100, $X = 100$ mM NaCl. CAP300, $X = 300$ mM NaCl. CAP500, $X = 500$ mM NaCl. CAP700, $X = 700$ mM NaCl. CAP1000, $X = 1000$ mM NaCl. Check conductivity of buffers using a conductivity meter.
9. 1.5 mL TPX microtubes: Diagenode, M-50050.
10. PicoGreen: Invitrogen, P11496.
11. 1× TE: 10 mM Tris, pH 8.0, 1 mM EDTA.

3 Methods

3.1 Expression and Purification of BioCAP ZF-CxxC Affinity Capture Protein

Recombinant KDM2B ZF-CxxC protein was expressed and purified from *E. coli*. The protocol is illustrated with appropriate stopping points indicated in Fig. 2.

1. Transform pNIC28-hKdm2b-CxxC-PHD (*see* **Note 1**) into 100 μL Rosetta2 *E. coli* (or any strain containing pRARE2

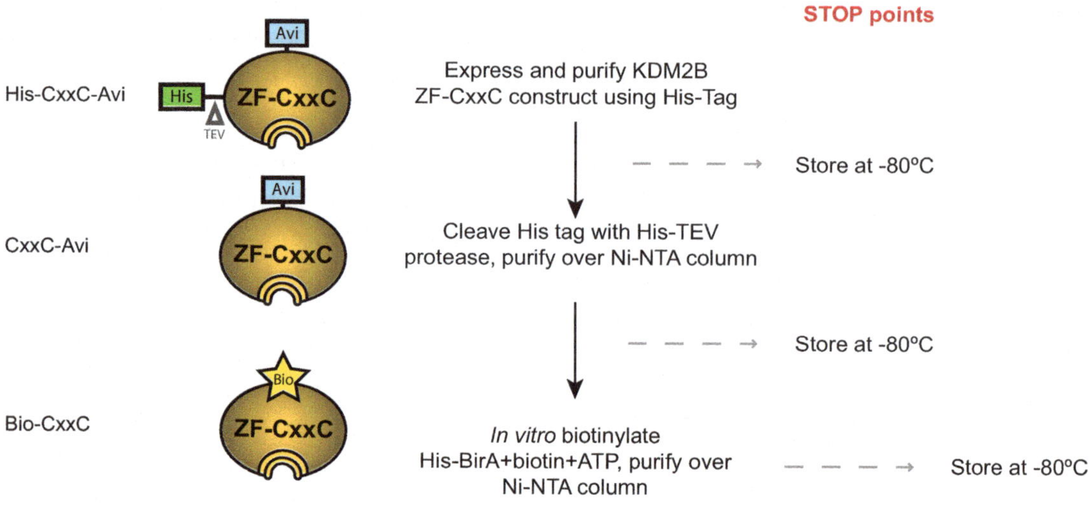

Fig. 2 Production of BioCAP protein. A schematic of the production of the affinity module used in BioCAP. Appropriate stopping points are indicated

plasmid) following a standard heat shock transformation protocol. Plate on agar plate containing kanamycin (50 μg/mL) and chloramphenicol (34 μg/mL) and incubate overnight at 37 °C. This plasmid encodes a bacterial expression cassette to produce a His-tagged human KDM2B ZF-CxxC domain with a C-terminal avi-tag and a TEV protease cleavage site immediately after the N-terminal His_6 tag.

2. Inoculate 100 mL 2× TY medium (+100 μL kanamycin and chloramphenicol 1000× stock solutions) with a single colony of Rosetta2 *E. coli* from the overnight transformation (**step 1**). Grow overnight at 37 °C.
3. Add 15 mL overnight culture to 650 mL 2×TY in a 2 L baffled Erlenmeyer flask (add 650 μL kanamycin and chloramphenicol 1000× stock solutions, and supplement with 162.5 μL of 1 M $ZnCl_2$ solution. Prepare six flasks in total, resulting in approximately 4 L of culture. Grow at 37 °C until the $OD_{600} = 0.6$, then cool to 30 °C and induce expression with the addition of 325 μL of 1 M IPTG (isopropyl β-D-1-thiogalactopyranoside) stock solution to each flask.
4. Grow at 30 °C for 5 h before harvesting by centrifugation at 10,000 rcf for 20 min at 4 °C.
5. Wash pellet once by resuspending in 20 mL PBS, transferring to a 50 mL Falcon tube, and centrifuging for 10 min at 3600 rcf. Remove supernatant. At this point the pellet can be stored at −80 °C until convenient to continue.

6. Add lysis buffer corresponding to 4× the volume of the cell pellet, and resuspend pellet gently without introducing air bubbles. Sonicate on ice in 20 mL aliquots at 60% amplitude (using Sonics Vibra-cell sonicator), alternating 30 s sonication with 30 s rest on ice.
7. Centrifuge the lysate at 48,000 rcf for 20 min. Filter the supernatant through a 0.22 μm syringe filter and set aside. This is the soluble protein fraction from *E. coli* that contains the recombinant His_6 tagged ZF-CxxC domain.
8. Charge IMAC sepharose beads with 0.2 M $NiSO_4$ following the manufacturer's instructions to generate Ni-NTA beads.
9. Prepare the Ni-NTA beads by transferring 4 mL of 50% bead slurry to a 15 mL Falcon tube and centrifuging at 850 rcf for 3 min. Remove the supernatant and resuspend the beads in 10 mL lysis buffer. Centrifuge again at 850 rcf for 3 min, remove supernatant, and repeat this wash once. Remove the supernatant.
10. Add the filtered lysate supernatant (from **step 7**) to the washed beads and rotate the suspension for 1 h at 4 °C to allow the His_6-tagged protein to bind to the beads.
11. Transfer the suspension to an empty 25 mL Bio-Rad column and collect the flowthrough. Wash the beads twice with 10 mL wash buffer. The protein is now bound to the Ni-NTA beads.
12. Elute protein from the column in batch by adding 1 mL of elution buffer gently to the bead bed in the Bio-Rad column. Collect the eluate in a 1.5 mL Eppendorf tube. Repeat the elution ten times.
13. Run 7.5 μL of each of the lysate supernatant, flowthrough, washes, and eluate fractions on a 12% SDS-PAGE gel. Pool the eluate fractions which contain significant amounts of CxxC protein, which is approximately 19 kDa in size. Check protein concentration by Bradford assay or NanoDrop.
14. Add the appropriate amount of TEV protease to pooled fractions and rotate overnight at 4 °C to cleave off the His_6-tag as per the manufacturer's recommedations.
15. Desalt the CxxC/TEV mixture on a HiPrep 26/10 desalting column (GE Healthcare) into wash buffer following the manufacturer's instructions to remove excess imidazole, and add the eluate to 300 μL prewashed Ni-NTA beads (prepare according to **step 8**) by washing 600 μL slurry in wash buffer). Rotate for 1 h at 4 °C to bind the His_6-tagged TEV protease and cleaved His_6-tag from the recombinant protein, then transfer to an empty 25 mL Bio-Rad column and collect flowthrough. Retain flowthrough as this now contains the CxxC protein without the His_6-tagged TEV protease or cleaved His_6

tag. Also elute bound TEV with 2 mL elution buffer and retain sample for SDS-PAGE analysis to verify efficient cleavage of the His_6-tag.

16. Desalt the CxxC protein (from the flowthrough in **step 15**) on a HiPrep 26/10 desalting column, into biotinylation buffer (*see* **Note 2**). Check the concentration of the protein by Bradford assay or NanoDrop spectrophotometry.
17. Add His_6-tagged BirA biotin ligase (add 2.5 μg for every 10 nmol CxxC substrate, *see* ref. 42 for expression and purification details; also *see* **Note 3**). Also add 500 μL BioMix. Rotate for 2 h at 4 °C, then add the same quantity of His_6-tagged BirA and 500 μL BioMix again. Allow the reaction to go to completion by rotating overnight at 4 °C.
18. Adjust the imidazole concentration of the reaction mixture to 20 mM imidazole by adding the appropriate amount of 1 M imidazole. Add 300 μL prewashed Ni-NTA beads to the reaction mixture (prepare according to **step 9** by washing 600 μL slurry in wash buffer), and rotate for 1 h at 4 °C to remove the His_6-tagged BirA. Transfer to an empty Bio-Rad column. Collect flowthrough (retain flowthrough as this contains the biotinylated CxxC protein). Elute bound BirA with 2 mL elution buffer and retain sample for SDS-PAGE analysis.
19. Desalt the CxxC protein into BC150 buffer on a HiPrep 26/10 desalting column, and check concentration by Bradford assay. Aliquot and store the biotinylated CxxC protein in volumes of 50 μL to avoid repeat freeze thaw cycles. To verify complete biotinylation, check the mass of the purified protein by mass spectrometry. The unmodified protein is 18,939 Da, and the biotinylated protein is 19,165 Da.

3.2 Prepare Genomic DNA

3.2.1 Genomic DNA Extraction Protocol from Tissue Samples

1. To prepare genomic DNA, add 950 μL extraction buffer to 25 mg tissue (without SDS or Proteinase K).
2. Homogenize tissue in a 1 mL Dounce homogenizer until tissue is dissociated into a cell suspension (some connective tissue may remain).
3. Add SDS and Proteinase K to homogenized tissue (and DTT if preparing genomic DNA from sperm, *see* **Note 4**).
4. Incubate at 50 °C for 3 h (or until the sample is homogeneous).
5. Add 10 μL of RNaseA/T1 (0.01 volumes, 2 mg/mL, 10,000 U/mL) and incubate at 37 °C for 1 h.
6. Split the extract into two 2 mL eppendorfs and add a further 400 μL extraction buffer to each (without SDS, Proteinase K, and DTT) to obtain a volume that is easy to extract by phenol–chloroform extraction.

7. To carry out a phenol–chloroform extraction, add an equal volume (900 μL) of phenol–chloroform–isoamyl alcohol (25:24:1 saturated with 10 mM Tris pH 8.0 and 1 mM EDTA) and vortex briefly. Separate phases by centrifugation for 5 min at 850 rcf in a benchtop centrifuge at room temperature. Transfer the upper aqueous phase into a fresh tube being careful not to disturb the interface between the upper and lower phases.
8. Repeat the phenol–chloroform extraction twice more.
9. Perform a final extraction with an equal volume of chloroform (900 μL), centrifuge as above and transfer the upper aqueous phase to a fresh tube. Each tube will have less than 900 μL due to loss of the sample during each transfer step.
10. To perform ethanol precipitation, split the upper phase from the final extraction into four 2 mL eppendorf tubes for ease of ethanol precipitation (around 400 μL extract per tube). Add 2.5 volumes ethanol (1 mL) and 0.1 volumes sodium acetate (3 M NaOAc, pH 5.2, 40 μL).
11. Place the samples at −80 °C for 30 min–1 h.
12. Centrifuge sample at 21,100 rcf in a benchtop centrifuge for 20 min at 4 °C.
13. Carefully decant the supernatant without disturbing the pellet and wash the pellet by adding 1 mL 70% ethanol per eppendorf tube.
14. Centrifuge sample at 21,100 rcf in a benchtop centrifuge for 10 min at 4 °C.
15. Carefully decant the supernatant without disturbing the pellet and pulse spin samples. Carefully remove the remaining liquid with a pipette taking care not to disturb the pellet.
16. Air-dry the pellet for 15 min at room temperature.
17. Resuspend each pellet in 50 μL MilliQ water. Warm the samples to 37 °C and gently flick the tubes if necessary to aid resuspension.
18. Combine the samples and determine the concentration using a NanoDrop 1000 spectrophotometer at 260 nm.
19. Store genomic DNA at −20 °C.

3.2.2 Alternative Purification of Genomic DNA

Alternatively, we have had success purifying genomic DNA using the 100/G Genomic Tip kit and routinely use this method for purification of genomic DNA from 80–100 mg of tissue (depending on the tissue-type). The manufacturer's instructions were followed for extraction of genomic DNA for tissue samples. A brief outline of the procedure is given below.

1. Homogenize 80–100 mg tissue in a 1 mL Dounce using Buffer G2 with RNaseA, final concentration 200 μg/mL.
2. Incubate the sample at 50 °C for 2 h with Proteinase K (1 mg/mL final concentration).
3. If particulate matter remains in the sample, centrifuge for 5000 rcf for 10 min at 4 °C. Discard the pellet.
4. Vortex for 10 s and apply the sample to a 100/G Genomic-tip column preequilibrated with 4 mL Buffer QBT.
5. Wash the column twice with 7.5 mL Buffer QC.
6. Elute genomic DNA with 5 mL of Buffer QF (warmed to 50 °C).
7. Precipitate genomic DNA by addition of 0.7 volumes of room temperature isopropanol. This is most easily done by aliquoting 1 mL of the sample into ten 2 mL eppendorf tubes and adding 700 μL isopropanol. Mix the samples by inversion and immediately centrifuge samples for 15 min at 21,100 rcf in a benchtop centrifuge at 4 °C.
8. Wash the DNA pellet with 1 mL 4 °C 70% ethanol.
9. Air-dry the pellet and resuspend in 50 μL MilliQ water.
10. Combine the samples and determine the concentration using a NanoDrop 1000 spectrophotometer at 260 nm.
11. Store genomic DNA at −20 °C.

3.3 Performing BioCAP

3.3.1 Preparation of Beads

1. For each BioCAP experiment, use 25 μL of packed NeutrAvidin Agarose Resin or NeutrAvidin-coated magnetic beads.
2. Pipette 50 μL beads (50% slurry) into a Protein LoBind tube and either pellet by centrifugation for 3 min at 850 rcf in a swing-bucket centrifuge by placing the centrifuge tubes in the top of a 15 mL falcon tube with the cap removed (for agarose beads), or immobilize by magnetization for magnetic beads.
3. Remove the supernatant by pipetting with a gel-loading tip and resuspend the beads in 1 mL of BC150 buffer by inversion.
4. Pellet the beads again, or immobilize using the magnet, and remove the supernatant.
5. Incubate the beads with 50 μL of 0.5 μg/μL biotinylated hKDM2B ZF-CxxC protein diluted in 425 μL BC150 buffer for 1 h at 4 °C, rotating the beads on a flywheel at 15 rpm.
6. Pellet the conjugated resin-CxxC protein, or immobilize on the magnet, and remove the supernatant.
7. Wash the beads three times with 1 mL CAP1000 buffer and once with 1 mL CAP100 buffer by inverting the tubes and rotating for 5 min at 4 °C on a flywheel (*see* **Note 5**). Between washes, remove the supernatant from the beads by centrifuging or immobilizing on a magnet.

3.3.2 Preparing Genomic DNA for Use in BioCAP

1. Typically 100 μL of approximately 350 ng/μL genomic DNA is used per BioCAP experiment.
2. Sonicate the genomic DNA to an average size of 150–250 bp. Sonication is performed in 1.5 mL TPX microtubes using a Diagenode Bioruptor. Sonication is performed on the high setting with 30 s on and 30 s off for 2 h. During sonication, spin the samples down at 10 min intervals by pulse centrifugation in a minicentrifuge.
3. Separate by gel electrophoresis 3 μL sonicated sample and 3 μL presonication sample on a 1% agarose gel to ensure the sample is sonicated to 150–350 bp in length (*see* Fig. 1b).
4. Dilute the sonicated DNA to approximately 17.5 μg/mL in CAP100 buffer: for example, 50 μL sample is diluted in 950 μL CAP100 buffer.
5. Five-hundred microliters of this material is used for the BioCAP procedure and the remaining volume is saved as an input control.

3.3.3 BioCAP

The BioCAP protocol is shown schematically in Fig. 1c.

1. For each BioCAP experiment, add 500 μL of diluted sonicated DNA, corresponding to approximately 8 μg of DNA, to the conjugated CxxC resin.
2. Incubate beads plus genomic DNA at 4 °C for 1 h, rotating on a flywheel at 15 rpm.
3. Collect the resin by centrifugation at 850 rcf for 3 min at 4 °C or by magnetization, and remove the unbound flowthrough (FT) material.
4. Wash the ZF-CxxC resin with bound DNA twice with 1 mL of CAP100 buffer by inverting the tube several times, and removing the supernatant following centrifugation at 850 rcf for 3 min at 4 °C or by magnetization.
5. The first BioCAP elution is performed by addition of 50 μL of CAP300 buffer to the ZF-CxxC resin with incubation at room temperature for 10 min with gentle agitiation (the tube should not be inverted due to the small volume of elution buffer). Following centrifugation or magnetization, the 50 μL elution fraction is carefully collected using a gel-loading tip.
6. The elution process is repeated using another 50 μL of CAP300 for 10 min and the 300 mM elution fractions are pooled (giving a total volume of 100 μL).
7. Perform subsequent elutions in the same manner using buffers CAP500, CAP700 and CAP1000 sequentially, with two elutions per buffer.

8. Each 100 μL elution fraction, and 100 μL of both the input and FT samples, should then be purified using a PCR purification column, eluting DNA in 50 μL distilled water.
9. For real-time qPCR analysis, BioCAP samples are typically diluted 10-fold and 5 μL used in a 15 μL quantitative PCR reaction volume.
10. For BioCAP sequencing, the BioCAP recovery should be verified for all elution fractions by qPCR at several loci and the amount of DNA in the CAP700 and CAP1000 elutions quantified using PicoGreen reagent (*see* Subheading 3.3.5). The combined CAP700 and CAP1000 elutions can then be used for library preparation for massively parallel sequencing.

3.3.4 Alternative Rapid Magnetic BioCAP

Mouse genomic DNA, sonicated to 150–350 bp, is incubated with the prepared magnetic CxxC resin (*see* Subheading 3.3.1) for 1 h in the same manner as a conventional BioCAP experiment (*see* Subheading 3.3.3). The magnetic CxxC resin and associated DNA are then collected using a magnetic microcentrifuge tube rack, allowing unbound FT material to be removed. Two 10 min washes are performed with 50 μL CAP500 (500 mM NaCl), followed by two 10 min elution steps with 50 μL CAP1000 (1 M NaCl). The CAP1000 elutions are combined for downstream analysis. The alternative magnetic BioCAP protocol may be preferred when rapid isolation of hypomethylated DNA is desired without interrogating the stepwise elution of increasingly hypomethylated DNA fractions. Equally, this method is convenient for rapid analysis of large numbers of samples, and is amenable to automated processing.

3.3.5 PicoGreen Quantification

To quantify DNA at low concentrations, PicoGreen quantification is used.

1. Determine the concentration of input DNA (i.e., sonicated DNA diluted to around 17.5 μg/mL in CAP100 buffer for BioCAP experiments, *see* Subheading 3.3.2) using a NanoDrop 1000 spectrophotometer at 260 nm.
2. Dilute the input DNA to 1 μg/mL in 1× TE and make a dilution series of 200 ng/mL, 100 ng/mL, 40 ng/mL, and 8 ng/mL of the 1 μg/mL input DNA using 1xTE. Mix 2 μL of each dilution with 2 μL of PicoGreen (diluted 1:200 in 1× TE). Create a calibration curve by measuring the fluorescence of the dilution series using a NanoDrop 3300. The concentration of the samples can then be inferred from this curve.
3. To calculate the concentration of the purified CAP elutions, dilute 1 μL of each samples tenfold in 1× TE. Then mix 2 μL of the tenfold diluted or undiluted samples with 2 μL diluted PicoGreen. Mix by flicking the tubes and pulse spin.

4. Detect fluorescence of the samples using a NanoDrop 3300 and infer the concentration of the samples from the calibration curve (from **step 2**). Average the two inferred DNA concentration values from the undiluted and tenfold diluted DNA samples to calculate an estimate of the DNA concentration for each sample.

4 Notes

1. AddGene, plasmid 49216: pNIC28-hKdm2b-CxxC-PHD.
2. Note that BirA activity is inhibited to some degree by NaCl, so this desalting step is crucial to ensure efficient biotinylation.
3. AddGene, plasmid 20857: pET21a-BirA.
4. Sperm heads are resilient to extraction by proteinase K and SDS alone as their membrane is enriched for disulfide bonds [43]. Therefore for genomic DNA extraction of sperm or testes tissue, the strong reducing agent dithiothreitol (DTT) is added to the gDNA extraction buffer to reduce these disulfide bonds.
5. This wash step is important, as it removes any bacterial CpG-containing DNA that might be bound to the CxxC domain from the protein expression and purification steps.

References

1. Klose RJ, Bird AP (2006) Genomic DNA methylation: The mark and its mediators. Trends Biochem Sci 31:89–97. https://doi.org/10.1016/j.tibs.2005.12.008
2. Cedar H, Bergman Y (2012) Programming of DNA methylation patterns. Annu Rev Biochem 81:97–117. https://doi.org/10.1146/annurev-biochem-052610
3. Jones PA (2012) Functions of DNA methylation: islands, start sites, gene bodies and beyond. Nat Rev Genet 13:484–492. https://doi.org/10.1038/nrg3230
4. Bergman Y, Cedar H (2013) DNA methylation dynamics in health and disease. Nat Struct Mol Biol 20:274–281. doi:nsmb.2518 [pii] \r10.1038/nsmb.2518
5. Seisenberger S, Peat JR, Hore T, Santos F, Dean W, Reik W Reprogramming DNA methylation in the mammalian life cycle: building and breaking epigenetic barriers. Philos Trans R Soc Lond 368(2013):20110330. https://doi.org/10.1098/rstb.2011.0330
6. Schübeler D (2015) Function and information content of DNA methylation. Nature 517:321–326. https://doi.org/10.1038/nature14192
7. Lister R, Pelizzola M, Dowen RH, Hawkins RD, Hon G, Tonti-Filippini J, Nery JR, Lee L, Ye Z, Ngo Q-M, Edsall L, Antosiewicz-Bourget J, Stewart R, Ruotti V, Millar H, Thomson J, Ren B, Ecker JR Human DNA methylomes at base resolution show widespread epigenomic differences. Nature 462(2009):315–322. https://doi.org/10.1038/nature08514
8. Stadler MB, Murr R, Burger L, Ivanek R, Lienert F, Schöler A, van Nimwegen E, Wirbelauer C, Oakeley EJ, Gaidatzis D, Tiwari VK, Schübeler D (2011) DNA-binding factors shape the mouse methylome at distal regulatory regions. Nature 480:490–495. https://doi.org/10.1038/nature10716
9. Jiang L, Zhang J, Wang JJ, Wang L, Zhang L, Li G, Yang X, Ma X, Sun X, Cai J, Zhang J, Huang X, Yu M, Wang X, Liu F, Wu CI, He C, Zhang B, Ci W, Liu J (2013) Sperm, but not oocyte, DNA methylome is inherited by zebrafish early embryos. Cell 153:773–784. https://doi.org/10.1016/j.cell.2013.04.041
10. Potok ME, Nix DA, Parnell TJ, Cairns BR (2013) Reprogramming the maternal zebrafish genome after fertilization to match the paternal

methylation pattern. Cell 153:759–772. https://doi.org/10.1016/j.cell.2013.04.030

11. Illingworth R, Kerr A, DeSousa D, Jørgensen H, Ellis P, Stalker J, Jackson D, Clee C, Plumb R, Rogers J, Humphray S, Cox T, Langford C, Bird A (2008) A novel CpG island set identifies tissue-specific methylation at developmental gene loci. PLoS Biol 6:0037–0051. https://doi.org/10.1371/journal.pbio.0060022
12. Conway KE, McConnell BB, Bowring CE, Donald CD, Warren ST, Vertino PM (2000) TMS1, a novel proapoptotic caspase recruitment domain protein, is a target of methylation-induced gene silencing in human breast cancers. Cancer Res 60:6236–6242. https://doi.org/10.1038/321209a0
13. Gardiner-Garden M, Frommer M (1987) CpG islands in vertebrate genomes. J Mol Biol 196:261–282. https://doi.org/10.1016/0022-2836(87)90689-9
14. Larsen F, Gundersen G, Lopez R, Prydz H (1992) CpG islands as gene markers in the human genome. Genomics 13:1095–1107. https://doi.org/10.1016/0888-7543(92)90024-M
15. Bird A, Taggart M, Frommer M, Miller OJ, Macleod D (1985) A fraction of the mouse genome that is derived from islands of non-methylated, CpG-rich DNA. Cell 40:91–99. https://doi.org/10.1016/0092-8674(85)90312-5
16. Cierpicki T, Risner LE, Grembecka J, Lukasik SM, Popovic R, Omonkowska M, Shultis DD, Zeleznik-Le NJ, Bushweller JH (2009) Structure of the MLL CXXC domain–DNA complex and its functional role in MLL-AF9 leukemia. Nat Struct Mol Biol 17:62–68. https://doi.org/10.1038/nsmb.1714
17. Song J, Rechkoblit O, Bestor TH, Patel DJ (2011) Structure of DNMT1-DNA complex reveals a role for autoinhibition in maintenance DNA methylation. Science 331:1036–1040. https://doi.org/10.1126/science.1195380
18. Xu C, Bian C, Lam R, Dong A, Min J (2011) The structural basis for selective binding of non-methylated CpG islands by the CFP1 CXXC domain. Nat Commun 2:227. https://doi.org/10.1038/ncomms1237
19. Song J, Teplova M, Ishibe-Murakami S, Patel DJ (2012) Structure-based mechanistic insights into DNMT1-mediated maintenance DNA methylation. Science 335:709–712. https://doi.org/10.1126/science.1214453
20. Blackledge NP, Zhou JC, Tolstorukov MY, Farcas AM, Park PJ, Klose RJ (2010) CpG islands recruit a histone H3 lysine 36 demethylase. Mol Cell 38:179–190. https://doi.org/10.1016/j.molcel.2010.04.009
21. Thomson JP, Skene PJ, Selfridge J, Clouaire T, Guy J, Webb S, Kerr ARW, Deaton A, Andrews R, James KD, Turner DJ, Illingworth R, Bird A (2010) CpG islands influence chromatin structure via the CpG-binding protein Cfp1. Nature 464:1082–1086. https://doi.org/10.1038/nature08924
22. Farcas AM, Blackledge NP, Sudbery I, Long HK, McGouran JF, Rose NR, Lee S, Sims D, Cerase A, Sheahan TW, Koseki H, Brockdorff N, Ponting CP, Kessler BM, Klose RJ (2012) KDM2B links the polycomb repressive complex 1 (PRC1) to recognition of CpG islands. elife 2012:1–26. https://doi.org/10.7554/eLife.00205
23. Boulard M, Edwards JR, Bestor TH (2015) FBXL10 protects polycomb-bound genes from hypermethylation. Nat Genet 47:1–9. https://doi.org/10.1038/ng.3272
24. Wu X, Johansen JV, Helin K (2013) Fbxl10/Kdm2b recruits polycomb repressive complex 1 to CpG islands and regulates H2A ubiquitylation. Mol Cell 49:1134–1146. https://doi.org/10.1016/j.molcel.2013.01.016
25. He J, Shen L, Wan M, Taranova O, Wu H, Zhang Y (2013) Kdm2b maintains murine embryonic stem cell status by recruiting PRC1 complex to CpG islands of developmental genes. Nat Cell Biol 15:373–384. https://doi.org/10.1038/ncb2702
26. Long HK, Blackledge NP, Klose RJ (2013) ZF-CxxC domain-containing proteins, CpG islands and the chromatin connection. Biochem Soc Trans 41:727–740. https://doi.org/10.1042/BST20130028
27. Blackledge NP, Farcas AM, Kondo T, King HW, McGouran JF, Hanssen LLP, Ito S, Cooper S, Kondo K, Koseki Y, Ishikura T, Long HK, Sheahan TW, Brockdorff N, Kessler BM, Koseki H, Klose RJ (2014) Variant PRC1 complex-dependent H2A ubiquitylation drives PRC2 recruitment and polycomb domain formation. Cell 157:1445–1459. https://doi.org/10.1016/j.cell.2014.05.004
28. Williams K, Christensen J, Helin K (2011) DNA methylation: TET proteins—guardians of CpG islands? EMBO Rep 13:28–35. https://doi.org/10.1038/embor.2011.233
29. Rasmussen KD, Helin K (2016) Role of TET enzymes in DNA methylation, development, and cancer. Genes Dev 30:733–750. https://doi.org/10.1101/gad.276568.115

30. Lienert F, Wirbelauer C, Som I, Dean A, Mohn F, Schübeler D (2011) Identification of genetic elements that autonomously determine DNA methylation states. Nat Genet 43:1091–1097. https://doi.org/10.1038/ng.946
31. Krebs AR, Dessus-Babus S, Burger L, Schubeler D (2014) High-throughput engineering of a mammalian genome reveals building principles of methylation states at CG rich regions. Elife 3:e04094. https://doi.org/10.7554/eLife.04094
32. Wachter E, Quante T, Merusi C, Arczewska A, Stewart F, Webb S, Bird A (2014) Synthetic CpG islands reveal DNA sequence determinants of chromatin structure. elife 3:e03397. https://doi.org/10.7554/eLife.03397
33. Long HK, King HW, Patient RK, Odom DT, Klose RJ (2016) Protection of CpG islands from DNA methylation is DNA-encoded and evolutionarily conserved. Nucleic Acids Res 44:gkw258. https://doi.org/10.1093/nar/gkw258
34. Van Vlodrop IJH, Niessen HEC, Derks S, Baldewijns MMLL, Van Criekinge W, Herman JG, Van Engeland M (2011) Analysis of promoter CpG island hypermethylation in cancer: location, location, location! Clin Cancer Res 17:4225–4231. https://doi.org/10.1158/1078-0432.CCR-10-3394
35. Laird PW (2010) Principles and challenges of genomewide DNA methylation analysis. Nat Rev Genet 11:191–203. https://doi.org/10.1038/nrg2732
36. Harris RA, Wang T, Coarfa C, Nagarajan RP, Hong C, Downey SL, Johnson BE, Fouse SD, Delaney A, Zhao Y, Olshen A, Ballinger T, Zhou X, Forsberg KJ, Gu J, Echipare L, O'Geen H, Lister R, Pelizzola M, Xi Y, Epstein CB, Bernstein BE, Hawkins RD, Ren B, Chung W-Y, Gu H, Bock C, Gnirke A, Zhang MQ, Haussler D, Ecker JR, Li W, Farnham PJ, a Waterland R, Meissner A, a Marra M, Hirst M, Milosavljevic A, Costello JF (2010) Comparison of sequencing-based methods to profile DNA methylation and identification of monoallelic epigenetic modifications. Nat Biotechnol 28:1097–1105. https://doi.org/10.1038/nbt.1682
37. Maunakea AK, Nagarajan RP, Bilenky M, Ballinger TJ, D'Souza C, Fouse SD, Johnson BE, Hong C, Nielsen C, Zhao Y, Turecki G, Delaney A, Varhol R, Thiessen N, Shchors K, Heine VM, Rowitch DH, Xing X, Fiore C, Schillebeeckx M, Jones SJM, Haussler D, Marra MA, Hirst M, Wang T, Costello JF (2010) Conserved role of intragenic DNA methylation in regulating alternative promoters. Nature 466:253–257. https://doi.org/10.1038/nature09165
38. Kriukienė E, Labrie V, Khare T, Urbanavičiūtė G, Lapinaitė A, Koncevičius K, Li D, Wang T, Pai S, Ptak C, Gordevičius J, Wang S-C, Petronis A, Klimašauskas S (2013) DNA unmethylome profiling by covalent capture of CpG sites. Nat Commun 4:2190. https://doi.org/10.1038/ncomms3190
39. Illingworth RS, Gruenewald-Schneider U, Webb S, Kerr ARW, James KD, Turner DJ, Smith C, Harrison DJ, Andrews R, Bird AP (2010) Orphan CpG islands identify numerous conserved promoters in the mammalian genome. PLoS Genet 6:e1001134. https://doi.org/10.1371/journal.pgen.1001134
40. Blackledge NP, Long HK, Zhou JC, Kriaucionis S, Patient R, Klose RJ (2012) Bio-CAP: a versatile and highly sensitive technique to purify and characterise regions of non-methylated DNA. Nucleic Acids Res 40:e32. https://doi.org/10.1093/nar/gkr1207
41. Long HK, Sims D, Heger A, Blackledge NP, Kutter C, Wright ML, Grützner F, Odom DT, Patient R, Ponting CP, Klose RJ (2013) Epigenetic conservation at gene regulatory elements revealed by non-methylated DNA profiling in seven vertebrates. elife 2013:1–19. https://doi.org/10.7554/eLife.00348
42. Howarth M, Ting AY (2008) Imaging proteins in live mammalian cells with biotin ligase and monovalent streptavidin. Nat Protoc 3:534–545. https://doi.org/10.1038/nprot.2008.20
43. Weyrich A (2012) Preparation of genomic DNA from mammalian sperm. Curr Protoc Mol Biol 1:2–4. https://doi.org/10.1002/0471142727.mb0213s98

Check for updates

Chapter 3

Prediction of CpG Islands as an Intrinsic Clustering Property Found in Many Eukaryotic DNA Sequences and Its Relation to DNA Methylation

Cristina Gómez-Martín, Ricardo Lebrón, José L. Oliver, and Michael Hackenberg

Abstract

The promoter region of around 70% of all genes in the human genome is overlapped by a CpG island (CGI). CGIs have known functions in the transcription initiation and outstanding compositional features like high G+C content and CpG ratios when compared to the bulk DNA. We have shown before that CGIs manifest as clusters of CpGs in mammalian genomes and can therefore be detected using clustering methods. These techniques have several advantages over sliding window approaches which apply compositional properties as thresholds. In this protocol we show how to determine local (CpG islands) and global (distance distribution) clustering properties of CG dinucleotides and how to generalize this analysis to any *k*-mer or combinations of it. In addition, we illustrate how to easily cross the output of a CpG island prediction algorithm with our methylation database to detect differentially methylated CGIs. The analysis is given in a step-by-step protocol and all necessary programs are implemented into a virtual machine or, alternatively, the software can be downloaded and easily installed.

Key words CpG islands, Clustering, DNA words, DNA methylation, Virtual machine

1 Introduction

Mammalian genomes are generally characterized by a global DNA methylation pattern of CpG dinucleotides [1]. Methylcytosine residues are deaminated to thymine and are therefore hotspots of spontaneous mutations [2]. Due to this reason, only about one fifth of the expected number of CpG dinucleotides can be found in a typical mammalian genome. Exceptions are short stretches of DNA called CpG islands which deviate significantly from the average genome background by being GC-rich, CpG-rich, and predominantly hypomethylated [3]. It is expected that most CpG islands are related to sites of transcription initiation by destabilizing nucleosomes and attracting proteins to form a transcriptionally

Tanya Vavouri and Miguel A. Peinado (eds.), *CpG Islands: Methods and Protocols*, Methods in Molecular Biology, vol. 1766, https://doi.org/10.1007/978-1-4939-7768-0_3, © Springer Science+Business Media, LLC, part of Springer Nature 2018

permissive, open chromatin state [3]. Given those properties, the distances between neighboring CpG dinucleotides are lower within CGIs than in the bulk genome sequences. We exploited this fact in our *CpGcluster* algorithm [4] which was the first method that detected CGIs as statistically significant clusters of CpG dinucleotides. Until the publication of our algorithm in 2006, many different algorithms had been developed to computationally detect CGIs, however, virtually all of them were based on the threshold criteria of Gardiner–Frommer [5]: GC-content $\geq$ 50%, CpG observed/expected (O/E) ratio $\geq$ 0.6 and length $\geq$ 200 bp. These thresholds were then applied in so called sliding window approaches [6]. Apart from a high parameter space, there are several other drawbacks of these approaches: (1) the thresholds need to be adjusted for each species individually, which hinders comparative genome analyses, (2) the main statistical properties of CGIs are used for the detection and therefore its distributions are artificially biased toward the thresholds [7], (3) the prediction can change notably if one of the thresholds gets changed.

On the contrary, methods based on the CpG clustering in chromosome sequences do not use ad hoc thresholds and the CGIs are predicted as "naturally occurring" clusters of CpGs in the DNA sequences. In concrete, *CpGcluster* is based only upon one parameter, the statistical significance as the distance threshold can be obtained directly from the DNA sequences. This adds the big advantage that stricter predictions are true subsets of predictions with a laxer *p*-value threshold. This is not true for sliding window approaches. Additionally, the theoretical framework can be easily extended to other DNA words (*k*-mers) [8].

In this protocol we first show that the distance distribution can be used to quantify the clustering and how this can be used to calculate the clusters of any arbitrary *k*-mer (like the 2-mer CG). Second, we demonstrate how to link the primary prediction to DNA methylation data. The chapter is structured like a bioinformatics protocol that can be easily followed by the user.

2 Preparation of the Protocol

The reader can follow the protocol in two ways. Either, the executables can be downloaded and installed on most operating systems, like Windows, Linux, and Mac (but preferable Linux). Alternatively, the user can download a virtual machine which can be executed on any operating system without the need to install or maintain the programs. More detailed information can be found under this link: http://bioinfo2.ugr.es/gCluster. The basic features of the programs are described in Table 1.

Note that all launch commands that we give in this protocol are for the usage with the virtual machine.

Table 1
Description of the programs needed to follow this protocol

Program	Description
prepareAssembly.py	(1) Obtains the genome sequences and (2) splits the assembly into canonical sequences (the reference sequences of the chromosomes) and alternative sequences (alternative assemblies, unassembled sequences, etc.). Normally, the predictions will be carried out on the canonical sequences only. It can work with local files, URLs or directly retrieve the corresponding files from UCSC
makeSeqObj.jar	Index the fasta input files in order to make them faster accessible. With an "assembly.fa" input file it generates an "assembly.zip" file which, in turn, is the input file for the programs described below
randomizer.jar	Randomizes DNA sequences preserving the dinucleotide frequencies. The theoretical framework does not contemplate the existence of N-runs, therefore: (1) contigs are merged, (2) the sequence is shuffled, (3) the N-runs are introduced again at the original position. Therefore the dinucleotides are not strictly conserved, because the last nucleotide of a contig and the first of the next contig form an "artificial" dinucleotide. However, the "error" is negligible even for highly fragmented assemblies (high number of contigs)
gCluster.jar	Determines the local clusters of a given DNA word (CpG islands if the DNA word is "CG") and its global clustering properties. It can work on both strands (for nonpalindromic words) and accepts any combination of DNA words (like CAG:CTG:CCG for the methylation context CHG)
GenomeCluster.pl	Determines the local clusters of genome elements identified by its chromosome coordinates
NGSmethDB_API_client.py	Takes a BED file as input and downloads the CpG methylation levels for the regions from our methylation database *NGSmethDB* [21–23]
calcDMIs.py	Takes the output files from *NGSmethDB_API_client.py* for two samples and calculates statistically significant differentially methylated CpG islands between them. The statistical significance is assessed by means of a Fisher exact test

2.1 Preparation of the Virtual Machine

The programs needed by the protocol can be used in a virtual machine, following these instructions:

1. Download and install VirtualBox (Windows, Mac OS X, Linux, and Solaris are supported). Install also the VirtualBox Extension Pack.
2. Download gClusterVM OVA file and import it on VirtualBox by double-clicking it.
3. Configures the virtual machine: allocated memory (minimum: 4 GB), CPUs assigned (minimum: 2) and folder shared for exchange files with the virtual machine (highly recommended).
4. Run the virtual machine.

More detailed information can be found under this link: http://bioinfo2.ugr.es/gCluster/manual.

2.2 Installation of the Standalone Executables

Another option is to install the standalone versions on your computer:

1. Download and install Java 8 or higher.
2. Download and install Python 3.4 or higher. Important: in Windows check the option "Add to the PATH" during the installation.
3. Download and extract the standalone executables.
4. Open a terminal to execute the commands of the protocol. If you are using windows, you can use CMD or PowerShell.

More details about how to use standalone executables can be found under this link: http://bioinfo2.ugr.es/gCluster/manual.

2.3 Preparation of the Input Files

In this section we will explain how to (1) obtain the sequences, (2) extract only the canonical sequences, (3) convert the assembly multi-fasta file to an indexed, easily and quickly accessible format, and (4) randomize the genome assembly so it can be used to assess the random expectation.

In this protocol we will give working examples for several species(assemblies): human (hg38), mouse (mm10), zebrafish (danRer10), *C. elegans* (ce11), *Arabidopsis thaliana* (tair10), tomato (sly2_50), 163 archaeas, and 124 bacterias (incluiding *E. coli*).

The sequences can be either downloaded from NCBI [9] for bacteria and archaea, Sol Genomics Network for tomato [10] or UCSC [11] for the rest.

To facilitate the download and processing of the genome assemblies we provide a small script that can either use local files, URLs or obtain the corresponding assemblies stored at UCSC.

Note that in general, fasta files from different sources can be used and not only those mentioned above.

2.3.1 Extract Canonical Sequences from Genome Assembly (from Local File)

```
prepareAssembly -i /home/gcluster/sequences/hg38.fa -o
/home/gcluster/sequences -l hg38
```

Explanation:

-i **<path>**: the multifasta genome assembly file.

-o **<output folder>**: the output folder where canonical (hg38_canonical.fa) and noncanonical (hg38_noncanonical.fa) multifasta files will be created.

-l **<label>**: label for output files.

2.3.2 Obtain the Genome Assembly and Extract Canonical Sequences (from UCSC)

```
prepareAssembly -u hg38 -o /home/gcluster/sequences -l
hg38
```

Explanation:

-u <UCSC assembly name>: the UCSC assembly name (hg38, mm10, canFam3, etc).

-o <output folder>: the output folder (see above).

-l <label>: label for output files.

2.3.3 Index the Fasta File

```
makeSeqObj /home/gcluster/sequences/hg38_canonical.fa
```

Explanation:

This script has only one parameter, the multifasta file of the genome assembly. The program indexes this file so that the sequences can be accessed quickly. The output indexed file will be placed in the directory of the input file. It will have the same name but with extension "zip". The indexed file will be the input for all subsequent analyses. Furthermore, it generates two more files: a file with the coordinates of the N-runs in the assembly (hg38_canonical.N) and a chromosome-size file (hg38_canonical.chromSize).

2.3.4 Randomize a Genome Assembly

```
randomizer /home/gcluster/sequences/hg38_canonical.zip
```

Explanation:

This command generates a genome assembly with a shuffled DNA sequence. Note that the randomization method preserves the frequencies of the dinucleotides [12].

3 The Distance as an Indicator of Clustering

We can consider both, local and global clustering of entities and both can be defined by the next-neighbor distance (*see* Subheading 3.1 on how we define the next-neighbour distance).

By means of the following command, the gCluster algorithm calculates all clustering properties:

```
mkdir /home/gcluster/results (generates the 'results'
output directory)
gCluster genome=/home/gcluster/sequences/hg38_canonical.
zip pattern=CG
```

Table 2
The output files generated by "gCluster"

Output file name	Description
cluster.txt	The clusters identified by its chromosomal coordinates and *p*-value adding basic compositional statistics like G+C content and *O/E* ratios of the pattern (the number of observed patterns divided by the number of expected pattern). In case of a compound pattern like CAG:CTG, the mean *O/E* ratio is calculated
CVnor.txt	The file holds the normalized coefficient of variation for each chromosome
*.distr files	The distance distribution, the observed and expected frequencies as a function of next-neighbor distance
log.txt	A log file
stat.txt	A basic statistic as a function of chromosome, i.e., pattern frequency, G+C content, lengths, and contig length (the sequence length minus the sum of all *N*'s)

```
output=/home/gcluster/results/hg38_CG writedistribu-
tion=true
chromStat=true
```

Explanation:

genome=<path>: the indexed multifasta file (*see* "Prepare the sequences" section above).

output=<output folder>: the output folder where the results will be written.

pattern=<*k*-mers>: the *k*-mers (DNA words) to be analyzed. In order to obtain CpG islands, pattern = CG will be used. However any other DNA word or combination of words can be used. For example, pattern = CAG:CTG would trigger the detection of CWG clusters, i.e., a methylation context always found in plants but also in some mammalian cell types [13].

writedistribution=<boolean>: write out the observed and expected distance distribution (default writedistribution=false).

chromStat=<boolean>: if true, the program writes out additional information of the chromosome sequences like G+C content, CpG frequencies (observed/expected ratios), lengths, etc. (default: chromStat=false).

Output

Table 2 gives an explanation of the produced output files.

3.1 The Distance Model

The next-neighbor distance between *k*-mers (DNA words) which can overlap each other, is much harder to define as the distance between solid, one dimensional entities like for example beads on a string. Below we illustrate this problem by means of the methylation contexts CCG and CWG which can be methylated in plants and also in embryonic stem cells. CCG can overlap itself if one of the copies is located on the direct and the other on the reverse strand (*k*-mers in red) while CAG and CTG (abbreviated CWG) are palindromes of each other and therefore cannot overlap each other.

```
5'- TCCGGTGCTACAGCTG -3' word start end strand
    ||||||||||||||||     CCG 2 4 +
3'- AGGCCACGATGTCGAC -5' CCG 3 5 -
```

Given this, the distance of overlapping words cannot be calculated as start (downstream word)–end (upstream word) as this can result in negative distances (in the example above it would be $3 - 4 = -1$.

On the other hand, if we define the distance as **start–start** we would have the situation that the minimal distance depends on the degree of overlap which two words can have. For example, using the coordinates of the illustration above:

CCG(+)-CCG(−): $3-2 = 1$ (the minimal distance possible).

CAG(+)-CTG(+): $14-11 = 3$ (the minimal distance of two, non-overlapping DNA words of length 3 nt).

Therefore we subtract the number of forced nonoverlapping bases and define the distance as:

$$d_{j,i} = (sc_j - sc_i) - nf_i.$$

Being sc_j and sc_i the start coordinate of downstream and upstream word, respectively, and nf_i the number of forced nonoverlapping bases of the upstream word. In the case of CAGCTG above, n_f would be 2 as CAG cannot be overlapped by CTG.

3.2 Local Clustering

The local clustering manifests itself in a higher local density which is simply defined as N/L, e.g., the number of entities (N) in a given space (length L here). The local density is inversely proportional to the average distance of the entities in the given region with length L, e.g., the higher the density, the shorter the distances. This means that if clustering exists, then the short distances should be overrepresented over a random estimate. Given that distances are integers, the random estimate is given by the Geometric distribution which is the discrete counterpart of the Poisson distribution.

gCluster allows to visualize the distance distribution both, for each chromosome individually and for the entire genome. However, given that compositional structure is mainly organized on a genome-scale [14, 15], gCluster uses by default the genome distribution, i.e., all distances are merged into one (genome wide) distance distribution in order to determine the distance threshold.

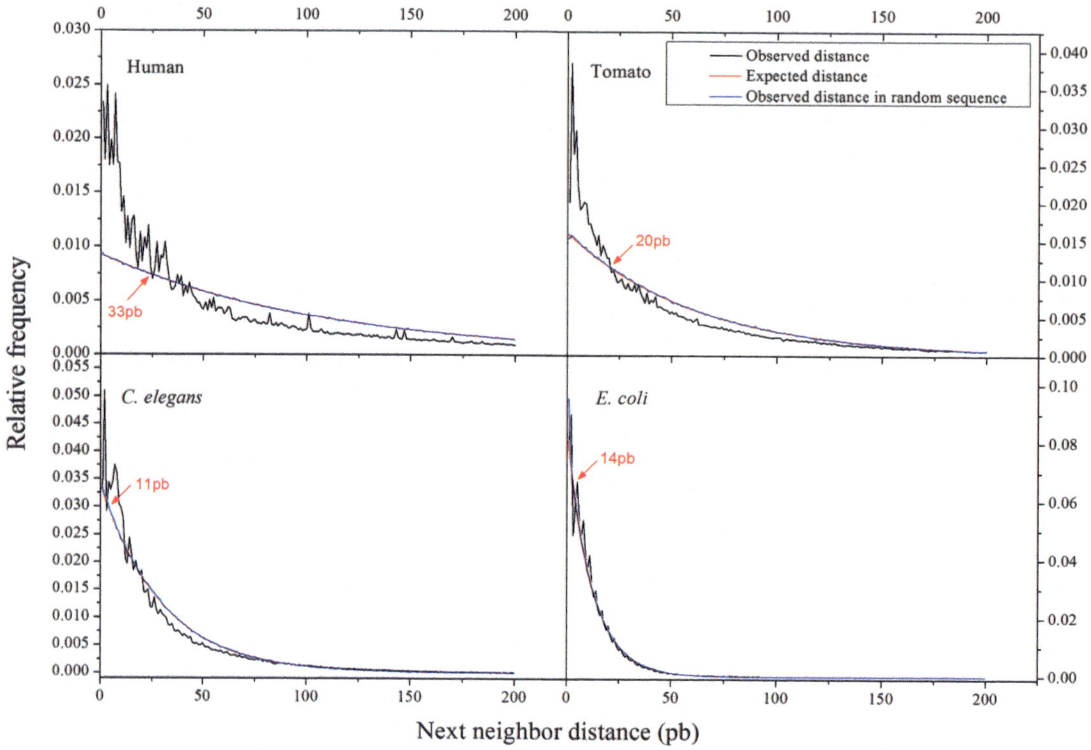

Fig. 1 Distance distributions (obtained from *.distr files) for four different species. Each plot represents the observed distances (black), the expected distances (red) and the observed distance frequencies after randomizing the chromosome sequence (blue). Arrows indicate the genome intersection

Figure 1 shows the distance distributions for four genomes: *Homo sapiens*, *C. elegans*, tomato, and *E. coli*. In the human and tomato genomes, short distances are clearly overrepresented (the black line is clearly above the random estimate) which is an indicator for clustering. In *C. elegans*, which lacks DNA methylation (at least in the CpG context) we observe a slight overrepresentation of short distances while *E. coli* does not show any clear overrepresentation of short distances. The Geometric distribution fits exactly the observed distribution in random sequences which proves that it models the randomness in an adequate way. The intersection between the observed and the expected distribution can be considered as the distance that separates clustered CpGs from those in the bulk DNA. We therefore recommend the usage of the genome intersection as distance threshold for cluster prediction (*see* Subheading 4).

3.3 Global Clustering of DNA Words

The global clustering of an entity in a one-dimensional space can be calculated by means of a coefficient of variation of the normalized distances [16, 17].

Briefly, in order to detect the clusters, (1) for each entity, the distance to the next-neighbour is calculated (*see* Subheading 3.1), (2) the distance distribution is normalized to 1 (by dividing each distance by the mean distance), (3) the coefficient of variation (CV) is calculated as the standard deviation divided by the mean, (4) the CV is corrected for certain biases of entities with high frequencies (please *see* [16, 17] for more details).

It can be shown that the CVcor is indicative of the clustering as:

- CVcor $=$ 1: the entity is randomly distributed (geometric distribution).
- CVcor $>$ 1: the entities are clustered, i.e., both short and long distances do exist.
- CVcor $<$ 1: the entities show repulsion, i.e., the distances tend to be "equidistant" (if only one distance would exist CVcor would be 0.

gCluster calculates the corrected CV by applying the distance model described above in all contigs of a chromosome. The CV can be calculated both, for each chromosome separately or at a genome level, i.e., for all distances 'glued together'. By default, gCluster uses always genome level analysis. Figure 2 shows the summary of the CVnor.txt files generated by the "gCluster command" (see above). It can be observed that the dinucleotide CG is clustered in all analyzed eukaryotic genomes (CVcor $>>$ 1), including C. elegans which does not show DNA methylation in its cytosines. Furthermore, the clustering of CpGs is unrelated to the O/E ratio. For example, *C. elegans* shows clustering (mean CVcor $=$ 1.33) but O/E ratios of 1 (the observed and expected numbers of CpGs are equal), while the human random sequences have O/E ratios of around 0.2 (as the frequency of dinucleotides was conserved in the randomization) but the CVnor coefficients are virtually 1 which indicates random distribution. Even for rather short DNA sequences like those for a typical bacteria (approx. 5 Mb), we observe only small fluctuation around CVcor $=$ 1 in the random sequences. For example, percentile 99 corresponds to a CVcor of 1.05 in the random sequences. This same value corresponds to percentile 45 in the natural sequences. This means that over half of all bacterial sequences show significant clustering of CpG dinucleotides.

4 Detect the Clusters of DNA Words

The statistically significant clusters are detected in two steps: (1) the patterns with Next Neighbour Distance (NND) $\leq d$ threshold are grouped together, and (2) the statistical significance is calculated by means of the negative binomial distribution.

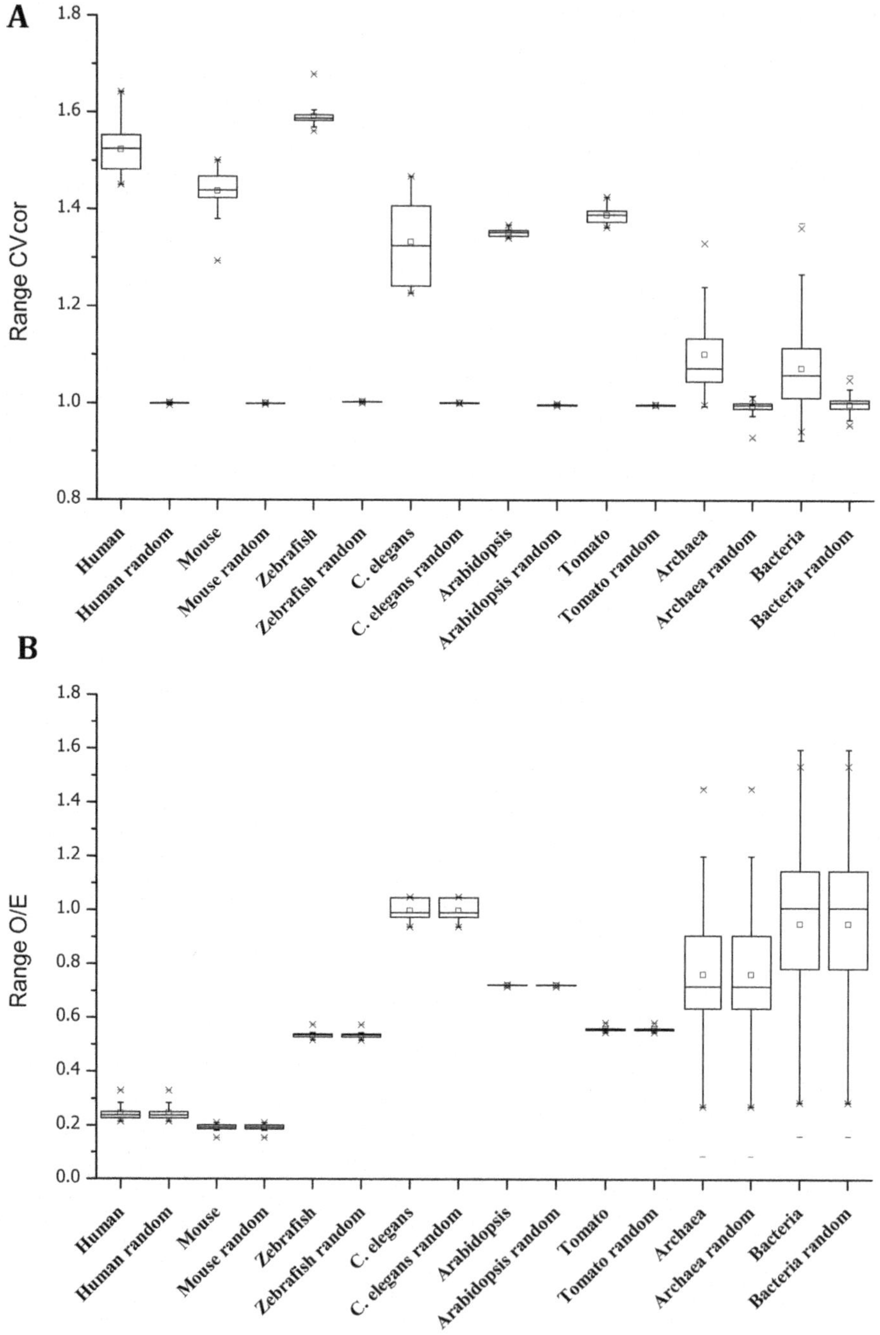

Fig. 2 (**a**) Global clustering (CVcor) for eight different species and its randomized assemblies (human, mouse, zebrafish, *C. elegans*, Arabidopsis, tomato and the collection of 163 archaea and 124 bacteria) and (**b**) the corresponding *O/E* ratios (ratio between the observed CpG frequency and the expected CpG frequency). Both measures are calculated for each chromosome and the plot shows the distribution among the different chromosomes by means of a box plot

4.1 Clustering

The drawing below illustrates the grouping of single CpGs (vertical ticks in middle track) into clusters (black boxes in bottom track).

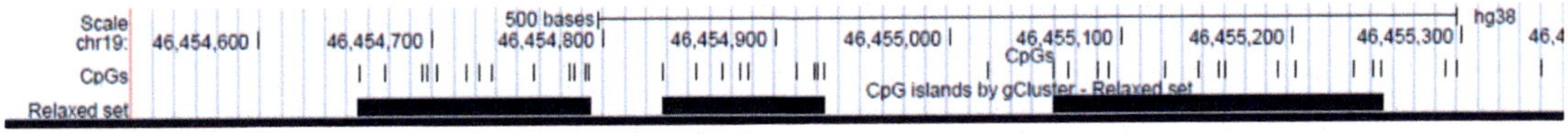

4.2 Statistical Significance

The negative binomial distribution gives the probability to obtain a certain number of "failures" (nonpattern) given N successes (pattern).

$$P_{N,\mathrm{p}}(n_{\mathrm{f}}) = \binom{(N-1)+n_{\mathrm{f}}-1}{N-1} p^{(N-1)}(1-p)^{n_{\mathrm{f}}}.$$

Being N the number of patterns found in the cluster, n_{f} the number of nonpattern found in the cluster and p the probability to find a pattern (either in the chromosome or genome) . Note that as each cluster starts trivially with a pattern, the number of "successes" is defined as $N-1$.

The probability to find a shorter or a cluster of the same length with the same number of patterns can then be calculated as the cumulative density function. Note that this probability corresponds to the p-value.

$$P_{N,\mathrm{p}}^{\mathrm{cum}}(x \le n_{\mathrm{f}}) = \sum_{x=0}^{n_{\mathrm{f}}} \binom{(N-1)+x-1}{N-1} p^{(N-1)}(1-p)^{x}.$$

4.3 CpG Islands

With the following command (same as above), the clusters are automatically determined and written to the cluster.txt file. The distance threshold is determined by the genome intersection (*see* Fig. 1).

```
gCluster genome=/home/gcluster/sequences/hg38_canonical.
zip pattern=CG
output=/home/gcluster/results/hg38_CG writedistribu-
tion=true
chromStat=true
```

Figure 3 shows the p-values in chromosome sequences vs. the corresponding random sequences in order to estimate the number of false positive predictions. The prediction on the random assembly yields 4640 CpG islands compared to 205,795 in the natural sequences. This means that for a relaxed p-value threshold of $1E{-}5$, 2.3% of the detected CGIs will be false positive predictions. The smallest p-value of a CGI in the random sequences is $4.18\ E^{-13}$. Therefore, for the strict threshold $1E^{-20}$ which we introduced before [7] we would not expect any false positive predictions. The number of false positive predictions is conditioned by the application of the

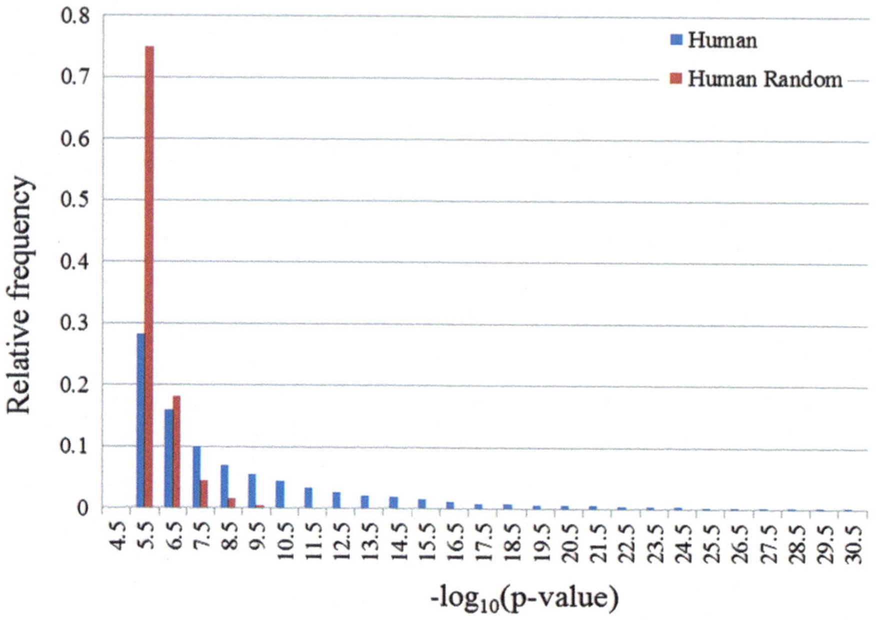

Fig. 3 Distribution of *p*-values (represented as −log10(*p*-value)) of human CpG islands and those from the corresponding randomized genome assembly

same distance threshold (genome intersection) to all chromosomes. G+C rich genome regions have shorter mean NND (next-neighbor distance) between CpGs compared to A+T rich genome regions. Thus, by applying the same threshold to all chromosomes, in G+C rich chromosomes the probability for false positive predictions increases. This could theoretically be prevented by using chromosome based statistics (or statistics of compositional superstructures [18]), i.e., each chromosome or superstructure having its own distance threshold. However, even so we recommend the usage of genome wide statistics as the compositional structure is mainly organized on a genome-scale [14, 15] which should include CpG islands. Therefore, the genome wide statistics will yield compositionally more similar CpG islands among the different chromosomes, what is biologically reasonable given that DNA methyltransferases will act equally on all chromosomes, while the usage of chromosome statistics (one threshold per chromosome) would lower considerably the number of false positive predictions.

4.4 Clusters of Other, Biologically Relevant k-mers

Like mentioned before, in plants cytosine can be methylated as well in other contexts like CWG (CAG, CTG), CCG, or CHH.

The clustering of these contexts can be calculated with the following command.

```
gCluster genome=/home/gcluster/sequences/ hg38_canoni-
cal.zip
pattern=CAG:CTG output=/home/gcluster/results/hg38_CWG
writedistribution=true chromStat=true
```

5 Cluster of Clusters: Any Biological Meaning?

As other genome elements, CpG islands detected by *gCluster* are known to be in turn clustered forming higher genome structures, which reveals a complex genome landscape dominated by hierarchical clustering [19]. The existence of such higher structures could throw light on the evolutionary origin of genome hierarchical clustering, but their structural or functional significance remains to be shown.

CGI clusters can be determined by the following command:

```
GenomeCluster start /home/gcluster/results/hg38_CG/clus-
ter.txt gi 1E-5
/home/gcluster/sequences/hg38_canonical.N 0
```

Explanation (the arguments must be given in this order):

Argument 1: the distance model <element; start; middle; end>.

Argument 2: the BED file within the chromosome coordinates of whatever genome element (e.g., CpG islands).

Argument 3: the distance threshold model <gi: genome intersection; ci: chromosome intersection; N (integer): percentile>.

Argument 4: the *p*-value threshold (by default $1E{-}5$ as in gCluster).

Argument 5: the file with the N-runs created by makeSeqObj.jar (*see* "Prepare the sequences" section above).

Argument 6: the maximum number of allowed Ns between two elements. It must be an integer greater or equal to 0 (default: 0).

Output files:

***_genomeIntersec_start_GenomeCluster.txt**: there are as many of these files as chromosomes in the assembly. They are tabular files with the columns described below.

Columns in outfile	Description
Chrom	Chromosome where the local cluster belongs
From	Start chromosome coordinate of the local cluster
To	End chromosome coordinate of the local cluster
Length	Length of the local cluster
Count	Number of genome elements within the local cluster
PValue	P-value of the local cluster
logPValue	Decimal logarithm of the *p*-value

6 Determine Differential Methylation of CpG Islands

6.1 Get DNA Methylation of CpG Islands

In order to calculate differentially methylation CGIs, we will have to obtain first the methylation values for the CpG islands from our database [20–22] and second, compare the methylation values between two samples.

Obtain the methylation values for the regions defined in a BED file:

```
NGSmethDB_API_client -i /home/gcluster/results/hg38_CG/
cluster.txt -o
/home/gcluster/results/methylationData
```

Explanation:

-i **<input bed file>**: in general, BED3 files will be accepted (only the first three columns will be considered), including the "cluster.txt" output files generated by gCluster.

-o **<output folder>**: output folder.

Output files:

<sample>.CG.meth.tsv: Contains the CG methylation data for all regions defined in the bed data input file. There is one <sample>.CG.meth.tsv file for each selected sample. Each line corresponds to one region.

Columns in outfile	Description
ID	chrom_start_end
refPatt	Number of CpGs within the region in the reference genome
indPatt	Number of CpGs in the region for the genotype of the sample
meanMR	Mean of CpG methylation ratio distribution
sdMR	Standard Deviation of CpG methylation ratio distribution
p10MR	Percentile 10 of CpG methylation ratio distribution
q1MR	Quartil 1 of CpG methylation ratio distribution
q2MR	Median of CpG methylation ratio distribution
q3MR	Quartil 3 of CpG methylation ratio distribution
p90MR	Percentile 90 of CpG methylation ratio distribution
totalMC	Total number of methylcytosines in the region
totalC	Total number of mapped reads within the region

<sample>.CHG.meth.tsv (plants only): Same as above, but for CHG.

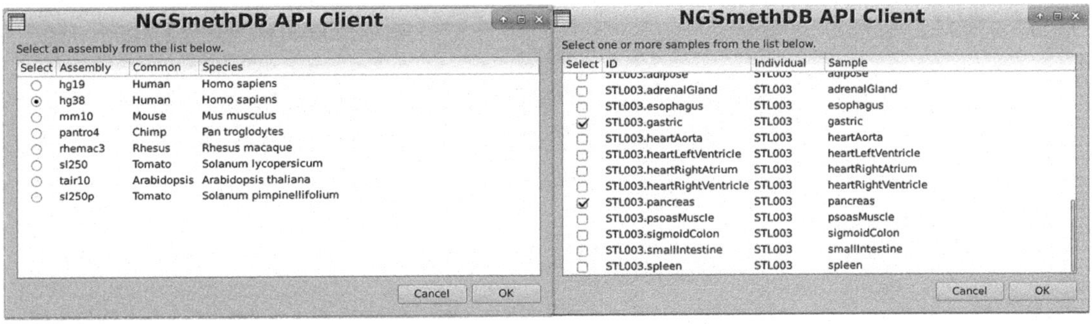

Fig. 4 NGSmeth_DB_API_client input display. Selection of the assembly (left) and of one or more samples (right)

6.2 Calculate Differential Methylation

```
calc_DMIs -a
/home/gcluster/results/methylationData/STL003.gastric.CG.
meth.tsv -b
/home/gcluster/results/methylationData/STL003.pancreas.
CG.meth.tsv -o
/home/gcluster/results/methylationData/STL003.gastric.
STL003.pancreas.DMI
s.txt -p 1E-5
```

Explanation

-a <sample1 file> and -b <sample2 file>: output methylation files from NGSmethDB_API_client.py (*see* Fig. 4).

-o <outfile>: output file. Default is "outputDMIs.txt".

-p <pvalue>: *p*-value threshold for DMIs. Default is $1E-5$.

Output file:

<outfile>: A file containing the differentially methylated CpG islands.

Columns in outfile	Description
ID	chrom_start_end
mC_Sample1	Number of methylcytosines for each cluster in sample 1
reads_Sample1	Number of reads for each cluster in sample 1
mRatio_Sample1	methRatio (number of methylcytosines at a given position/total number of reads mapped to that given position) for each cluster in sample 1
mC_Sample2	Number of methylcytosines for each cluster in sample 2
reads_Sample2	Number of reads for each cluster in sample 2
mRatio_Sample2	methRatio for each cluster in sample 2

diffMeth	Absolute difference between methRatio_Sample1 and methRatio_Sample2
pvalue	Statistical significance assessed by means of the Fisher exact test

7 Notes

1. Out of memory
   "gCluster" might throw an "out of memory" exception in rare cases (if too many patterns are used or the genome assemblies are very big). In such a case, you can increase the memory on the Java virtual machine. The following command would assign 5Gb to the virtual machine.

```
gCluster MaxMem=5000m genome=/home/gcluster/genomes/
hg38.fa
pattern=CAG:CTG output=/home/gcluster/results/hg38_CWG
genomeDist=false
```

If memory usage exceeds the memory allocated to the virtual machine, the process will use swap (memory on disk), which is much slower than the RAM. If the RAM and the swap are fully used, the process will fail and it is possible that the virtual machine not respond.

References

1. Suzuki MM, Bird A (2008) DNA methylation landscapes: provocative insights from epigenomics. Nat Rev Genet 9:465–476. https://doi.org/10.1038/nrg2341
2. Duncan BK, Miller JH (1980) Mutagenic deamination of cytosine residues in DNA. Nature 287:560–561. https://doi.org/10.1038/287560a0
3. Deaton AM, Bird A (2011) CpG islands and the regulation of transcription. Genes Dev 25:1010–1022. https://doi.org/10.1101/gad.2037511
4. Hackenberg M, Previti C, Luque-escamilla PL et al (2006) CpGcluster: a distance-based algorithm for CpG-island detection. BMC Bioinformatics 13:1–13. https://doi.org/10.1186/1471-2105-7-446
5. Gardiner-Garden M, Frommer M (1987) CpG islands in vertebrate genomes. J Mol Biol 196:261–282
6. Takai D, Jones PA (2002) Comprehensive analysis of CpG islands in human chromosomes 21 and 22. Proc Natl Acad Sci U S A 99:3740–3745. https://doi.org/10.1073/pnas.052410099
7. Hackenberg M, Barturen G, Carpena P et al (2010) Prediction of CpG-island function: CpG clustering vs. sliding-window methods. BMC Genomics 11:327. https://doi.org/10.1186/1471-2164-11-327
8. Hackenberg M, Carpena P, Bernaola-galván P et al (2011) WordCluster : detecting clusters of DNA words and genomic elements. Algorithms Mol Biol 6:2. https://doi.org/10.1186/1748-7188-6-2
9. Pruitt KD, Tatusova T, Brown GR, Maglott DR (2012) NCBI reference sequences (RefSeq): current status, new features and genome annotation policy. Nucleic Acids Res 40:D130–D135. https://doi.org/10.1093/nar/gkr1079
10. Fernandez-Pozo N, Menda N, Edwards JD et al (2015) The Sol Genomics Network (SGN)—from genotype to phenotype to

breeding. Nucleic Acids Res 43: D1036–D1041. https://doi.org/10.1093/nar/gku1195

11. Kent WJ, Sugnet CW, Furey TS et al (2002) The human genome browser at UCSC. Genome Res 12:996–1006. https://doi.org/10.1101/gr.229102
12. Altschul SF, Erickson BW (1985) Significance of nucleotide sequence alignments: a method for random sequence permutation that preserves dinucleotide and codon usage. Mol Biol Evol 2:526–538
13. Lister R, Pelizzola M, Dowen RH et al (2009) Human DNA methylomes at base resolution show widespread epigenomic differences. Nature 462:315–322. https://doi.org/10.1038/nature08514
14. Grantham R, Gautier C, Gouy M et al (1980) Codon catalog usage and the genome hypothesis. Nucleic Acids Res 8:197. https://doi.org/10.1093/nar/8.1.197-c
15. Bernardi G (1993) Genome organization and species formation in vertebrates. J Mol Evol 37 (4):331–337
16. Bernaola-Galván P, Oliver JL, Hackenberg M et al (2012) Segmentation of time series with long-range fractal correlations. Eur Phys J B. https://doi.org/10.1140/epjb/e2012-20969-5
17. Hackenberg M, Rueda A, Carpena P et al (2012) Clustering of DNA words and biological function: a proof of principle. J Theor Biol 297:127–136. https://doi.org/10.1016/j.jtbi.2011.12.024
18. Carpena P, Oliver JL, Hackenberg M et al (2011) High-level organization of isochores into gigantic superstructures in the human genome. Phys Rev E Stat Nonlin Soft Matter Phys 83:31908
19. Dios F, Barturen G, Lebrón R et al (2014) DNA clustering and genome complexity. Comput Biol Chem 53:71–78. https://doi.org/10.1016/j.compbiolchem.2014.08.011
20. Oliver L, Hackenberg M, Barturen G, De GD (2011) NGSmethDB: a database for next-generation sequencing single-cytosine-resolution DNA methylation data. Nucleic Acids Res 39:75–79. https://doi.org/10.1093/nar/gkq942
21. Hackenberg M, Barturen G, Oliver JL (2011) NGSmethDB: a database for next-generation sequencing single-cytosine-resolution DNA methylation data. Nucleic Acids Res 39: D75–D79. https://doi.org/10.1093/nar/gkq942
22. Lebrón R, Gómez-Martín C, Carpena P et al (2016) NGSmethDB 2017: enhanced methylomes and differential methylation. Nucleic Acids Res 45:gkw996. https://doi.org/10.1093/nar/gkw996
23. Geisen S, Barturen G, Alganza M et al (2014) NGSmethDB: an updated genome resource for high quality, single-cytosine resolution methylomes. Nucleic Acids Res 42:53–59. https://doi.org/10.1093/nar/gkt1202

Check for updates

Chapter 4

CpG Islands in Cancer: Heads, Tails, and Sides

Humberto J. Ferreira and Manel Esteller

Abstract

DNA methylation is a dynamic epigenetic mark that characterizes different cellular developmental stages, including tissue-specific profiles. This CpG dinucleotide modification cooperates in the regulation of the output of the cellular genetic content, in both healthy and pathological conditions. According to endogenous and exogenous stimuli, DNA methylation is involved in gene transcription, alternative splicing, imprinting, X-chromosome inactivation, and control of transposable elements. When these dinucleotides are organized in dense regions are called CpG islands (CGIs), being commonly known as transcriptional regulatory regions frequently associated with the promoter region of several genes. In cancer, promoter DNA hypermethylation events sustained the mechanistic hypothesis of epigenetic transcriptional silencing of an increasing number of tumor suppressor genes. CGI hypomethylation-mediated reactivation of oncogenes was also documented in several cancer types. In this chapter, we aim to summarize the functional consequences of the differential DNA methylation at CpG dinucleotides in cancer, focused in CGIs. Interestingly, cancer methylome is being recently explored, looking for biomarkers for diagnosis, prognosis, and predictors of drug response.

Key words CpG islands, DNA methylation, Cancer

1 DNA Methylation in the Epigenetic Context

Biologically, the molecular accessibility to the genetic information in a certain cell is regulated by different epigenetic layers that control essential processes such as transcription, replication, recombination, and DNA repair [1]. For instance, the chromatin status in a particular gene expression regulatory region directs its accessibility to transcription factors. Accordingly, the variability of these profiles across the genome are responsible for tissue-specific expression profiles, in addition to controlled expression profiles during cell differentiation and development, defining lineage patterns and specific cell-type identities [2, 3]. Adult cell renewal ability, telomere length, alternative transcription start site (TSS) establishment, and different splicing events are also attributed to specific chromatin profiles [4, 5].

Tanya Vavouri and Miguel A. Peinado (eds.), *CpG Islands: Methods and Protocols*, Methods in Molecular Biology, vol. 1766, https://doi.org/10.1007/978-1-4939-7768-0_4, © Springer Science+Business Media, LLC, part of Springer Nature 2018

High levels of chromatin compaction, characterized by DNA methylation and specific histone modifications, guarantee the repression of chromosomal loci containing transposable elements and endogenous retroviruses. Hence, deleterious translocations to other genomic loci or insertional mutagenesis are prevented, ensuring genomic stability in healthy cells [6, 7]. On the other hand, X-chromosome inactivation [8, 9] and genomic imprinting [10, 11] represent important epigenetic mechanisms that regulate allele-specific gene expression.

Epigenetics is as regulatory layer above genetics, controlling the outcome of the genetic content of a particular cell, comprising DNA methylation that appears as a broadly studied epigenetic modification in mammals.

2 Genomic and Functional Relevance of CpG Islands

DNA methylation events occur commonly in CpG dinucleotides by the addition of a methyl group from S-adenosylmethionine to the 5′ carbon of the cytosine ring (5-methylcytosine, 5mC) [12]. Curiously, the frequency of CpG dinucleotides in human genome is only 25% of their expected occurrence with a concomitant genome-wide decrease of cytosines and guanines to about 40% of all nucleotides. These lower percentages could be explained by spontaneous methylcytosine–thymine mutations with their subsequent depletion during evolution [13, 14]. Moreover, the DNA methylation profile through the genome is not arbitrary. According to a recent hypothesis, a relationship between the DNA methylation of the individual CpG dinucleotides is established within the region where they are located. Large and dense clusters of CpGs are usually hypomethylated, while clusters with a sparse distribution or less number of CpGs are typically hypermethylated. This hypothesis takes into account the positive feedback that a methylated or unmethylated CpG exerts at nearby CpGs, by long-range methylation or short-range demethylation, respectively [15].

In mammals, the statistically CpG dinucleotide underrepresentation contrasts with genomic loci of large repetitive sequences (centromeres, telomeres, and endoparasitic sequences) and short and dense extensions of CpGs, normally associated to gene promoters, called CpG islands (CGIs). Their assignment is based on prediction algorithms [16]. Usually, a CGI is defined as a genomic region longer than 200 bases, presenting at least a content of 50% of guanines and cytosines and an observed-to-expected CpG ratio greater than 0.6 [17]. Taking into consideration these parameters, even if 72% of all mammalian gene promoters enclose a high CpG density [18], just 50% of them are considered to be linked with at least one CGI [19]. The human genome contains around 29,000 CGIs [20, 21]. CGIs are genomically spread in a nonrandom pattern, preferentially in the promoter regions, close to TSSs or in the first exons [20, 22].

2.1 Regulation of Transcription

In 1975, DNA methylation was associated to gene expression [23, 24]. The impact of DNA methylation dynamic alterations were subsequently highlighted in mammalian development [25], being related to the repression of both coding and noncoding genes [26, 27]. Around 75% of the cytosines belonging to CpG dinucleotides are methylated, whereas those clustered in CGIs are generally protected from DNA methylation [16, 28]. CGIs are coupled to stable and inherited patterns of transcriptional activity and may adopt different DNA methylation profiles. Promoter CGIs hypermethylation is genome-wide correlated to gene repression [29, 30]. This silencing is explained by the spatial impediment of the transcription factor binding sites for their accessibility [31], usually based on the attachment of methyl CpG-binding domain (MBD)-containing family of proteins to methylated CpGs (Fig. 1a) [40]. Moreover, while the methylation of nucleosome-bound DNA increases their rigidity and compaction, methylation of nucleosome-free DNA reduces the affinity of DNA to enroll into nucleosomes, shifting nucleosome positioning around TSSs [41]. Across mammals, DNA methylation adopt tissue-specific patterns [42]. Accordingly, around 5% of the CGIs are differentially methylated, being essential for tissue-specific expression profiles and embryonic development [43].

In human genome, some of the CGIs not associated to promoter regions of known protein-coding or noncoding genes (in intergenic or intragenic loci) were found to overlap enhancers. Associated with transcription regulators, their hypermethylation is linked to the loss of enhancer marks and concomitant transcriptional silencing of the target genes [44]. Although the majority of the epigenetic studies concerning DNA methylation are focused in CGIs, their upstream regions up to 2 Kb, termed *CGI shores*, are also a focus of increasing interest. They present less density of CpG dinucleotides but are more susceptible to hypermethylation and hypomethylation events, correlating strongly with transcriptional activity [45, 46].

DNA methylation is intimately linked with gene silencing, allowing the conception of successful correlations in normal and unhealthy tissues. In terms of a cause–effect relationship, the molecular mechanism behind the coexistence of these events is still an unsolved issue. The hypotheses for this open question consider that DNA methylation is able to inhibit transcription factor binding, leading to gene silencing; or that gene repression caused by chromatin modifications are consequently reflected and stabilized by DNA methylation. A third hypothesis consider a more complex panorama where both regulatory mechanisms are relatively independent, functioning on the same target loci (promoters), but having no causal relationship [47].

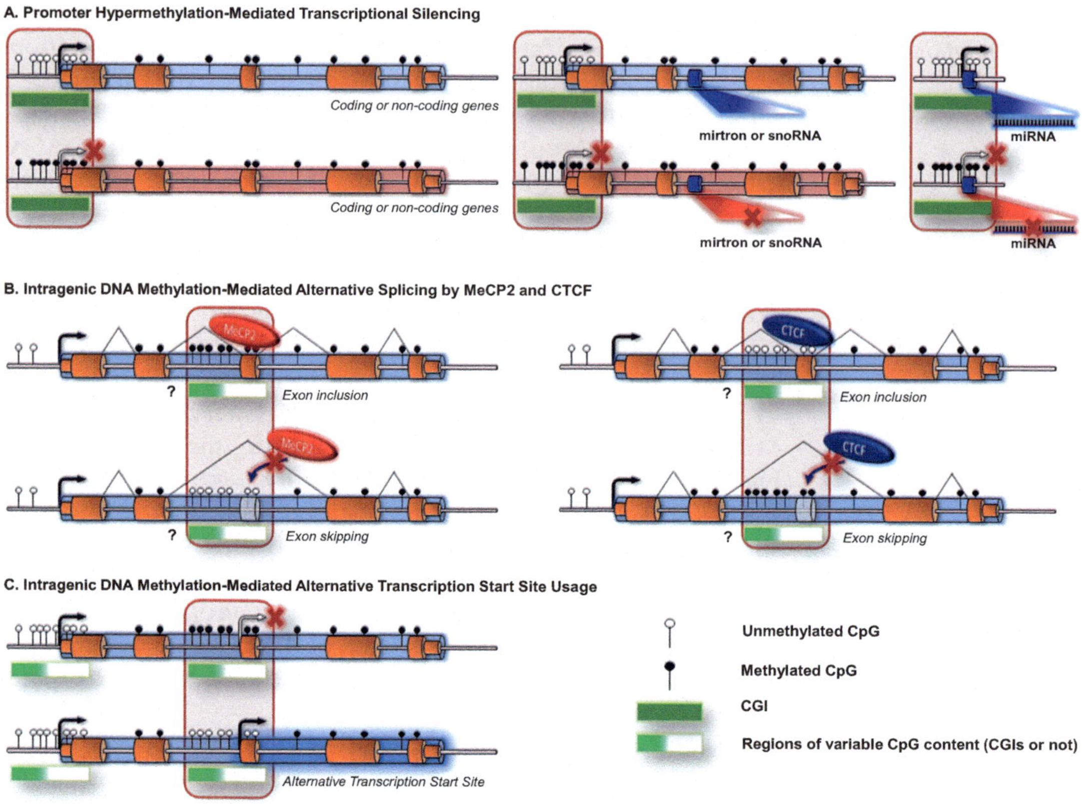

Fig. 1 Model of canonical and noncanonical functions of CpG islands in gene regulation. (**a**) Differentially methylated CGIs in the promoter region of coding and noncoding genes are associated with specific expression profiles. CGIs can be associated with the transcription of more than one gene in both sense and antisense orientations. Intronically, they could also encode distinct RNA entities, such as miRNAs (mirtrons) and small nucleolar RNAs (snoRNAs) [32–35]. (**b**) The intragenic CpG methylation can mediate alternative splicing by regulating MeCP2 and CTCF binding. MeCP2 and CTCF specifically bind to methylated and unmethylated DNA, respectively, promoting exon inclusion by decreasing the elongation rate of Pol II [36]. Some evidences point out that the methylation of CGIs could be associated with alternative splicing events [37], but the existence of a role of intragenic CGIs in alternative splicing is still unknown [38]. (**c**) Model showing the intragenic methylation-mediated activation of alternative TSSs [39]. Note that in all the represented models, density and distribution of CpG sites are merely illustrative; sparse CpGs are typically hypermethylated [15]

2.2 Alternative Splicing

Alternative splicing is a transcriptomic mechanism that generates multiple transcripts from a unique gene, increasing the coding capacity of the human genome. Since the large majority of the human genes undergo alternative splicing, they are responsible for transcriptomic and proteomic variability. This transcriptional mechanism is differentially synchronized across different tissues and developmental stages, controlling mRNA stability, translational activity, and subcellular localization [48, 49]. Abnormal splicing events result in developmental defects and illustrate disorders such as neurodegenerative diseases and cancer [50].

The fact that splicing can occur cotranscriptionally [51] points out that this mechanism could be customized by epigenetic changes. In certain cellular contexts, the chromatin structure can influence splicing by shifting the elongation rate of Pol II, increasing or decreasing the inclusion of alternative exons [52]. In turn, DNA looping is responsible for alternative promoter selection and pre-mRNA splicing [53, 54]. Moreover, in addition to nucleosome occupancy, DNA methylation is able to distinguish exons from introns, as exons tend to present higher methylation levels. Curiously, human brain present one-third of all the intragenic CGIs methylated [55], suggesting a role in the prevention of transcription from ambiguous alternative internal promoters [56]. Likewise, constitutively spliced exons have higher levels of DNA methylation than the alternative ones, indicating that DNA methylation may promote exon inclusion [57]. Accordingly, the hypomethylating drug 5-Aza-2′-deoxycytidine treatment leads to exon-skipping events [58].

Three known proteins are responsible to mediate the intragenic DNA methylation code into (alternative) splicing patterns, namely MeCP2, CTCF, and HP1. While MeCP2 specifically binds to methylated DNA, methylation of CpG dinucleotides within a CTCF binding site inhibits its attachment. Hence, since DNA binding of these two factors facilitate exon inclusion by decreasing the elongation rate of Pol II, DNA methylation in their binding loci results in exon inclusion or skipping, respectively (Fig. 1b) [36]. Alternatively, an enrichment in TET-catalyzed 5-hydroxymethylcytosine (5hmC) and 5-carboxycytosine (5caC) promote an enhanced CTCF binding, leading to alternative exon inclusion [59]. At chromatin level, HP1 protein is indirectly recruited to alternative exons, upon histone H3K9 trimethylation (H3K9me3), induced by DNA methylation. In turn, HP1 recruits splicing factors ultimately transferred onto the mRNA precursor [36]. Notably, HP1 binding can either enhance or hamper exon recognition, depending on its genomic location, upstream (intron) or in the alternative spliced exon itself, respectively [60]. For example, the *Drosophila* HP1 homolog Rhino attaches to a nuclear complex that suppresses piRNA precursor splicing [61]. In addition, ChIP-Seq analysis established a significant association between HP1α and CTCF close to alternative spliced exons, emphasizing a putative DNA binding motif for HP1α [62].

Highlighting their possible role in cell differentiation, differentially methylated regulatory regions are known to be in charge of differentially spliced transcripts. They characterize different developmental stages and tissues, being also involved in some pathological conditions. For instance, it was shown that DNA methylation blocks CTCF binding to *CD45* exon 5, preventing RNA polymerase II pausing and exon 5 inclusion [63]. Since *MeCP2* mutations are characteristic of Rett syndrome, studies conducted in a mouse model of this disorder showed an abnormal alternative splicing profile. However, in vitro assays failed to demonstrate the MeCP2-dependent increase in exon inclusion upon DNA methylation [64]. Conversely, it was demonstrated that knockdown of MeCP2 or drug-induced hypomethylation, inhibiting exonic MeCP2 binding, results in abnormal exclusion of alternative exons, highlighting the significance of MeCP2 recruitment and intragenic DNA methylation in exon genomic structure [57].

The large majority of methylated CGIs are located in intergenic and intragenic loci. Interestingly, the last ones were demonstrated to be associated to tissue-specific transcripts derived from an alternative TSS usage (Fig. 1c) [55]. Moreover, high-resolution genome-wide analysis established a strong association between DNA methylation and alternative splicing, in a tissue-specific manner. These transcriptional events were coupled to different genomic regions comprising those related to active promoters, CTCF binding peaks and transcription factor binding peak motifs. In a slighter level, the methylation of the CpG sites overlapping CGIs was also associated with alternative splicing [37].

2.3 Transposable Elements, Imprinting, and X-Chromosome Inactivation

Genomically, the deleterious effect of parasitic sequence reactivation, including translocations, chromosome instability, and gene disruption, is blocked by their considerable DNA methylation [65, 66]. Evolutionarily, the introduction of transposable elements in the eukaryotic genome overlapped with the emergence of epigenetic suppressive mechanisms able to stop their mobility and restrict their expression [67]. Nevertheless, the permissive nature in terms of DNA methylation changes in some of these regions may have evolved to establish tissue-specific regulatory networks, where they lack DNA methylation and exhibit enhancer activity [68].

Biologically, DNA methylation is also involved in the epigenetic silencing of other genomic regions, by mechanisms that are thought to be coevolutional, namely X-chromosome inactivation and genomic imprinting [69, 70]. The last mechanism reveals the meticulous molecular nature of DNA methylation, by comprising the hypermethylation-mediated transcriptional silencing of one of the parental alleles across different chromosomal locations [71]. Across all autosomes, more than one thousand of CpG sites were found to be differentially methylated between men and women. These differentially methylated loci were found to be enriched in imprinted genes, concentrated within CGI shores and associated with differential transcriptional activity [72]. Normal genomic imprinting is accurately regulated sustaining an allele-specific maternal or paternal transcription, often associated with allele-specific methylation at the corresponding CGIs [73]. In healthy cells, maternal allele-specific methylation at the promoter-associated CGI of the *KCNQ1OT1* antisense noncoding RNA controls its transcriptional activity, being responsible for its exclusive paternal expression. In turn, this noncoding RNA abrogates the expression of several other genes that are paternally imprinted, around the imprinting control region 2 (ICR2) [74]. Interestingly, studies conducted in mice suggested that *Kcnq1ot1* transcript is implicated in guiding and preserving the DNA methylation, repressing ubiquitously imprinted genes at all developmental stages [75].

Generally X-inactivation establishes a comparable gene expression pattern among both sexes. Exceptionally, a subset of human X-linked genes escapes this epigenetic mechanism, allowing a biallelic expression with comparative higher level of particular transcripts in females. The assessment of the DNA methylation in human neuronal cells focused in these genes showed a promoter CpG methylation decrease and, by contrast, a substantial increase in intragenic non-CpG methylation [76]. Apart from brain, the escape from X-inactivation was also highlighted by tissue-specific unmethylated CGIs [77]. The genomic loci of the genes that escape X-inactivation are not coated by the noncoding RNA *Xist*, an essential controller of this mechanism [78]. This transcript is expressed exclusively from the inactive X chromosome, being stably repressed on the active one in females and in the unpaired X chromosome in males [79]. Importantly, X-inactivation maintenance is guaranteed by the cooperation between *Xist* RNA expression, hypoacetylation of histone H4, and hypermethylation of CGIs [80]. Curiously, the transcriptional activity of *Xist* promoter is itself regulated by its CGI-associated DNA methylation dependent on two other noncoding RNAs with antagonistic effects, *Tsix* promoting DNA methylation and *Ftx* correlating with its loss [81, 82]. It was also observed that Mbd2 protein binds to the methylated CGI, contributing to DNA methylation-directed silencing of the *Xist* transcript in mouse fibroblasts [83].

3 Epigenetic Maintenance of CpG Methylation Status

Mammalian DNA methylation is an epigenetic mark inherited over the mitotic and meiotic cell divisions [12]. It is established and maintained through the functional combination of three main DNA methyltransferase enzymes called DNMT1, DNMT3A, and DNMT3B. Importantly, cell differentiation and mammalian development are supported by their unique and overlapping functions. DNMT3A and DNMT3B are responsible for the de novo methylation, being essential for mammalian development at very early stages [84, 85]. Nevertheless, these events are not exclusive of cell differentiation and development. For instance, during endotoxin tolerance, TNFα promoter hypermethylation-mediated transcriptional silencing is mediated by the H3K9 histone methyltransferase G9a that by H3K9 dimethylation mediate HP1 binding that in turn recruits

DNMT3A/B [86]. By contrast, DNMT1 guarantees DNA methylation maintenance, binding preferentially to the hemimethylated double-strands and restoring the DNA methylation status at CpG dinucleotides after DNA replication. Thus, this enzyme is responsible for copying the methylation patterns from the old strands to the new ones [87, 88]. All the above mentioned DNMTs have an essential function, since their lack results in impaired embryonic development or lethality [85, 89].

3.1 Neither Head Nor Tail: 5-Hydroxymethylation and Non-CpG Methylation

The discovery of 5hmC across the genome introduced new reflections about DNA methylation, considered as a quite invariable DNA modification. 5hmC is regarded as an intermediary in DNA demethylation and as a signal for transcriptional regulation [90, 91]. Thus, DNA methylation reversibility and methylome shaping brought up new issues in cell differentiation, embryonic development, and cancer fields [92]. In specific tissues, the 5mC to 5hmC frequent dynamic modification is the outcome of 2-oxoglutarate- and Fe(II)-dependent oxygenases TET1, TET2, and TET3 activity [92, 93]. These proteins were first described in human myeloid malignancies [94–96]. The current scenario suggests that while Tet1 and Tet2 perform crucial roles in pluripotency, Tet3, not present in embryonic stem cells, but upregulated in their differentiation, is probably an additional mediator of hydroxymethylation in differentiated cells, simultaneously with Tet1 or Tet2 [97, 98].

5hmC enrichment plays a fundamental role in development, assuring specific gene expression patterns, being also associated with alternative exon inclusion [59]. Genome-wide analysis, in human fetal and adult cerebellum, showed that 5hmC is intragenically enriched at 5′-UTRs and exons, being depleted at introns and intergenic regions. Suggesting a possible role in transcriptional regulation, this modification is strongly enriched at CGIs and CGI shores. Notably, the most differentially hydroxymethylated regions during human cerebellum development are associated with CGI shores [99]. DNA hydroxymethylation was also shown to play an important role during cardiac maturation. In cardiomyocites, 5-hmC is enriched at gene bodies and specific distal regulatory regions. Overall, CG regions containing this epigenetic mark in neonatal cardiomyocites become demethylated during development. For instance, the mouse *Myh7* gene was suggested to be regulated either by its gene body or by its associated enhancer hydroxymethylation. Accordingly, *Myh7* expression was associated to the decrease of 5-hmC in the first mentioned locus, during cardiac maturation; and to its increase on its intergenic enhancer region, in hypertrophic cardiomyocites [100].

CGI hydroxymethylation plays a role in the homeostatic maintenance of their methylation status. For instance, allele-specific p16 CGI methylation is escorted by focal de novo methylation, hydroxymethylation, and demethylation [101]. Moreover, the antifibrotic activity of the bone morphogenic protein 7 (BMP7) in a kidney fibrosis model was suggested to arise primarily from *Tet3* expression reactivation. The hypothesis suggests that Tet3-derived hydroxymethylation of the epigenetically repressed *Rasal1* CGI leads to its demethylation and associated transcriptional re-expression [102]. Similar results were replicated in cardiac fibrosis [103]. In hepatic tissue, differentially DNA hydroxymethylation was found to characterize chronic alcohol consumption and aging. While a non-CGI region located about 1 kb upstream of the *leptin receptor* gene (*Lepr*) showed a reduced level of hydroxymethylation and related transcriptional repression, in aging; the CGI overlapping the promoter region of the glucocorticoid receptor gene, *Nr3c1*, was documented to increase its hydroxymethylation and transcriptional activity, upon chronic alcohol consumption in young mice [104].

In mammals, the DNA methylation was the first epigenetic mark described in CpG dinucleotides, being their location continuously depicted at single-base resolution, in different cellular contexts. The dynamic balance between the DNA methylation and the functional effects at the correspondent regulatory regions is bringing new pieces to fill the transcriptional regulation puzzle [105]. Recently, unknown pieces were exposed by the existence of DNA methylation in a non-CpG context. CpH (H = Adenine/Cytosine/Thymine) dinucleotides were discovered in embryonic stem cells, disappearing upon induced differentiation and restored in induced pluripotent stem cells [106]. Additionally, neuronal maturation involves de novo CpH methylation established by DNMT3A, suggesting its significance as a key epigenetic player in the nervous system [107].

4 Disruption of CpG Island Methylation in Cancer

During cancer progression, genetic and epigenetic acquired changes cooperate to create variability in cell populations. The permanent cellular Darwinian evolutionary course is responsible for selecting cells with advantageous features [108–110]. The epigenetic landscape of a cancer cell is usually characterized by a genome-wide DNA hypomethylation simultaneous to specific focal hypermethylation events in promoter regions, altered histone tail modifications patterns and abnormal expression of chromatin-modifying enzymes [32, 111]. Demethylation and de novo methylation could be functionally understood as equivalent to gain- and loss-of-function genetic mutations, respectively. These events have been described in an increasing number of genes, according to the original tissue and tumor stage, emphasizing their functional significance (Fig. 2). Genome-wide studies highlighted a new

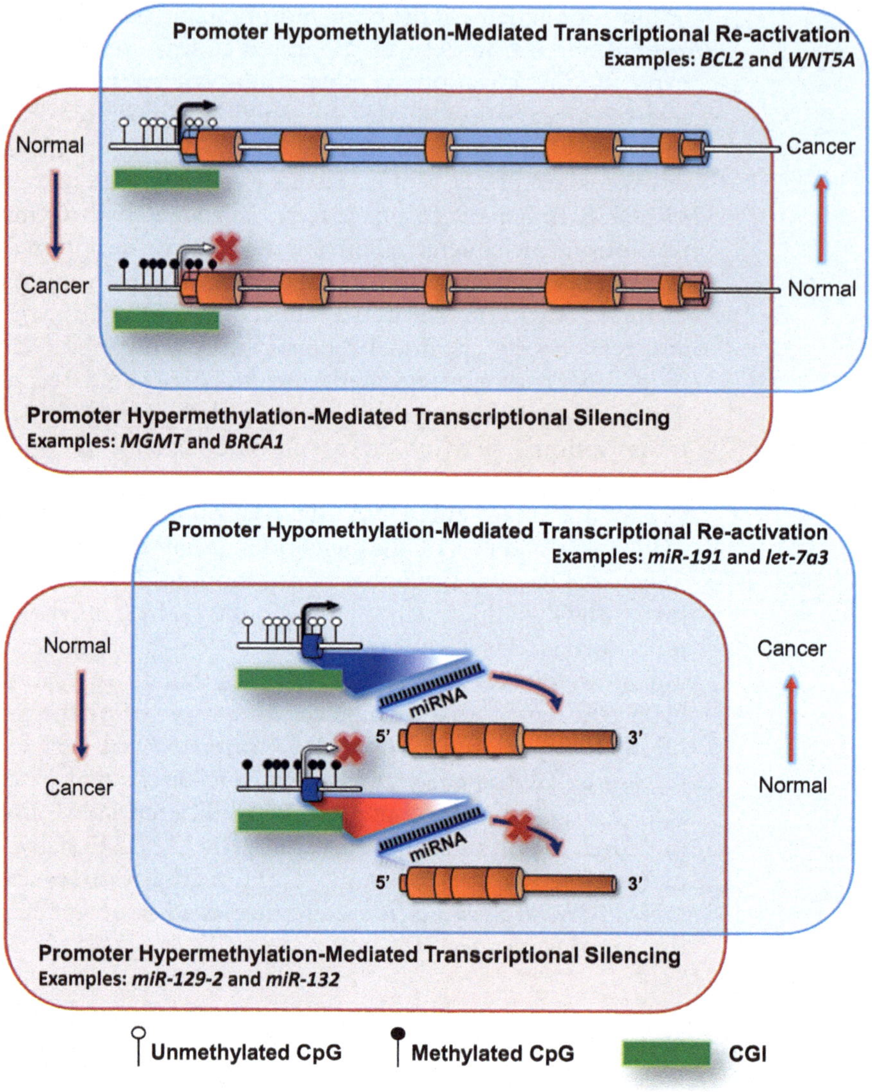

Fig. 2 Canonical CpG island methylation-mediated gene transcriptional regulation in cancer. Model showing the promoter hypomethylation-mediated transcriptional re-activation and promoter hypermethylation-mediated transcriptional silencing of both coding and noncoding genes, in normal and cancer tissues. Differentially methylated CGIs are responsible for the activation and silencing of cancer-related genes. Note that density of CpGs at CGIs is merely illustrative. DNA methylation outside the differentially methylated regions is not represented

generation of biomarkers for diagnosis, prognosis, and predictors of drug response [112].

Cancer transcriptional epigenetic reprogramming has been approached mainly in the promoter region of specific target genes, highlighting their CGI hypermethylation and gain of nucleosomes [113]. However, the knowledge about these epigenetic mechanisms at distal regulatory elements, such as enhancers and insulators, is limited. Importantly, enhancers direct the expression of distal positioned genes through chromatin looping, while insulators are genetic barrier elements that restrain the invasion of neighboring genomic condensed chromatin. In cancer, the epigenetic silencing or activation of these regulatory elements occur in parallel with nucleosome occupancy changes. This epigenetic plasticity is characterized by an aberrant gain of nucleosomes at enhancer and insulator-associated nucleosome-depleted regions, simultaneous to DNA methylation susceptibility, contributing to an abnormal epigenetic architecture. In contrast, a landscape characterized by loss of nucleosomes is coupled to DNA hypomethylation and epigenetic activation of these distal regulatory elements [114]. Curiously, CGIs overlapping with enhancer regulatory regions are more prone to DNA methylation changes than typical promoter CGIs, suggesting a stronger role in cancer [44]. Among all the identified enhancers, a subset of super-enhancers, defined as large clusters of transcriptional enhancers, was also identified. These regulators play an important role in shaping the transcriptional landscape that directs cell identity and disease, including cancer [44, 115]. Abnormal DNA methylation at these nonproximal promoter regions was described in cancer. Human tumors undergo a shift in super-enhancer DNA methylation profiles that are translated into transcriptional repression or re-expression of the corresponding target genes. According to the hypothesis that transcription factor binding affects the DNA methylome of regulatory regions, it was observed an association between DNA hypomethylation and transcription factor occupancy [116].

DNA methylation is not restricted to transcriptional regulation. Methylated CpGs increase the potential for cytosine to thymine transition mutations, being responsible for more than 30% of all known disease-related point mutations [117].

4.1 DNA Hypomethylation

The DNA hypomethylation was one of the first epigenetic anomalies discovered in cancer, distinguishing human tumors from the correspondent healthy tissue [118]. Interestingly, a global DNA hypomethylation was shown in transgenic mice bearing the hypomorphic *DNA methyltransferase 1* (*Dnmt1*) allele. It was also verified that DNA hypomethylation induced chromosome instability and tumor development [119, 120]. These genome-wide events are known to occur in repetitive sequences, CpG-poor promoters, CpG-rich promoters and introns [121, 122]. LINE-1 and latent viral sequences integrated in the genome, typically silenced by DNA methylation, can be reactivated by DNA hypomethylation, promoting cancer progression [123, 124]. Genomic instability derived from retrotransposon activation is illustrated by deletions, translocations and chromosomal rearrangements [125]. Importantly, aberrant transcriptional changes can be derived from mutagenic insertions in regulatory regions or within genes, by changing

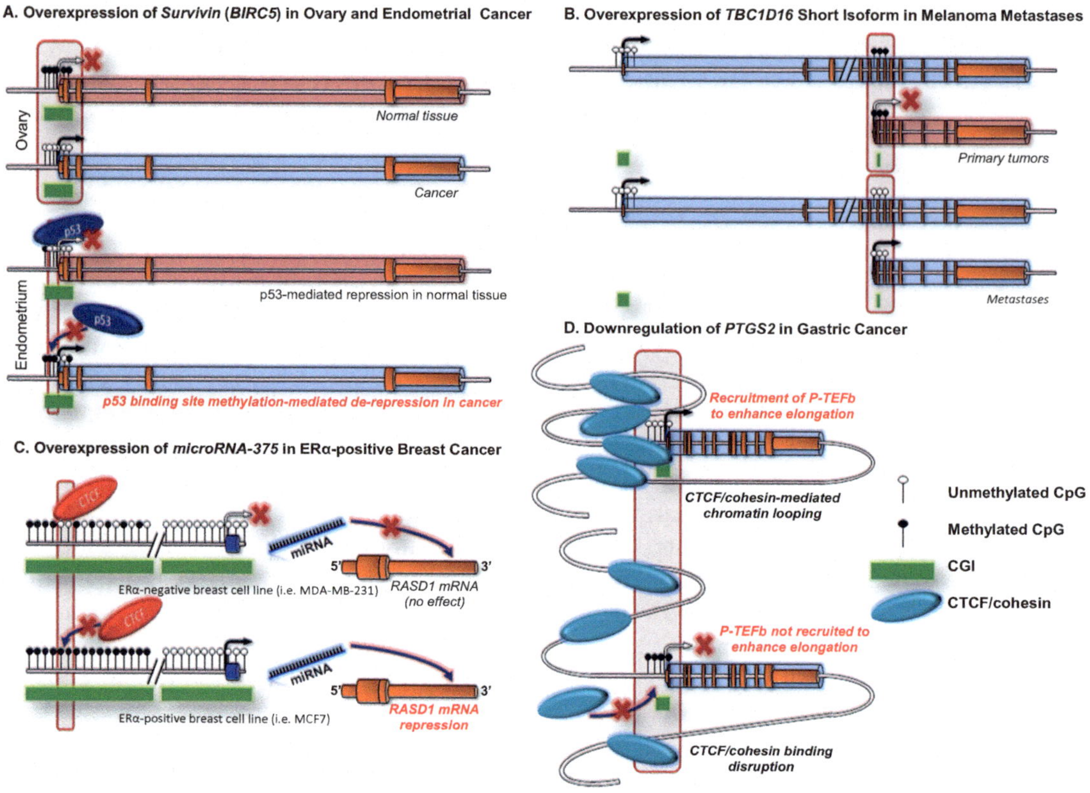

Fig. 3 Noncanonical examples of CpG island methylation-mediated gene transcriptional regulation in cancer. (**a**) *Survivin* (*BIRC5*) is overexpressed in both ovary and endometrial cancer. In ovary cancer, the standard demethylation of the CGI overlapping the promoter region of *BIRC5* leads to its overexpression [131]. However, the promoter CGI methylation at specific p53 binding sites blocks its attachment, leading to the derepression of *BIRC5* in endometrial cancer [132]. (**b**) The epigenetic activation of a cryptic *TBC1D16* transcript, through the DNA hypomethylation of an intragenic CGI contributes to melanoma progression, keeping the expression of the larger transcript [39]. (**c**) *miR-375* is overexpressed in the ERα-positive breast cancer cell line MCF7. It is suggested that this happens through DNA hypermethylation at specific CTCF binding sites overlapping one of its promoter CGIs. The inhibition of CTCF union would permit the expression of the miRNA, leading to RASD1 silencing. By contrast, the ERα-negative breast cancer cell line MDA-MB-231 shows lower expression of *miR-375* and hypomethylated CTCF binding sites [133]. (**d**) The CTCF/cohesion-mediated high-order chromatin structure leads to the recruitment of P-TEFb, allowing *PTGS2* transcription. In gastric cancer, promoter CGI methylation comprises CTCF binding sites, leading to the disruption of chromatin looping, repressing *PTGS2* expression [134]. Note that density of CpGs at CGIs is merely illustrative. DNA methylation outside the differentially methylated regions is not represented

their alternative splicing, or creating new exons or polyadenylation sites [67, 126]. Several studies underline the reactivation of transposable elements leading to singular patterns of transcriptional activity, loss of heterozygosity and aneuploidy [127–130]. Promoter CGI hypomethylation-dependent reactivation of potential oncogenic transcripts, commonly silenced in normal cells, can be observed in several cancer types (Fig. 3a, *upper panel* and Table 1).

Table 1
Hypomethylated CpG islands in cancer (non-exhaustive list of examples)

Gene	CpG island location	Functional consequence	Cancer type/ reference
ABCB1	Promoter	Upregulation	Leukemia [135]
ANK1/MIR486	Promoter	Upregulation	Pancreas [33]
BCL2	Promoter	Upregulation	Leukemia [136]
BIRC5	Promoter	Upregulation	Ovary [131]
BNIP3	Promoter	Upregulation	Breast [137]; Prostate [138]
CAV1	CGI shore/promoter	Upregulation	Breast (ERα-negative) [139]
CYP2W1	Promoter	Upregulation	Colorectal [140]
DNMT3B	Promoter	Upregulation	Gliomas [141]; Breast [142]
HK2	CGI shore/promoter	Upregulation	Liver [143]
MAL	Promoter	Upregulation	Ovary [144]
MIR191	Promoter	Upregulation	Liver [145]
MIR519D	miRNA cluster CGI located 17.6 kb upstream	Upregulation	Liver [146]
MIRLET7A3	Promoter	Upregulation	Lung [147]
MUC1	Promoter	Upregulation	Breast [148]
POMC	Promoter	Upregulation	Thymus [149]
PRDM16 (alternative isorform MEL1S)	Gene body	Upregulation	Leukemia [150]
PROM1	Alternative transcripts promoter	Upregulation (alternative transcripts)	Gliomas [151, 152]
R3HDM1/MIR128A	Promoter	Upregulation (miRNA and host gene)	Leukemia [153]
ROS1	Promoter	Upregulation	Gliomas [154]
RRAS	Promoter	Upregulation	Stomach [155]
SHH	Promoter	Upregulation	Colorectal [156]
SNCG	Promoter	Upregulation	Breast [157]; Ovary [157, 158]
SOX2	Promoter	Upregulation	Glioblastoma [159]
ST3GAL2	Promoter	Upregulation	Prostate [160]

(continued)

Table 1 (continued)

Gene	CpG island location	Functional consequence	Cancer type/ reference
TBC1D16	Alternative transcript promoter	Upregulation (alternative transcript)	Skin [39]
WT1/WT1-AS	Promoter	Upregulation	Leukemia [34]
WNT5A	Promoter	Upregulation	Prostate [161]

The epigenetic re-expression of oncogenes include both coding and noncoding genes such as the *related RAS viral oncogene homolog* (*RRAS*) in gastric cancer [155], the *wingless-type MMTV integration site family, member 5A* (*WNT5A*) in prostate cancer [161], the *miR-191* in hepatocellular carcinoma [145], and the *miR-128a* in T-cell leukemias (Fig. 2) [153].

Recently, it was demonstrated that the intragenic CGI hypomethylation-mediated reactivation of a shorter isoform of the *TBC1 domain family, member 16* (*TBC1D16*) exacerbate melanoma growth and metastasis, being connected to a poor clinical outcome (Fig. 3b) [39]. Imprinting is another genomic mechanism affected by DNA hypomethylation, triggering the erroneous transcription of maternal or paternal imprinted loci. In cancer, several studies have addressed the loss of imprinting in genomic regions with different CpG dinucleotide content, unsurprisingly affecting CGIs. For instance, loss of imprinting with consequent upregulation of the *IGF2* gene was studied in prostate [162], colorectal [163], and Wilms tumors [164]. Particularly, in breast cancer it was noticed that one of its intragenic CGIs is affected [165]. Loss of imprinting in an intronic CGI of *RB1* locus in hepatocellular carcinoma was also documented to increase the expression of an alternative transcript and to deregulate *RB1* expression [166].

CGI shores overlapping tissue-specific differentially methylated regions are associated with epigenetic reprogramming and cancer. Their hypomethylation have been associated with transcriptional activation of *caveolin-1* (*CAV1*) in highly aggressive breast cancer [139] and *hexokinase 2* (*HK2*) in hepatocellular carcinoma. This last study revealed an additional opposing regulatory mechanism of *HK2* expression in some patients. Despite the loss of methylation in the CGI shore, the CGI itself, unmethylated in healthy tissue, can experience DNA methylation through a CpG island methylator phenotype, maintaining the expression of HK2 repressed [143].

4.2 DNA Hypermethylation

In cancer, the nonrandom hypermethylation fingerprint in the promoter region of tumor suppressor genes is responsible for their inactivation (Table 2). It affects one or both alleles, depending

Table 2
Hypermethylated CpG Islands in cancer (non-exhaustive list of examples)

Gene	CpG island location	Functional consequence	Cancer type/ reference
AR	Promoter	Downregulation	Prostate [167]
BIRC5	Promoter	Upregulation (by DNA methylation inhibition of p53-mediated repression)	Endometrium [132]
BMP6	Promoter	Downregulation	Prostate [168]
BRCA1	Promoter	Downregulation	Breast, Ovary [169]
CADM1	Promoter	Downregulation	Colorectal [170]
CDH1	Promoter	Downregulation	Breast, Prostate [171]
CDH11	Promoter	Downregulation	Skin and Head and Neck metastases [172]; Bladder [173]
CDKN2A	Promoter	Downregulation	Stomach [174]; Liver [175]; Esophagus [176]
DAPK1	Promoter	Downregulation	Leukemia [177]; Bladder [178]; Head and Neck [179]
ERG	Promoter	Downregulation (Specific isoform)	Leukemia [180]
MGMT	Promoter	Downregulation	Colon, Gliomas [181]; Esophagus [176]
MIR129-2	Promoter	Downregulation	Endometrium [182]
MIR132	Promoter	Downregulation	Prostate [183]
MIR375	Promoter	Upregulation (by DNA methylation inhibition of CTCF-mediated repression)	Breast [133]
MIR9	Promoter	Downregulation	Stomach [184]
MLH1	Promoter	Downregulation	Colorectal [185]; Endometrium [186]; Stomach [187]
NSD1	Promoter	Downregulation	Neuroblastoma and Glioma [188]
PTGS2	Promoter	Downregulation (by DNA methylation disruption of CTCF/cohesin-mediated high-order chromatin structure)	Stomach [134]

(continued)

Table 2 (continued)

Gene	CpG island location	Functional consequence	Cancer type/ reference
RASSF1 (Isoform A)	Promoter	Downregulation	Lung [189]
RASSF5	Promoter	Downregulation	Lung [190]; Esophagus [191]
RB1	Promoter	Downregulation	Retinoblastoma [192]
SEPT9	Promoter of an alternative transcript	Downregulation	Colon [193]
SFRP1,2,5	Promoter	Downregulation	Colon [194]
SLFN11	Promoter	Downregulation	Ovary, Lung [195]
SNORD123/ LOC100505806/ SEMA5A	Promoter	Downregulation	Colorectal [35]
TFPI2	Promoter	Downregulation	Stomach [196]
TIMP3	Promoter	Downregulation	Stomach [197]; Endometrium [198]

on the inactivation (mutation) status of each allele [199]. As a result, the main cellular pathways such as cell cycle control, DNA repair, apoptosis, angiogenesis, cell adhesion, and metastases could be compromised. The inactivation of particular genes confers thereby a proliferative advantage, resulting in clonal selection [200]. According to the Knudson's two-hit hypothesis, this failure is functionally correspondent to a loss of function gene mutation with a similar or even higher frequency [199, 201]. Interestingly the first described tumor suppressor gene, the *retinoblastoma 1 gene* (*RB1*) [202], can be inactivated by DNA hypermethylation [192]. Other examples involve the promoter DNA hypermethylation of *O-6-methylguanine-DNA methyltransferase* (*MGMT*) in colorectal cancer and gliomas [181]; *MLH1* in colorectal [185], endometrial [186], and gastric [187] cancers; and *BRCA1* in breast and ovarian cancer (Fig. 2) [169]. CGI shore hypermethylation was also identified in pancreatic adenocarcinomas [203] and bladder cancer [204].

The repression of genes holding CGIs in their promoter region is not always correlated with their DNA hypermethylation. Genomically, when insulators are present between enhancers and promoters, they can operate as blocking elements, preventing the expression of the target genes [205]. The transcriptional repressor CTCF is known to bind insulators and to inhibit enhancer activities, being also documented in CGIs [206]. The existence of an inverse correlation among DNA methylation and CTCF occupancy at two crucial positions within its recognition sequence, suggest its mediated transcriptional regulation [207]. In ERα-positive breast cancer cell lines, it was demonstrated that *miR-375* upregulation is linked to DNA hypermethylation in a distal part of an upstream CGI. It was observed that methylation of a particular locus inhibits CTCF binding, increasing the expression of this noncoding RNA (Fig. 3c) [133]. Moreover, DNA hypermethylation of a CTCF-binding site located 6.3 kb downstream of the transcriptional start site, in a CGI located mostly in the third intron of *GAD1* gene leads to its overexpression. The methylation-dependent inhibition of CTCF binding at this locus prevents the formation of a chromatin loop with the 5′ untranslated region and the recruitment of the polycomb repressive complex 2 to the promoter, allowing transcription [208]. Oppositely, the CGI hypermethylation-related silencing of *COX-2* gene, found in gastric cancer patients with better clinical prognosis [209], was suggested to be an outcome of the methylation-dependent dissociation of CTCF/cohesin complex from this locus. The hypothesis considers that DNA methylation disrupts chromatin looping interactions, abolishing the enrichment of transcriptional factors (Fig. 3d) [134]. In endometrial cancer, the promoter CGI methylation at specific p53 binding sites blocks its attachment, leading to the expression of *Survivin* (Fig. 3a, *bottom panel*) [132].

The hypermethylation of CGIs was associated to an increased expression of *DNMT1*, *DNMT3A*, and *DNMT3B* that characterizes several cancer types [210]. Curiously, polymorphisms in *DNMT3B* gene promoter region are linked to its increased activity and associated to an earlier onset of hereditary non-polyposis colorectal cancer [211] and an increased risk of lung [212], breast [213], and prostate [214] cancers. In cancer, there are not only changes in the activity of DNMTs, but also in their recruitment to specific genes. For instance, the promyelocytic leukemia-retinoic acid receptor alpha fusion oncoprotein (PML-RARA) interacts with DNMT1 and DNMT3A, leading to the promoter DNA methylation-dependent repression of the *retinoic acid receptor β2 gene* (*RARβ2*), in leukemia [215].

4.3 DNA Hydroxymethylation

Cancer is characterized by several aberrant DNA hypermethylation events, conferring growth advantages to malignant cells, for example, by an associated repression of tumor suppressor genes. The origin of these events, not always associated to the activity of DNA methyltransferases, is being the subject of different hypotheses.

In hematological malignancies, it was shown that the presence of a defective TET2 activity, by miRNA-mediated silencing, mutations or deletions, is associated with an inefficient conversion of 5mC to 5hmC [217–219]. Loss and redistribution of 5hmC genome-wide was also linked to several solid tumors including prostate, breast, colorectal cancers, esophageal squamous cell carcinoma, hepatocellular carcinoma, and cholangiocarcinoma [219–221]. A recent study showed that the 5-methylcytosine oxidation catalyzed by TET enzymes is decreased under hypoxic conditions, with a subsequent DNA hypermethylation. In normoxic conditions, their induced knockout decreases 5hmC levels. However, it was demonstrated that in hypoxia, 5hmC reduction arises independently of *TET* expression or other hypoxia-associated alterations. Accordingly, it was also noticed an association between the hypermethylation of tumor suppressor promoter regions and patient hypoxic tumor tissues, suggesting a selective advantage of this environment. In vitro studies confirmed a global 5hmC loss under hypoxic conditions, covering enhancers, promoters and CGIs. For instance, a decline in 5hmC levels was found near the TSSs of the tumor suppressor genes *NSD1*, *FOXA1*, and *CDKN2A* [222]. Curiously, in addition to their location at promoter regions, these genomic sites overlap or are close to CGIs that are suggested as transcriptional regulatory regions of several tumor suppressor genes [222, 223].

5 Epigenetic Diagnosis and Pharmacoepigenomics

Different epigenetic landscapes define singular cell types under healthy and pathological conditions, being responsible for diverse responses to equal endogenous or exogenous stimulus. Personalized medicine based on these cause-effect relationships is particularly important to predict drug response and achieve better clinical outcomes. In cancer, the characterization of both irreversible genetic and dynamic epigenetic modifications provides extremely relevant information for therapy selection, improving clinical outcomes.

Cancer of unknown primary has a particularly poor prognosis being dependent on the uncovering of the original primary tumor site. This prediction allows a later personalized tumor type-specific therapy with an overall survival improvement, when compared with patients treated empirically. Using a DNA methylation based assay and profiling more than ten thousand cancer samples, the diagnosis of this set of cancer samples was significantly improved. The method was based on the similarities among the specific methylation signatures of each case sample compared to those from known primary tumors, predicting the tissue of origin in 87% of the cases [224].

The methylome of a cancer cell is able to discriminate cells from different tissues. Moreover, the establishment of correlations between DNA methylation and the transcriptional activity of certain genes opens a therapeutic opportunity for personalized medicine. In glioblastoma and colorectal cancer patients, the epigenetic silencing of the *MGMT* DNA-repair gene, by hypermethylation of the CGI overlapping its promoter region, predicts the clinical better response to alkylating agents [225–227]. In the same regard, one-third of triple-negative breast cancer tumors show a promoter CGI hypermethylation associated repression of *BRCA1* gene. Interestingly, it was shown that these patients have a higher sensitivity to adjuvant chemotherapy, presenting a better survival contrasted with patients with *BRCA1*-unmethylated triple-negative tumors [228]. In contrast, promoter CGI hypermethylation of the BRCA1 interactor SRBC gene leads to its epigenetic silencing, being associated with a worse clinical outcome of colorectal cancer patients treated with oxaliplatin [229]. In addition, CGI hypermethylation-associated silencing of the putative DNA/RNA helicase SLFN11 was shown to predict resistance to cisplatin/carboplatin drugs in human cancer. Its silencing correlated with a poor outcome in ovarian or lung cancer patients upon platinum-based treatment, representing a response predictive biomarker [195].

References

1. Esteller M (2008) Epigenetics in cancer. N Engl J Med 358(11):1148–1159
2. Arney KL, Fisher AG (2004) Epigenetic aspects of differentiation. J Cell Sci 117 (Pt 19):4355–4363
3. Elango N, Yi SV (2008) DNA methylation and structural and functional bimodality of vertebrate promoters. Mol Biol Evol 25 (8):1602–1608
4. Gonzalo S, Jaco I, Fraga MF, Chen T, Li E, Esteller M, Blasco MA (2006) DNA methyltransferases control telomere length and telomere recombination in mammalian cells. Nat Cell Biol 8(4):416–424
5. Schwartz S, Meshorer E, Ast G (2009) Chromatin organization marks exon-intron structure. Nat Struct Mol Biol 16(9):990–995
6. Hollister JD, Gaut BS (2009) Epigenetic silencing of transposable elements: a trade-off between reduced transposition and deleterious effects on neighboring gene expression. Genome Res 19(8):1419–1428
7. Lee YC (2015) The role of piRNA-mediated epigenetic silencing in the population dynamics of transposable elements in Drosophila melanogaster. PLoS Genet 11(6):e1005269
8. Lyon MF (1962) Sex chromatin and gene action in the mammalian X-chromosome. Am J Hum Genet 14:135–148
9. Wutz A (2011) Gene silencing in X-chromosome inactivation: advances in understanding facultative heterochromatin formation. Nat Rev Genet 12(8):542–553
10. Reik W, Collick A, Norris ML, Barton SC, Surani MA (1987) Genomic imprinting determines methylation of parental alleles in transgenic mice. Nature 328(6127):248–251
11. Collick A, Reik W, Barton SC, Surani AH (1988) CpG methylation of an X-linked transgene is determined by somatic events postfertilization and not germline imprinting. Development 104(2):235–244
12. Bird A (2002) DNA methylation patterns and epigenetic memory. Genes Dev 16(1):6–21
13. Bird AP (1986) CpG-rich islands and the function of DNA methylation. Nature 321 (6067):209–213
14. Strichman-Almashanu LZ, Lee RS, Onyango PO, Perlman E, Flam F, Frieman MB, Feinberg AP (2002) A genome-wide screen for normally methylated human CpG islands that can identify novel imprinted genes. Genome Res 12(4):543–554

15. Lovkvist C, Dodd IB, Sneppen K, Haerter JO (2016) DNA methylation in human epigenomes depends on local topology of CpG sites. Nucleic Acids Res 44(11):5125–5132
16. Illingworth RS, Bird AP (2009) CpG islands—a rough guide. FEBS Lett 583 (11):1713–1720
17. Wu H, Caffo B, Jaffee HA, Irizarry RA, Feinberg AP (2010) Redefining CpG islands using hidden Markov models. Biostatistics 11 (3):499–514
18. Saxonov S, Berg P, Brutlag DL (2006) A genome-wide analysis of CpG dinucleotides in the human genome distinguishes two distinct classes of promoters. Proc Natl Acad Sci U S A 103(5):1412–1417
19. Ioshikhes IP, Zhang MQ (2000) Large-scale human promoter mapping using CpG islands. Nat Genet 26(1):61–63
20. Venter JC, Adams MD, Myers EW, Li PW, Mural RJ, Sutton GG, Smith HO, Yandell M, Evans CA, Holt RA, Gocayne JD, Amanatides P, Ballew RM, Huson DH, Wortman JR, Zhang Q, Kodira CD, Zheng XH, Chen L, Skupski M, Subramanian G, Thomas PD, Zhang J, Gabor Miklos GL, Nelson C, Broder S, Clark AG, Nadeau J, McKusick VA, Zinder N, Levine AJ, Roberts RJ, Simon M, Slayman C, Hunkapiller M, Bolanos R, Delcher A, Dew I, Fasulo D, Flanigan M, Florea L, Halpern A, Hannenhalli S, Kravitz S, Levy S, Mobarry C, Reinert K, Remington K, Abu-Threideh J, Beasley E, Biddick K, Bonazzi V, Brandon R, Cargill M, Chandramouliswaran I, Charlab R, Chaturvedi K, Deng Z, Di Francesco V, Dunn P, Eilbeck K, Evangelista C, Gabrielian AE, Gan W, Ge W, Gong F, Gu Z, Guan P, Heiman TJ, Higgins ME, Ji RR, Ke Z, Ketchum KA, Lai Z, Lei Y, Li Z, Li J, Liang Y, Lin X, Lu F, Merkulov GV, Milshina N, Moore HM, Naik AK, Narayan VA, Neelam B, Nusskern D, Rusch DB, Salzberg S, Shao W, Shue B, Sun J, Wang Z, Wang A, Wang X, Wang J, Wei M, Wides R, Xiao C, Yan C, Yao A, Ye J, Zhan M, Zhang W, Zhang H, Zhao Q, Zheng L, Zhong F, Zhong W, Zhu S, Zhao S, Gilbert D, Baumhueter S, Spier G, Carter C, Cravchik A, Woodage T, Ali F, An H, Awe A, Baldwin D, Baden H, Barnstead M, Barrow I, Beeson K, Busam D, Carver A, Center A, Cheng ML, Curry L, Danaher S, Davenport L, Desilets R, Dietz S, Dodson K, Doup L, Ferriera S, Garg N, Gluecksmann A, Hart B, Haynes J, Haynes C, Heiner C, Hladun S, Hostin D, Houck J, Howland T, Ibegwam C, Johnson J, Kalush F, Kline L, Koduru S, Love A, Mann F, May D, McCawley S, McIntosh T, McMullen I, Moy M, Moy L, Murphy B, Nelson K, Pfannkoch C, Pratts E, Puri V, Qureshi H, Reardon M, Rodriguez R, Rogers YH, Romblad D, Ruhfel B, Scott R, Sitter C, Smallwood M, Stewart E, Strong R, Suh E, Thomas R, Tint NN, Tse S, Vech C, Wang G, Wetter J, Williams S, Williams M, Windsor S, Winn-Deen E, Wolfe K, Zaveri J, Zaveri K, Abril JF, Guigo R, Campbell MJ, Sjolander KV, Karlak B, Kejariwal A, Mi H, Lazareva B, Hatton T, Narechania A, Diemer K, Muruganujan A, Guo N, Sato S, Bafna V, Istrail S, Lippert R, Schwartz R, Walenz B, Yooseph S, Allen D, Basu A, Baxendale J, Blick L, Caminha M, Carnes-Stine J, Caulk P, Chiang YH, Coyne M, Dahlke C, Mays A, Dombroski M, Donnelly M, Ely D, Esparham S, Fosler C, Gire H, Glanowski S, Glasser K, Glodek A, Gorokhov M, Graham K, Gropman B, Harris M, Heil J, Henderson S, Hoover J, Jennings D, Jordan C, Jordan J, Kasha J, Kagan L, Kraft C, Levitsky A, Lewis M, Liu X, Lopez J, Ma D, Majoros W, McDaniel J, Murphy S, Newman M, Nguyen T, Nguyen N, Nodell M, Pan S, Peck J, Peterson M, Rowe W, Sanders R, Scott J, Simpson M, Smith T, Sprague A, Stockwell T, Turner R, Venter E, Wang M, Wen M, Wu D, Wu M, Xia A, Zandieh A, Zhu X (2001) The sequence of the human genome. Science 291(5507):1304–1351
21. Lander ES, Linton LM, Birren B, Nusbaum C, Zody MC, Baldwin J, Devon K, Dewar K, Doyle M, FitzHugh W, Funke R, Gage D, Harris K, Heaford A, Howland J, Kann L, Lehoczky J, LeVine R, McEwan P, McKernan K, Meldrim J, Mesirov JP, Miranda C, Morris W, Naylor J, Raymond C, Rosetti M, Santos R, Sheridan A, Sougnez C, Stange-Thomann N, Stojanovic N, Subramanian A, Wyman D, Rogers J, Sulston J, Ainscough R, Beck S, Bentley D, Burton J, Clee C, Carter N, Coulson A, Deadman R, Deloukas P, Dunham A, Dunham I, Durbin R, French L, Grafham D, Gregory S, Hubbard T, Humphray S, Hunt A, Jones M, Lloyd C, McMurray A, Matthews L, Mercer S, Milne S, Mullikin JC, Mungall A, Plumb R, Ross M, Shownkeen R, Sims S, Waterston RH, Wilson RK, Hillier LDW, McPherson JD, Marra MA, Mardis ER, Fulton LA, Chinwalla AT, Pepin KH, Gish WR, Chissoe SL, Wendl MC, Delehaunty KD, Miner TL, Delehaunty A, Kramer JB, Cook LL, Fulton RS, Johnson DL, Minx PJ, Clifton

SW, Hawkins T, Branscomb E, Predki P, Richardson P, Wenning S, Slezak T, Doggett N, Cheng J-F, Olsen A, Lucas S, Elkin C, Uberbacher E, Frazier M, Gibbs RA, Muzny DM, Scherer SE, Bouck JB, Sodergren EJ, Worley KC, Rives CM, Gorrell JH, Metzker ML, Naylor SL, Kucherlapati RS, Nelson DL, Weinstock GM, Sakaki Y, Fujiyama A, Hattori M, Yada T, Toyoda A, Itoh T, Kawagoe C, Watanabe H, Totoki Y, Taylor T, Weissenbach J, Heilig R, Saurin W, Artiguenave F, Brottier P, Bruls T, Pelletier E, Robert C, Wincker P, Rosenthal A, Platzer M, Nyakatura G, Taudien S, Rump A, Smith DR, Doucette-Stamm L, Rubenfield M, Weinstock K, Lee HM, Dubois JA, Yang H, Yu J, Wang J, Huang G, Jun G, Hood L, Lee R, Madan A, Qin S, Davis RW, Federspiel NA, Pia Abola A, Proctor MJ, Roe BA, Chen F, Pan H, Ramser J, Lehrach H, Richard R, Richard McCombie W, de la Bastide M, Dedhia N, Blöcker H, Hornischer K, Nordsiek G, Richa A, Aravind L, Bailey JA, Bateman A, Batzoglou S, Birney E, Bork P, Brown DG, Burge CB, Cerutti L, Chen H-C, Church D, Clamp M, Copley RR, Doerks T, Eddy SR, Eichler EE, Furey TS, Galagan J, Gilbert JGR, Harmon C, Hayashizaki Y, Haussler D, Hermjakob H, Hokamp K, Jang W, Steven Johnson L, Jones TA, Kasif S, Kaspryzk A, Kennedy S, James Kent W, Kitts P, Koonin EV, Korf I, Kulp D, Lancet D, Lowe TM, McLysaght A, Mikkelsen T, Moran JV, Mulder N, Pollara VJ, Ponting CP, Schuler G, Schultz J, Slater G, Smit AFA, Stupka E, Szustakowki J, Thierry-Mieg D, Thierry-Mieg J, Wagner L, Wallis J, Wheeler R, Williams A, Wolf YI, Wolfe KH, Yang SP, Yeh R-F, Collins F, Guyer MS, Peterson J, Felsenfeld A, Wetterstrand KA, Myers RM, Schmutz J, Dickson M, Grimwood J, Cox DR, Olson MV, Kaul R, Raymond C, Shimizu N, Kawasaki K, Minoshima S, Evans GA, Athanasiou M, Schultz R, Patrinos A, Morgan MJ (2001) Initial sequencing and analysis of the human genome. Nature 409(6822):860–921

22. Vavouri T, Lehner B (2012) Human genes with CpG island promoters have a distinct transcription-associated chromatin organization. Genome Biol 13(11):R110
23. Riggs AD (1975) X Inactivation, differentiation, and DNA methylation. Cytogenet Cell Genet 14(1):9–25
24. Holliday R, Pugh JE (1975) DNA modification mechanisms and gene activity during development. Science 187(4173):226–232
25. Kitamura E, Igarashi J, Morohashi A, Hida N, Oinuma T, Nemoto N, Song F, Ghosh S, Held WA, Yoshida-Noro C, Nagase H (2007) Analysis of tissue-specific differentially methylated regions (TDMs) in humans. Genomics 89(3):326–337
26. Lee HS, Chen ZJ (2001) Protein-coding genes are epigenetically regulated in Arabidopsis polyploids. Proc Natl Acad Sci U S A 98(12):6753–6758
27. Suzuki H, Maruyama R, Yamamoto E, Kai M (2012) DNA methylation and microRNA dysregulation in cancer. Mol Oncol 6 (6):567–578
28. Kanwal R, Gupta S (2012) Epigenetic modifications in cancer. Clin Genet 81 (4):303–311
29. Kang SH, Bang YJ, Im YH, Yang HK, Lee DA, Lee HY, Lee HS, Kim NK, Kim SJ (1999) Transcriptional repression of the transforming growth factor-beta type I receptor gene by DNA methylation results in the development of TGF-beta resistance in human gastric cancer. Oncogene 18 (51):7280–7286
30. Baylin SB (2005) DNA methylation and gene silencing in cancer. Nat Clin Pract Oncol 2 (Suppl 1):S4–11
31. Rodriguez C, Borgel J, Court F, Cathala G, Forne T, Piette J (2010) CTCF is a DNA methylation-sensitive positive regulator of the INK/ARF locus. Biochem Biophys Res Commun 392(2):129–134
32. Portela A, Esteller M (2010) Epigenetic modifications and human disease. Nat Biotechnol 28(10):1057–1068
33. Omura N, Mizuma M, MacGregor A, Hong SM, Ayars M, Almario JA, Borges M, Kanda M, Li A, Vincent A, Maitra A, Goggins M (2016) Overexpression of ankyrin1 promotes pancreatic cancer cell growth. Oncotarget 7(23):34977–34987
34. McCarty G, Loeb DM (2015) Hypoxia-sensitive epigenetic regulation of an antisense-oriented lncRNA controls WT1 expression in myeloid leukemia cells. PLoS One 10(3):e0119837
35. Ferreira HJ, Heyn H, Moutinho C, Esteller M (2012) CpG island hypermethylation-associated silencing of small nucleolar RNAs in human cancer. RNA Biol 9(6):881–890
36. Lev Maor G, Yearim A, Ast G (2015) The alternative role of DNA methylation in splicing regulation. Trends Genet 31(5):274–280

37. Gutierrez-Arcelus M, Ongen H, Lappalainen T, Montgomery SB, Buil A, Yurovsky A, Bryois J, Padioleau I, Romano L, Planchon A, Falconnet E, Bielser D, Gagnebin M, Giger T, Borel C, Letourneau A, Makrythanasis P, Guipponi M, Gehrig C, Antonarakis SE, Dermitzakis ET (2015) Tissue-specific effects of genetic and epigenetic variation on gene regulation and splicing. PLoS Genet 11(1): e1004958
38. Deaton AM, Webb S, Kerr AR, Illingworth RS, Guy J, Andrews R, Bird A (2011) Cell type-specific DNA methylation at intragenic CpG islands in the immune system. Genome Res 21(7):1074–1086
39. Vizoso M, Ferreira HJ, Lopez-Serra P, Carmona FJ, Martinez-Cardus A, Girotti MR, Villanueva A, Guil S, Moutinho C, Liz J, Portela A, Heyn H, Moran S, Vidal A, Martinez-Iniesta M, Manzano JL, Fernandez-Figueras MT, Elez E, Munoz-Couselo E, Botella-Estrada R, Berrocal A, Ponten F, Oord J, Gallagher WM, Frederick DT, Flaherty KT, McDermott U, Lorigan P, Marais R, Esteller M (2015) Epigenetic activation of a cryptic TBC1D16 transcript enhances melanoma progression by targeting EGFR. Nat Med 21(7):741–750
40. Wade PA (2001) Methyl CpG-binding proteins and transcriptional repression. BioEssays 23(12):1131–1137
41. Perez A, Castellazzi CL, Battistini F, Collinet K, Flores O, Deniz O, Ruiz ML, Torrents D, Eritja R, Soler-Lopez M, Orozco M (2012) Impact of methylation on the physical properties of DNA. Biophys J 102 (9):2140–2148
42. Gama-Sosa MA, Midgett RM, Slagel VA, Githens S, Kuo KC, Gehrke CW, Ehrlich M (1983) Tissue-specific differences in DNA methylation in various mammals. Biochim Biophys Acta 740(2):212–219
43. Song F, Smith JF, Kimura MT, Morrow AD, Matsuyama T, Nagase H, Held WA (2005) Association of tissue-specific differentially methylated regions (TDMs) with differential gene expression. Proc Natl Acad Sci U S A 102(9):3336–3341
44. Bae MG, Kim JY, Choi JK (2016) Frequent hypermethylation of orphan CpG islands with enhancer activity in cancer. BMC Med Genomics 9(Suppl 1):38
45. Doi A, Park IH, Wen B, Murakami P, Aryee MJ, Irizarry R, Herb B, Ladd-Acosta C, Rho J, Loewer S, Miller J, Schlaeger T, Daley GQ, Feinberg AP (2009) Differential methylation of tissue- and cancer-specific CpG island shores distinguishes human induced pluripotent stem cells, embryonic stem cells and fibroblasts. Nat Genet 41 (12):1350–1353
46. Irizarry RA, Ladd-Acosta C, Wen B, Wu Z, Montano C, Onyango P, Cui H, Gabo K, Rongione M, Webster M, Ji H, Potash JB, Sabunciyan S, Feinberg AP (2009) The human colon cancer methylome shows similar hypo- and hypermethylation at conserved tissue-specific CpG island shores. Nat Genet 41(2):178–186
47. Medvedeva YA, Khamis AM, Kulakovskiy IV, Ba-Alawi W, Bhuyan MS, Kawaji H, Lassmann T, Harbers M, Forrest AR, Bajic VB (2014) Effects of cytosine methylation on transcription factor binding sites. BMC Genomics 15:119
48. Wang ET, Sandberg R, Luo S, Khrebtukova I, Zhang L, Mayr C, Kingsmore SF, Schroth GP, Burge CB (2008) Alternative isoform regulation in human tissue transcriptomes. Nature 456(7221):470–476
49. Mockenhaupt S, Makeyev EV (2015) Non-coding functions of alternative pre-mRNA splicing in development. Semin Cell Dev Biol 47-48:32–39
50. Cooper TA, Wan L, Dreyfuss G (2009) RNA and disease. Cell 136(4):777–793
51. Listerman I, Sapra AK, Neugebauer KM (2006) Cotranscriptional coupling of splicing factor recruitment and precursor messenger RNA splicing in mammalian cells. Nat Struct Mol Biol 13(9):815–822
52. Luco RF, Allo M, Schor IE, Kornblihtt AR, Misteli T (2011) Epigenetics in alternative pre-mRNA splicing. Cell 144(1):16–26
53. Monahan K, Rudnick ND, Kehayova PD, Pauli F, Newberry KM, Myers RM, Maniatis T (2012) Role of CCCTC binding factor (CTCF) and cohesin in the generation of single-cell diversity of protocadherin-alpha gene expression. Proc Natl Acad Sci U S A 109(23):9125–9130
54. Guo Y, Monahan K, Wu H, Gertz J, Varley KE, Li W, Myers RM, Maniatis T, Wu Q (2012) CTCF/cohesin-mediated DNA looping is required for protocadherin alpha promoter choice. Proc Natl Acad Sci U S A 109 (51):21081–21086
55. Maunakea AK, Nagarajan RP, Bilenky M, Ballinger TJ, D'Souza C, Fouse SD, Johnson BE, Hong C, Nielsen C, Zhao Y, Turecki G, Delaney A, Varhol R, Thiessen N, Shchors K, Heine VM, Rowitch DH, Xing X, Fiore C, Schillebeeckx M, Jones SJ, Haussler D, Marra MA, Hirst M, Wang T,

Costello JF (2010) Conserved role of intragenic DNA methylation in regulating alternative promoters. Nature 466(7303):253–257

56. Suzuki MM, Bird A (2008) DNA methylation landscapes: provocative insights from epigenomics. Nat Rev Genet 9(6):465–476
57. Maunakea AK, Chepelev I, Cui K, Zhao K (2013) Intragenic DNA methylation modulates alternative splicing by recruiting MeCP2 to promote exon recognition. Cell Res 23 (11):1256–1269
58. Ding XL, Yang X, Liang G, Wang K (2016) Isoform switching and exon skipping induced by the DNA methylation inhibitor 5-Aza-2-′-deoxycytidine. Sci Rep 6:24545
59. Marina RJ, Sturgill D, Bailly MA, Thenoz M, Varma G, Prigge MF, Nanan KK, Shukla S, Haque N, Oberdoerffer S (2016) TET-catalyzed oxidation of intragenic 5-methylcytosine regulates CTCF-dependent alternative splicing. EMBO J 35(3):335–355
60. Yearim A, Gelfman S, Shayevitch R, Melcer S, Glaich O, Mallm JP, Nissim-Rafinia M, Cohen AH, Rippe K, Meshorer E, Ast G (2015) HP1 is involved in regulating the global impact of DNA methylation on alternative splicing. Cell Rep 10(7):1122–1134
61. Zhang Z, Wang J, Schultz N, Zhang F, Parhad SS, Tu S, Vreven T, Zamore PD, Weng Z, Theurkauf WE (2014) The HP1 homolog rhino anchors a nuclear complex that suppresses piRNA precursor splicing. Cell 157 (6):1353–1363
62. Agirre E, Bellora N, Allo M, Pages A, Bertucci P, Kornblihtt AR, Eyras E (2015) A chromatin code for alternative splicing involving a putative association between CTCF and HP1alpha proteins. BMC Biol 13:31
63. Shukla S, Kavak E, Gregory M, Imashimizu M, Shutinoski B, Kashlev M, Oberdoerffer P, Sandberg R, Oberdoerffer S (2011) CTCF-promoted RNA polymerase II pausing links DNA methylation to splicing. Nature 479(7371):74–79
64. Young JI, Hong EP, Castle JC, Crespo-Barreto J, Bowman AB, Rose MF, Kang D, Richman R, Johnson JM, Berget S, Zoghbi HY (2005) Regulation of RNA splicing by the methylation-dependent transcriptional repressor methyl-CpG binding protein 2. Proc Natl Acad Sci U S A 102 (49):17551–17558
65. Walsh CP, Chaillet JR, Bestor TH (1998) Transcription of IAP endogenous retroviruses is constrained by cytosine methylation. Nat Genet 20(2):116–117
66. Liang G, Chan MF, Tomigahara Y, Tsai YC, Gonzales FA, Li E, Laird PW, Jones PA (2002) Cooperativity between DNA methyltransferases in the maintenance methylation of repetitive elements. Mol Cell Biol 22 (2):480–491
67. Slotkin RK, Martienssen R (2007) Transposable elements and the epigenetic regulation of the genome. Nat Rev Genet 8(4):272–285
68. Xie M, Hong C, Zhang B, Lowdon RF, Xing X, Li D, Zhou X, Lee HJ, Maire CL, Ligon KL, Gascard P, Sigaroudinia M, Tlsty TD, Kadlecek T, Weiss A, O'Geen H, Farnham PJ, Madden PA, Mungall AJ, Tam A, Kamoh B, Cho S, Moore R, Hirst M, Marra MA, Costello JF, Wang T (2013) DNA hypomethylation within specific transposable element families associates with tissue-specific enhancer landscape. Nat Genet 45 (7):836–841
69. Reik W, Lewis A (2005) Co-evolution of X-chromosome inactivation and imprinting in mammals. Nat Rev Genet 6(5):403–410
70. Paulsen M, Ferguson-Smith AC (2001) DNA methylation in genomic imprinting, development, and disease. J Pathol 195(1):97–110
71. Kacem S, Feil R (2009) Chromatin mechanisms in genomic imprinting. Mamm Genome 20(9-10):544–556
72. Singmann P, Shem-Tov D, Wahl S, Grallert H, Fiorito G, Shin SY, Schramm K, Wolf P, Kunze S, Baran Y, Guarrera S, Vineis P, Krogh V, Panico S, Tumino R, Kretschmer A, Gieger C, Peters A, Prokisch H, Relton CL, Matullo G, Illig T, Waldenberger M, Halperin E (2015) Characterization of whole-genome autosomal differences of DNA methylation between men and women. Epigenetics Chromatin 8:43
73. Li E, Beard C, Jaenisch R (1993) Role for DNA methylation in genomic imprinting. Nature 366(6453):362–365
74. Du M, Zhou W, Beatty LG, Weksberg R, Sadowski PD (2004) The KCNQ1OT1 promoter, a key regulator of genomic imprinting in human chromosome 11p15.5. Genomics 84(2):288–300
75. Mohammad F, Pandey GK, Mondal T, Enroth S, Redrup L, Gyllensten U, Kanduri C (2012) Long noncoding RNA-mediated maintenance of DNA methylation and transcriptional gene silencing. Development 139 (15):2792–2803
76. Lister R, Mukamel EA, Nery JR, Urich M, Puddifoot CA, Johnson ND, Lucero J, Huang Y, Dwork AJ, Schultz MD, Yu M,

Tonti-Filippini J, Heyn H, Hu S, Wu JC, Rao A, Esteller M, He C, Haghighi FG, Sejnowski TJ, Behrens MM, Ecker JR (2013) Global epigenomic reconfiguration during mammalian brain development. Science 341 (6146):1237905

77. Cotton AM, Lam L, Affleck JG, Wilson IM, Penaherrera MS, McFadden DE, Kobor MS, Lam WL, Robinson WP, Brown CJ (2011) Chromosome-wide DNA methylation analysis predicts human tissue-specific X inactivation. Hum Genet 130(2):187–201
78. Murakami K, Ohhira T, Oshiro E, Qi D, Oshimura M, Kugoh H (2009) Identification of the chromatin regions coated by non-coding Xist RNA. Cytogenet Genome Res 125(1):19–25
79. Brown CJ, Ballabio A, Rupert JL, Lafreniere RG, Grompe M, Tonlorenzi R, Willard HF (1991) A gene from the region of the human X inactivation centre is expressed exclusively from the inactive X chromosome. Nature 349 (6304):38–44
80. Csankovszki G, Nagy A, Jaenisch R (2001) Synergism of Xist RNA, DNA methylation, and histone hypoacetylation in maintaining X chromosome inactivation. J Cell Biol 153 (4):773–784
81. Navarro P, Page DR, Avner P, Rougeulle C (2006) Tsix-mediated epigenetic switch of a CTCF-flanked region of the Xist promoter determines the Xist transcription program. Genes Dev 20(20):2787–2792
82. Chureau C, Chantalat S, Romito A, Galvani A, Duret L, Avner P, Rougeulle C (2011) Ftx is a non-coding RNA which affects Xist expression and chromatin structure within the X-inactivation center region. Hum Mol Genet 20(4):705–718
83. Barr H, Hermann A, Berger J, Tsai HH, Adie K, Prokhortchouk A, Hendrich B, Bird A (2007) Mbd2 contributes to DNA methylation-directed repression of the Xist gene. Mol Cell Biol 27(10):3750–3757
84. Jin B, Ernst J, Tiedemann RL, Xu H, Sureshchandra S, Kellis M, Dalton S, Liu C, Choi JH, Robertson KD (2012) Linking DNA methyltransferases to epigenetic marks and nucleosome structure genome-wide in human tumor cells. Cell Rep 2(5):1411–1424
85. Okano M, Bell DW, Haber DA, Li E (1999) DNA methyltransferases Dnmt3a and Dnmt3b are essential for de novo methylation and mammalian development. Cell 99 (3):247–257
86. El Gazzar M, Yoza BK, Chen X, Hu J, Hawkins GA, McCall CE (2008) G9a and HP1 couple histone and DNA methylation to TNFalpha transcription silencing during endotoxin tolerance. J Biol Chem 283 (47):32198–32208
87. Song J, Teplova M, Ishibe-Murakami S, Patel DJ (2012) Structure-based mechanistic insights into DNMT1-mediated maintenance DNA methylation. Science 335 (6069):709–712
88. Pathania R, Ramachandran S, Elangovan S, Padia R, Yang P, Cinghu S, Veeranan-Karmegam R, Arjunan P, Gnana-Prakasam JP, Sadanand F, Pei L, Chang CS, Choi JH, Shi H, Manicassamy S, Prasad PD, Sharma S, Ganapathy V, Jothi R, Thangaraju M (2015) DNMT1 is essential for mammary and cancer stem cell maintenance and tumorigenesis. Nat Commun 6:6910
89. Li E, Bestor TH, Jaenisch R (1992) Targeted mutation of the DNA methyltransferase gene results in embryonic lethality. Cell 69 (6):915–926
90. Song CX, He C (2013) Potential functional roles of DNA demethylation intermediates. Trends Biochem Sci 38(10):480–484
91. Guibert S, Weber M (2013) Functions of DNA methylation and hydroxymethylation in mammalian development. Curr Top Dev Biol 104:47–83
92. Pastor WA, Aravind L, Rao A (2013) TETonic shift: biological roles of TET proteins in DNA demethylation and transcription. Nat Rev Mol Cell Biol 14(6):341–356
93. Williams K, Christensen J, Helin K (2011) DNA methylation: TET proteins-guardians of CpG islands? EMBO Rep 13(1):28–35
94. Lorsbach RB, Moore J, Mathew S, Raimondi SC, Mukatira ST, Downing JR (2003) TET1, a member of a novel protein family, is fused to MLL in acute myeloid leukemia containing the t(10,11)(q22;q23). Leukemia 17 (3):637–641
95. Tefferi A, Levine RL, Lim KH, Abdel-Wahab O, Lasho TL, Patel J, Finke CM, Mullally A, Li CY, Pardanani A, Gilliland DG (2009) Frequent TET2 mutations in systemic mastocytosis: clinical, KITD816V and FIP1L1-PDGFRA correlates. Leukemia 23 (5):900–904
96. Abdel-Wahab O, Mullally A, Hedvat C, Garcia-Manero G, Patel J, Wadleigh M, Malinge S, Yao J, Kilpivaara O, Bhat R, Huberman K, Thomas S, Dolgalev I, Heguy A, Paietta E, Le Beau MM, Beran M, Tallman MS, Ebert BL, Kantarjian HM, Stone RM, Gilliland DG, Crispino JD, Levine RL (2009) Genetic characterization of TET1,

TET2, and TET3 alterations in myeloid malignancies. Blood 114(1):144–147

97. Koh KP, Yabuuchi A, Rao S, Huang Y, Cunniff K, Nardone J, Laiho A, Tahiliani M, Sommer CA, Mostoslavsky G, Lahesmaa R, Orkin SH, Rodig SJ, Daley GQ, Rao A (2011) Tet1 and Tet2 regulate 5-hydroxymethylcytosine production and cell lineage specification in mouse embryonic stem cells. Cell Stem Cell 8(2):200–213
98. Walter J (2011) An epigenetic Tet a Tet with pluripotency. Cell Stem Cell 8(2):121–122
99. Wang T, Pan Q, Lin L, Szulwach KE, Song CX, He C, Wu H, Warren ST, Jin P, Duan R, Li X (2012) Genome-wide DNA hydroxymethylation changes are associated with neurodevelopmental genes in the developing human cerebellum. Hum Mol Genet 21 (26):5500–5510
100. Greco CM, Kunderfranco P, Rubino M, Larcher V, Carullo P, Anselmo A, Kurz K, Carell T, Angius A, Latronico MV, Papait R, Condorelli G (2016) DNA hydroxymethylation controls cardiomyocyte gene expression in development and hypertrophy. Nat Commun 7:12418
101. Qin S, Li Q, Zhou J, Liu ZJ, Su N, Wilson J, Lu ZM, Deng D (2014) Homeostatic maintenance of allele-specific p16 methylation in cancer cells accompanied by dynamic focal methylation and hydroxymethylation. PLoS One 9(5):e97785
102. Tampe B, Tampe D, Muller CA, Sugimoto H, LeBleu V, Xu X, Muller GA, Zeisberg EM, Kalluri R, Zeisberg M (2014) Tet3-mediated hydroxymethylation of epigenetically silenced genes contributes to bone morphogenic protein 7-induced reversal of kidney fibrosis. J Am Soc Nephrol 25(5):905–912
103. Xu X, Tan X, Tampe B, Nyamsuren G, Liu X, Maier LS, Sossalla S, Kalluri R, Zeisberg M, Hasenfuss G, Zeisberg EM (2015) Epigenetic balance of aberrant Rasal1 promoter methylation and hydroxymethylation regulates cardiac fibrosis. Cardiovasc Res 105(3):279–291
104. Tammen SA, Park LK, Dolnikowski GG, Ausman LM, Friso S, Choi SW (2015) Hepatic DNA hydroxymethylation is site-specifically altered by chronic alcohol consumption and aging. Eur J Nutr 56(2):535–544
105. Lister R, Ecker JR (2009) Finding the fifth base: genome-wide sequencing of cytosine methylation. Genome Res 19(6):959–966
106. Lister R, Pelizzola M, Dowen RH, Hawkins RD, Hon G, Tonti-Filippini J, Nery JR, Lee L, Ye Z, Ngo QM, Edsall L, Antosiewicz-Bourget J, Stewart R, Ruotti V, Millar AH, Thomson JA, Ren B, Ecker JR (2009) Human DNA methylomes at base resolution show widespread epigenomic differences. Nature 462(7271):315–322
107. Guo JU, Su Y, Shin JH, Shin J, Li H, Xie B, Zhong C, Hu S, Le T, Fan G, Zhu H, Chang Q, Gao Y, Ming GL, Song H (2014) Distribution, recognition and regulation of non-CpG methylation in the adult mammalian brain. Nat Neurosci 17(2):215–222
108. Goelz SE, Vogelstein B, Hamilton SR, Feinberg AP (1985) Hypomethylation of DNA from benign and malignant human colon neoplasms. Science 228(4696):187–190
109. Hansen KD, Timp W, Bravo HC, Sabunciyan S, Langmead B, McDonald OG, Wen B, Wu H, Liu Y, Diep D, Briem E, Zhang K, Irizarry RA, Feinberg AP (2011) Increased methylation variation in epigenetic domains across cancer types. Nat Genet 43 (8):768–775
110. Issa JP (2011) Epigenetic variation and cellular Darwinism. Nat Genet 43(8):724–726
111. Sharma S, Kelly TK, Jones PA (2010) Epigenetics in cancer. Carcinogenesis 31(1):27–36
112. Heyn H, Esteller M (2012) DNA methylation profiling in the clinic: applications and challenges. Nat Rev Genet 13(10):679–692
113. Lin JC, Jeong S, Liang G, Takai D, Fatemi M, Tsai YC, Egger G, Gal-Yam EN, Jones PA (2007) Role of nucleosomal occupancy in the epigenetic silencing of the MLH1 CpG island. Cancer Cell 12(5):432–444
114. Taberlay PC, Statham AL, Kelly TK, Clark SJ, Jones PA (2014) Reconfiguration of nucleosome-depleted regions at distal regulatory elements accompanies DNA methylation of enhancers and insulators in cancer. Genome Res 24(9):1421–1432
115. Hnisz D, Abraham BJ, Lee TI, Lau A, Saint-Andre V, Sigova AA, Hoke HA, Young RA (2013) Super-enhancers in the control of cell identity and disease. Cell 155(4):934–947
116. Heyn H, Vidal E, Ferreira HJ, Vizoso M, Sayols S, Gomez A, Moran S, Boque-Sastre R, Guil S, Martinez-Cardus A, Lin CY, Royo R, Sanchez-Mut JV, Martinez R, Gut M, Torrents D, Orozco M, Gut I, Young RA, Esteller M (2016) Epigenomic analysis detects aberrant super-enhancer DNA methylation in human cancer. Genome Biol 17 (1):11
117. Rideout WM III, Coetzee GA, Olumi AF, Jones PA (1990) 5-Methylcytosine as an endogenous mutagen in the human LDL receptor and p53 genes. Science 249 (4974):1288–1290

118. Feinberg AP, Vogelstein B (1983) Hypomethylation distinguishes genes of some human cancers from their normal counterparts. Nature 301(5895):89–92
119. Eden A, Gaudet F, Waghmare A, Jaenisch R (2003) Chromosomal instability and tumors promoted by DNA hypomethylation. Science 300(5618):455
120. Gaudet F, Hodgson JG, Eden A, Jackson-Grusby L, Dausman J, Gray JW, Leonhardt H, Jaenisch R (2003) Induction of tumors in mice by genomic hypomethylation. Science 300(5618):489–492
121. Kaneda A, Tsukamoto T, Takamura-Enya T, Watanabe N, Kaminishi M, Sugimura T, Tatematsu M, Ushijima T (2004) Frequent hypomethylation in multiple promoter CpG islands is associated with global hypomethylation, but not with frequent promoter hypermethylation. Cancer Sci 95(1):58–64
122. Esteller M (2007) Cancer epigenomics: DNA methylomes and histone-modification maps. Nat Rev Genet 8(4):286–298
123. Badal V, Chuang LS, Tan EH, Badal S, Villa LL, Wheeler CM, Li BF, Bernard HU (2003) CpG methylation of human papillomavirus type 16 DNA in cervical cancer cell lines and in clinical specimens: genomic hypomethylation correlates with carcinogenic progression. J Virol 77(11):6227–6234
124. Hur K, Cejas P, Feliu J, Moreno-Rubio J, Burgos E, Boland CR, Goel A (2014) Hypomethylation of long interspersed nuclear element-1 (LINE-1) leads to activation of proto-oncogenes in human colorectal cancer metastasis. Gut 63(4):635–646
125. Collier LS, Largaespada DA (2007) Transposable elements and the dynamic somatic genome. Genome Biol 8(Suppl 1):S5
126. Chenais B (2015) Transposable elements in cancer and other human diseases. Curr Cancer Drug Targets 15(3):227–242
127. Bourc'his D, Bestor TH (2004) Meiotic catastrophe and retrotransposon reactivation in male germ cells lacking Dnmt3L. Nature 431(7004):96–99
128. Lee E, Iskow R, Yang L, Gokcumen O, Haseley P, Luquette LJ III, Lohr JG, Harris CC, Ding L, Wilson RK, Wheeler DA, Gibbs RA, Kucherlapati R, Lee C, Kharchenko PV, Park PJ (2012) Landscape of somatic retrotransposition in human cancers. Science 337 (6097):967–971
129. Carreira PE, Richardson SR, Faulkner GJ (2014) L1 retrotransposons, cancer stem cells and oncogenesis. FEBS J 281(1):63–73
130. Helman E, Lawrence MS, Stewart C, Sougnez C, Getz G, Meyerson M (2014) Somatic retrotransposition in human cancer revealed by whole-genome and exome sequencing. Genome Res 24(7):1053–1063
131. Hattori M, Sakamoto H, Satoh K, Yamamoto T (2001) DNA demethylase is expressed in ovarian cancers and the expression correlates with demethylation of CpG sites in the promoter region of c-erbB-2 and survivin genes. Cancer Lett 169(2):155–164
132. Nabilsi NH, Broaddus RR, Loose DS (2009) DNA methylation inhibits p53-mediated survivin repression. Oncogene 28 (19):2046–2050
133. de Souza Rocha Simonini P, Breiling A, Gupta N, Malekpour M, Youns M, Omranipour R, Malekpour F, Volinia S, Croce CM, Najmabadi H, Diederichs S, Sahin O, Mayer D, Lyko F, Hoheisel JD, Riazalhosseini Y (2010) Epigenetically deregulated microRNA-375 is involved in a positive feedback loop with estrogen receptor alpha in breast cancer cells. Cancer Res 70 (22):9175–9184
134. Kang JY, Song SH, Yun J, Jeon MS, Kim HP, Han SW, Kim TY (2015) Disruption of CTCF/cohesin-mediated high-order chromatin structures by DNA methylation downregulates PTGS2 expression. Oncogene 34 (45):5677–5684
135. Nakayama M, Wada M, Harada T, Nagayama J, Kusaba H, Ohshima K, Kozuru M, Komatsu H, Ueda R, Kuwano M (1998) Hypomethylation status of CpG sites at the promoter region and overexpression of the human MDR1 gene in acute myeloid leukemias. Blood 92(11):4296–4307
136. Hanada M, Delia D, Aiello A, Stadtmauer E, Reed JC (1993) bcl-2 gene hypomethylation and high-level expression in B-cell chronic lymphocytic leukemia. Blood 82 (6):1820–1828
137. Naushad SM, Prayaga A, Digumarti RR, Gottumukkala SR, Kutala VK (2012) Bcl-2/adenovirus E1B 19 kDa-interacting protein 3 (BNIP3) expression is epigenetically regulated by one-carbon metabolism in invasive duct cell carcinoma of breast. Mol Cell Biochem 361(1-2):189–195
138. Divyya S, Naushad SM, Murthy PV, Reddy Ch R, Kutala VK (2013) GCPII modulates oxidative stress and prostate cancer susceptibility through changes in methylation of RASSF1, BNIP3, GSTP1 and Ec-SOD. Mol Biol Rep 40(10):5541–5550

139. Rao X, Evans J, Chae H, Pilrose J, Kim S, Yan P, Huang RL, Lai HC, Lin H, Liu Y, Miller D, Rhee JK, Huang YW, Gu F, Gray JW, Huang TM, Nephew KP (2013) CpG island shore methylation regulates caveolin-1 expression in breast cancer. Oncogene 32 (38):4519–4528

140. Gomez A, Karlgren M, Edler D, Bernal ML, Mkrtchian S, Ingelman-Sundberg M (2007) Expression of CYP2W1 in colon tumors: regulation by gene methylation. Pharmacogenomics 8(10):1315–1325

141. Rajendran G, Shanmuganandam K, Bendre A, Muzumdar D, Goel A, Shiras A (2011) Epigenetic regulation of DNA methyltransferases: DNMT1 and DNMT3B in gliomas. J Neuro-Oncol 104(2):483–494

142. Naghitorabi M, Mohammadi Asl J, Mir Mohammad Sadeghi H, Rabbani M, Jafarian-Dehkordi A, Javanmard HS (2013) Quantitative evaluation of DNMT3B promoter methylation in breast cancer patients using differential high resolution melting analysis. Res Pharm Sci 8(3):167–175

143. Lee HG, Kim H, Son T, Jeong Y, Kim SU, Dong SM, Park YN, Lee JD, Lee JM, Park JH (2016) Regulation of HK2 expression through alterations in CpG methylation of the HK2 promoter during progression of hepatocellular carcinoma. Oncotarget 7 (27):41798–41810

144. Lee PS, Teaberry VS, Bland AE, Huang Z, Whitaker RS, Baba T, Fujii S, Secord AA, Berchuck A, Murphy SK (2010) Elevated MAL expression is accompanied by promoter hypomethylation and platinum resistance in epithelial ovarian cancer. Int J Cancer 126 (6):1378–1389

145. He Y, Cui Y, Wang W, Gu J, Guo S, Ma K, Luo X (2011) Hypomethylation of the hsa--miR-191 locus causes high expression of hsa--mir-191 and promotes the epithelial-to-mesenchymal transition in hepatocellular carcinoma. Neoplasia 13(9):841–853

146. Fornari F, Milazzo M, Chieco P, Negrini M, Marasco E, Capranico G, Mantovani V, Marinello J, Sabbioni S, Callegari E, Cescon M, Ravaioli M, Croce CM, Bolondi L, Gramantieri L (2012) In hepatocellular carcinoma miR-519d is up-regulated by p53 and DNA hypomethylation and targets CDKN1A/p21, PTEN, AKT3 and TIMP2. J Pathol 227(3):275–285

147. Brueckner B, Stresemann C, Kuner R, Mund C, Musch T, Meister M, Sultmann H, Lyko F (2007) The human let-7a-3 locus contains an epigenetically regulated microRNA gene with oncogenic function. Cancer Res 67(4):1419–1423

148. Zrihan-Licht S, Weiss M, Keydar I, Wreschner DH (1995) DNA methylation status of the MUC1 gene coding for a breast-cancer-associated protein. Int J Cancer 62(3):245–251

149. Ye L, Li X, Kong X, Wang W, Bi Y, Hu L, Cui B, Li X, Ning G (2005) Hypomethylation in the promoter region of POMC gene correlates with ectopic overexpression in thymic carcinoids. J Endocrinol 185(2):337–343

150. Yoshida M, Nosaka K, Yasunaga J, Nishikata I, Morishita K, Matsuoka M (2004) Aberrant expression of the MEL1S gene identified in association with hypomethylation in adult T-cell leukemia cells. Blood 103(7):2753–2760

151. Tabu K, Sasai K, Kimura T, Wang L, Aoyanagi E, Kohsaka S, Tanino M, Nishihara H, Tanaka S (2008) Promoter hypomethylation regulates CD133 expression in human gliomas. Cell Res 18 (10):1037–1046

152. Gopisetty G, Xu J, Sampath D, Colman H, Puduvalli VK (2013) Epigenetic regulation of CD133/PROM1 expression in glioma stem cells by Sp1/myc and promoter methylation. Oncogene 32(26):3119–3129

153. Yamada N, Noguchi S, Kumazaki M, Shinohara H, Miki K, Naoe T, Akao Y (2014) Epigenetic regulation of microRNA-128a expression contributes to the apoptosis-resistance of human T-cell leukaemia jurkat cells by modulating expression of fas-associated protein with death domain (FADD). Biochim Biophys Acta 1843 (3):590–602

154. Jun HJ, Woolfenden S, Coven S, Lane K, Bronson R, Housman D, Charest A (2009) Epigenetic regulation of c-ROS receptor tyrosine kinase expression in malignant gliomas. Cancer Res 69(6):2180–2184

155. Nishigaki M, Aoyagi K, Danjoh I, Fukaya M, Yanagihara K, Sakamoto H, Yoshida T, Sasaki H (2005) Discovery of aberrant expression of R-RAS by cancer-linked DNA hypomethylation in gastric cancer using microarrays. Cancer Res 65(6):2115–2124

156. Fu X, Deng H, Zhao L, Li J, Zhou Y, Zhang Y (2010) Distinct expression patterns of hedgehog ligands between cultured and primary colorectal cancers are associated with aberrant methylation of their promoters. Mol Cell Biochem 337(1-2):185–192

157. Gupta A, Godwin AK, Vanderveer L, Lu A, Liu J (2003) Hypomethylation of the synuclein gamma gene CpG island promotes its

aberrant expression in breast carcinoma and ovarian carcinoma. Cancer Res 63 (3):664–673

158. Czekierdowski A, Czekierdowska S, Wielgos M, Smolen A, Kaminski P, Kotarski J (2006) The role of CpG islands hypomethylation and abnormal expression of neuronal protein synuclein-gamma (SNCG) in ovarian cancer. Neuro Endocrinol Lett 27 (3):381–386
159. Alonso MM, Diez-Valle R, Manterola L, Rubio A, Liu D, Cortes-Santiago N, Urquiza L, Jauregi P, Lopez de Munain A, Sampron N, Aramburu A, Tejada-Solis S, Vicente C, Odero MD, Bandres E, Garcia-Foncillas J, Idoate MA, Lang FF, Fueyo J, Gomez-Manzano C (2011) Genetic and epigenetic modifications of Sox2 contribute to the invasive phenotype of malignant gliomas. PLoS One 6(11):e26740
160. Hatano K, Miyamoto Y, Mori M, Nimura K, Nakai Y, Nonomura N, Kaneda Y (2012) Androgen-regulated transcriptional control of sialyltransferases in prostate cancer cells. PLoS One 7(2):e31234
161. Wang Q, Williamson M, Bott S, Brookman-Amissah N, Freeman A, Nariculam J, Hubank MJ, Ahmed A, Masters JR (2007) Hypomethylation of WNT5A, CRIP1 and S100P in prostate cancer. Oncogene 26 (45):6560–6565
162. Bhusari S, Yang B, Kueck J, Huang W, Jarrard DF (2011) Insulin-like growth factor-2 (IGF2) loss of imprinting marks a field defect within human prostates containing cancer. Prostate 71(15):1621–1630
163. Baba Y, Nosho K, Shima K, Huttenhower C, Tanaka N, Hazra A, Giovannucci EL, Fuchs CS, Ogino S (2010) Hypomethylation of the IGF2 DMR in colorectal tumors, detected by bisulfite pyrosequencing, is associated with poor prognosis. Gastroenterology 139 (6):1855–1864
164. Ogawa O, Eccles MR, Szeto J, McNoe LA, Yun K, Maw MA, Smith PJ, Reeve AE (1993) Relaxation of insulin-like growth factor II gene imprinting implicated in Wilms' tumour. Nature 362(6422):749–751
165. Shetty PJ, Movva S, Pasupuleti N, Vedicherlla B, Vattam KK, Venkatasubramanian S, Ahuja YR, Hasan Q (2011) Regulation of IGF2 transcript and protein expression by altered methylation in breast cancer. J Cancer Res Clin Oncol 137 (2):339–345
166. Anwar SL, Krech T, Hasemeier B, Schipper E, Schweitzer N, Vogel A, Kreipe H, Lehmann U (2014) Deregulation of RB1 expression by loss of imprinting in human hepatocellular carcinoma. J Pathol 233(4):392–401
167. Jarrard DF, Kinoshita H, Shi Y, Sandefur C, Hoff D, Meisner LF, Chang C, Herman JG, Isaacs WB, Nassif N (1998) Methylation of the androgen receptor promoter CpG island is associated with loss of androgen receptor expression in prostate cancer cells. Cancer Res 58(23):5310–5314
168. Tamada H, Kitazawa R, Gohji K, Kitazawa S (2001) Epigenetic regulation of human bone morphogenetic protein 6 gene expression in prostate cancer. J Bone Miner Res 16 (3):487–496
169. Esteller M, Silva JM, Dominguez G, Bonilla F, Matias-Guiu X, Lerma E, Bussaglia E, Prat J, Harkes IC, Repasky EA, Gabrielson E, Schutte M, Baylin SB, Herman JG (2000) Promoter hypermethylation and BRCA1 inactivation in sporadic breast and ovarian tumors. J Natl Cancer Inst 92 (7):564–569
170. Chen K, Wang G, Peng L, Liu S, Fu X, Zhou Y, Yu H, Li A, Li J, Zhang S, Bai Y, Zhang Y (2011) CADM1/TSLC1 inactivation by promoter hypermethylation is a frequent event in colorectal carcinogenesis and correlates with late stages of the disease. Int J Cancer 128(2):266–273
171. Graff JR, Herman JG, Lapidus RG, Chopra H, Xu R, Jarrard DF, Isaacs WB, Pitha PM, Davidson NE, Baylin SB (1995) E-cadherin expression is silenced by DNA hypermethylation in human breast and prostate carcinomas. Cancer Res 55 (22):5195–5199
172. Carmona FJ, Villanueva A, Vidal A, Munoz C, Puertas S, Penin RM, Goma M, Lujambio A, Piulats JM, Mesia R, Sanchez-Cespedes M, Manos M, Condom E, Eccles SA, Esteller M (2012) Epigenetic disruption of cadherin-11 in human cancer metastasis. J Pathol 228(2):230–240
173. Lin YL, Gui SL, Ma JG (2015) Aberrant methylation of CDH11 predicts a poor outcome for patients with bladder cancer. Oncol Lett 10(2):647–652
174. Mino A, Onoda N, Yashiro M, Aya M, Fujiwara I, Kubo N, Sawada T, Ohira M, Kato Y, Hirakawa K (2006) Frequent p16 CpG island hypermethylation in primary remnant gastric cancer suggesting an independent carcinogenic pathway. Oncol Rep 15 (3):615–620
175. Zhang YJ, Ahsan H, Chen Y, Lunn RM, Wang LY, Chen SY, Lee PH, Chen CJ, Santella RM (2002) High frequency of promoter hypermethylation of RASSF1A and p16 and

its relationship to aflatoxin B1-DNA adduct levels in human hepatocellular carcinoma. Mol Carcinog 35(2):85–92

176. Wang J, Sasco AJ, Fu C, Xue H, Guo G, Hua Z, Zhou Q, Jiang Q, Xu B (2008) Aberrant DNA methylation of P16, MGMT, and hMLH1 genes in combination with MTHFR C677T genetic polymorphism in esophageal squamous cell carcinoma. Cancer Epidemiol Biomark Prev 17(1):118–125
177. Raval A, Tanner SM, Byrd JC, Angerman EB, Perko JD, Chen SS, Hackanson B, Grever MR, Lucas DM, Matkovic JJ, Lin TS, Kipps TJ, Murray F, Weisenburger D, Sanger W, Lynch J, Watson P, Jansen M, Yoshinaga Y, Rosenquist R, de Jong PJ, Coggill P, Beck S, Lynch H, de la Chapelle A, Plass C (2007) Downregulation of death-associated protein kinase 1 (DAPK1) in chronic lymphocytic leukemia. Cell 129(5):879–890
178. Tada Y, Wada M, Taguchi K, Mochida Y, Kinugawa N, Tsuneyoshi M, Naito S, Kuwano M (2002) The association of death-associated protein kinase hypermethylation with early recurrence in superficial bladder cancers. Cancer Res 62(14):4048–4053
179. Rosas SL, Koch W, da Costa Carvalho MG, Wu L, Califano J, Westra W, Jen J, Sidransky D (2001) Promoter hypermethylation patterns of p16, O6-methylguanine-DNA-methyltransferase, and death-associated protein kinase in tumors and saliva of head and neck cancer patients. Cancer Res 61 (3):939–942
180. Bohne A, Schlee C, Mossner M, Thibaut J, Heesch S, Thiel E, Hofmann WK, Baldus CD (2009) Epigenetic control of differential expression of specific ERG isoforms in acute T-lymphoblastic leukemia. Leuk Res 33 (6):817–822
181. Esteller M, Hamilton SR, Burger PC, Baylin SB, Herman JG (1999) Inactivation of the DNA repair gene O6-methylguanine-DNA methyltransferase by promoter hypermethylation is a common event in primary human neoplasia. Cancer Res 59(4):793–797
182. Huang YW, Liu JC, Deatherage DE, Luo J, Mutch DG, Goodfellow PJ, Miller DS, Huang TH (2009) Epigenetic repression of microRNA-129-2 leads to overexpression of SOX4 oncogene in endometrial cancer. Cancer Res 69(23):9038–9046
183. Formosa A, Lena AM, Markert EK, Cortelli S, Miano R, Mauriello A, Croce N, Vandesompele J, Mestdagh P, Finazzi-Agro E, Levine AJ, Melino G, Bernardini S, Candi E (2013) DNA methylation silences miR-132 in prostate cancer. Oncogene 32 (1):127–134
184. Li Y, Xu Z, Li B, Zhang Z, Luo H, Wang Y, Lu Z, Wu X (2014) Epigenetic silencing of miRNA-9 is correlated with promoter-proximal CpG island hypermethylation in gastric cancer in vitro and in vivo. Int J Oncol 45(6):2576–2586
185. Herman JG, Umar A, Polyak K, Graff JR, Ahuja N, Issa JP, Markowitz S, Willson JK, Hamilton SR, Kinzler KW, Kane MF, Kolodner RD, Vogelstein B, Kunkel TA, Baylin SB (1998) Incidence and functional consequences of hMLH1 promoter hypermethylation in colorectal carcinoma. Proc Natl Acad Sci U S A 95(12):6870–6875
186. Esteller M, Levine R, Baylin SB, Ellenson LH, Herman JG (1998) MLH1 promoter hypermethylation is associated with the microsatellite instability phenotype in sporadic endometrial carcinomas. Oncogene 17 (18):2413–2417
187. Leung SY, Yuen ST, Chung LP, Chu KM, Chan AS, Ho JC (1999) hMLH1 promoter methylation and lack of hMLH1 expression in sporadic gastric carcinomas with high-frequency microsatellite instability. Cancer Res 59(1):159–164
188. Berdasco M, Ropero S, Setien F, Fraga MF, Lapunzina P, Losson R, Alaminos M, Cheung NK, Rahman N, Esteller M (2009) Epigenetic inactivation of the Sotos overgrowth syndrome gene histone methyltransferase NSD1 in human neuroblastoma and glioma. Proc Natl Acad Sci U S A 106 (51):21830–21835
189. Dammann R, Li C, Yoon JH, Chin PL, Bates S, Pfeifer GP (2000) Epigenetic inactivation of a RAS association domain family protein from the lung tumour suppressor locus 3p21.3. Nat Genet 25(3):315–319
190. Irimia M, Fraga MF, Sanchez-Cespedes M, Esteller M (2004) CpG island promoter hypermethylation of the Ras-effector gene NORE1A occurs in the context of a wild-type K-ras in lung cancer. Oncogene 23 (53):8695–8699
191. Guo W, Wang C, Guo Y, Shen S, Guo X, Kuang G, Dong Z (2015) RASSF5A, a candidate tumor suppressor, is epigenetically inactivated in esophageal squamous cell carcinoma. Clin Exp Metastasis 32(1):83–98
192. Sakai T, Toguchida J, Ohtani N, Yandell DW, Rapaport JM, Dryja TP (1991) Allele-specific hypermethylation of the retinoblastoma tumor-suppressor gene. Am J Hum Genet 48(5):880–888

193. Wasserkort R, Kalmar A, Valcz G, Spisak S, Krispin M, Toth K, Tulassay Z, Sledziewski AZ, Molnar B (2013) Aberrant septin 9 DNA methylation in colorectal cancer is restricted to a single CpG island. BMC Cancer 13:398
194. Qi J, Zhu YQ, Luo J, Tao WH (2006) Hypermethylation and expression regulation of secreted frizzled-related protein genes in colorectal tumor. World J Gastroenterol 12 (44):7113–7117
195. Nogales V, Reinhold WC, Varma S, Martinez-Cardus A, Moutinho C, Moran S, Heyn H, Sebio A, Barnadas A, Pommier Y, Esteller M (2016) Epigenetic inactivation of the putative DNA/RNA helicase SLFN11 in human cancer confers resistance to platinum drugs. Oncotarget 7(3):3084–3097
196. Takada H, Wakabayashi N, Dohi O, Yasui K, Sakakura C, Mitsufuji S, Taniwaki M, Yoshikawa T (2010) Tissue factor pathway inhibitor 2 (TFPI2) is frequently silenced by aberrant promoter hypermethylation in gastric cancer. Cancer Genet Cytogenet 197 (1):16–24
197. Guan Z, Zhang J, Song S, Dai D (2013) Promoter methylation and expression of TIMP3 gene in gastric cancer. Diagn Pathol 8:110
198. Catasus L, Pons C, Munoz J, Espinosa I, Prat J (2013) Promoter hypermethylation contributes to TIMP3 down-regulation in high stage endometrioid endometrial carcinomas. Histopathology 62(4):632–641
199. Wajed SA, Laird PW, DeMeester TR (2001) DNA methylation: an alternative pathway to cancer. Ann Surg 234(1):10–20
200. Esteller M (2007) Epigenetic gene silencing in cancer: the DNA hypermethylome. Hum Mol Genet 16(Spec 1):R50–R59
201. Baylin SB, Herman JG (2000) DNA hypermethylation in tumorigenesis: epigenetics joins genetics. Trends Genet 16(4):168–174
202. Murphree AL, Benedict WF (1984) Retinoblastoma: clues to human oncogenesis. Science 223(4640):1028–1033
203. Zhao Y, Sun J, Zhang H, Guo S, Gu J, Wang W, Tang N, Zhou X, Yu J (2014) High-frequency aberrantly methylated targets in pancreatic adenocarcinoma identified via global DNA methylation analysis using methylCap-seq. Clin Epigenetics 6(1):18
204. Dudziec E, Miah S, Choudhry HM, Owen HC, Blizard S, Glover M, Hamdy FC, Catto JW (2011) Hypermethylation of CpG islands and shores around specific microRNAs and mirtrons is associated with the phenotype and presence of bladder cancer. Clin Cancer Res 17(6):1287–1296
205. Cuddapah S, Jothi R, Schones DE, Roh TY, Cui K, Zhao K (2009) Global analysis of the insulator binding protein CTCF in chromatin barrier regions reveals demarcation of active and repressive domains. Genome Res 19 (1):24–32
206. Mukhopadhyay R, Yu W, Whitehead J, Xu J, Lezcano M, Pack S, Kanduri C, Kanduri M, Ginjala V, Vostrov A, Quitschke W, Chernukhin I, Klenova E, Lobanenkov V, Ohlsson R (2004) The binding sites for the chromatin insulator protein CTCF map to DNA methylation-free domains genome-wide. Genome Res 14(8):1594–1602
207. Wang H, Maurano MT, Qu H, Varley KE, Gertz J, Pauli F, Lee K, Canfield T, Weaver M, Sandstrom R, Thurman RE, Kaul R, Myers RM, Stamatoyannopoulos JA (2012) Widespread plasticity in CTCF occupancy linked to DNA methylation. Genome Res 22(9):1680–1688
208. Yan H, Tang G, Wang H, Hao L, He T, Sun X, Ting AH, Deng A, Sun S (2016) DNA methylation reactivates GAD1 expression in cancer by preventing CTCF-mediated polycomb repressive complex 2 recruitment. Oncogene 35(30):3995–4008
209. de Maat MF, van de Velde CJ, Umetani N, de Heer P, Putter H, van Hoesel AQ, Meijer GA, van Grieken NC, Kuppen PJ, Bilchik AJ, Tollenaar RA, Hoon DS (2007) Epigenetic silencing of cyclooxygenase-2 affects clinical outcome in gastric cancer. J Clin Oncol 25 (31):4887–4894
210. Miremadi A, Oestergaard MZ, Pharoah PD, Caldas C (2007) Cancer genetics of epigenetic genes. Hum Mol Genet 16(Spec 1): R28–R49
211. Jones JS, Amos CI, Pande M, Gu X, Chen J, Campos IM, Wei Q, Rodriguez-Bigas M, Lynch PM, Frazier ML (2006) DNMT3b polymorphism and hereditary nonpolyposis colorectal cancer age of onset. Cancer Epidemiol Biomark Prev 15(5):886–891
212. Shen H, Wang L, Spitz MR, Hong WK, Mao L, Wei Q (2002) A novel polymorphism in human cytosine DNA-methyltransferase-3B promoter is associated with an increased risk of lung cancer. Cancer Res 62 (17):4992–4995
213. Montgomery KG, Liu MC, Eccles DM, Campbell IG (2004) The DNMT3B C→T promoter polymorphism and risk of breast cancer in a British population: a case-control study. Breast Cancer Res 6(4):R390–R394

214. Singal R, Das PM, Manoharan M, Reis IM, Schlesselman JJ (2005) Polymorphisms in the DNA methyltransferase 3b gene and prostate cancer risk. Oncol Rep 14(2):569–573
215. Di Croce L, Raker VA, Corsaro M, Fazi F, Fanelli M, Faretta M, Fuks F, Lo Coco F, Kouzarides T, Nervi C, Minucci S, Pelicci PG (2002) Methyltransferase recruitment and DNA hypermethylation of target promoters by an oncogenic transcription factor. Science 295(5557):1079–1082
216. Coutinho DF, Monte-Mor BC, Vianna DT, Rouxinol ST, Batalha AB, Bueno AP, Boulhosa AM, Fernandez TS, Pombo-de-Oliveira MS, Gutiyama LM, Abdelhay E, Zalcberg IR (2015) TET2 expression level and 5-hydroxymethylcytosine are decreased in refractory cytopenia of childhood. Leuk Res 39(10):1103–1108
217. Delhommeau F, Dupont S, Della Valle V, James C, Trannoy S, Masse A, Kosmider O, Le Couedic JP, Robert F, Alberdi A, Lecluse Y, Plo I, Dreyfus FJ, Marzac C, Casadevall N, Lacombe C, Romana SP, Dessen P, Soulier J, Viguie F, Fontenay M, Vainchenker W, Bernard OA (2009) Mutation in TET2 in myeloid cancers. N Engl J Med 360(22):2289–2301
218. Langemeijer SM, Kuiper RP, Berends M, Knops R, Aslanyan MG, Massop M, Stevens-Linders E, van Hoogen P, van Kessel AG, Raymakers RA, Kamping EJ, Verhoef GE, Verburgh E, Hagemeijer A, Vandenberghe P, de Witte T, van der Reijden BA, Jansen JH (2009) Acquired mutations in TET2 are common in myelodysplastic syndromes. Nat Genet 41(7):838–842
219. Haffner MC, Chaux A, Meeker AK, Esopi DM, Gerber J, Pellakuru LG, Toubaji A, Argani P, Iacobuzio-Donahue C, Nelson WG, Netto GJ, De Marzo AM, Yegnasubramanian S (2011) Global 5-hydroxymethylcytosine content is significantly reduced in tissue stem/progenitor cell compartments and in human cancers. Oncotarget 2(8):627–637
220. Murata A, Baba Y, Ishimoto T, Miyake K, Kosumi K, Harada K, Kurashige J, Iwagami S, Sakamoto Y, Miyamoto Y, Yoshida N, Yamamoto M, Oda S, Watanabe M, Nakao M, Baba H (2015) TET family proteins and 5-hydroxymethylcytosine in esophageal squamous cell carcinoma. Oncotarget 6 (27):23372–23382
221. Udali S, Guarini P, Moruzzi S, Ruzzenente A, Tammen SA, Guglielmi A, Conci S, Pattini P, Olivieri O, Corrocher R, Choi SW, Friso S (2015) Global DNA methylation and hydroxymethylation differ in hepatocellular carcinoma and cholangiocarcinoma and relate to survival rate. Hepatology 62(2):496–504
222. Thienpont B, Steinbacher J, Zhao H, D'Anna F, Kuchnio A, Ploumakis A, Ghesquiere B, Van Dyck L, Boeckx B, Schoonjans L, Hermans E, Amant F, Kristensen VN, Koh KP, Mazzone M, Coleman ML, Carell T, Carmeliet P, Lambrechts D (2016) Tumour hypoxia causes DNA hypermethylation by reducing TET activity. Nature 537 (7618):63–68
223. Kent WJ, Sugnet CW, Furey TS, Roskin KM, Pringle TH, Zahler AM, Haussler D (2002) The human genome browser at UCSC. Genome Res 12(6):996–1006
224. Moran S, Martinez-Cardus A, Sayols S, Musulen E, Balana C, Estival-Gonzalez A, Moutinho C, Heyn H, Diaz-Lagares A, de Moura MC, Stella GM, Comoglio PM, Ruiz-Miro M, Matias-Guiu X, Pazo-Cid R, Anton A, Lopez-Lopez R, Soler G, Longo F, Guerra I, Fernandez S, Assenov Y, Plass C, Morales R, Carles J, Bowtell D, Mileshkin L, Sia D, Tothill R, Tabernero J, Llovet JM, Esteller M (2016) Epigenetic profiling to classify cancer of unknown primary: a multicentre, retrospective analysis. Lancet Oncol 17(10):1386–1395
225. Esteller M, Garcia-Foncillas J, Andion E, Goodman SN, Hidalgo OF, Vanaclocha V, Baylin SB, Herman JG (2000) Inactivation of the DNA-repair gene MGMT and the clinical response of gliomas to alkylating agents. N Engl J Med 343(19):1350–1354
226. Hegi ME, Diserens AC, Gorlia T, Hamou MF, de Tribolet N, Weller M, Kros JM, Hainfellner JA, Mason W, Mariani L, Bromberg JE, Hau P, Mirimanoff RO, Cairncross JG, Janzer RC, Stupp R (2005) MGMT gene silencing and benefit from temozolomide in glioblastoma. N Engl J Med 352 (10):997–1003
227. Amatu A, Sartore-Bianchi A, Moutinho C, Belotti A, Bencardino K, Chirico G, Cassingena A, Rusconi F, Esposito A, Nichelatti M, Esteller M, Siena S (2013) Promoter CpG island hypermethylation of the DNA repair enzyme MGMT predicts clinical response to dacarbazine in a phase II study for metastatic colorectal cancer. Clin Cancer Res 19(8):2265–2272
228. Xu Y, Diao L, Chen Y, Liu Y, Wang C, Ouyang T, Li J, Wang T, Fan Z, Fan T, Lin B, Deng D, Narod SA, Xie Y (2013)

Promoter methylation of BRCA1 in triple-negative breast cancer predicts sensitivity to adjuvant chemotherapy. Ann Oncol 24 (6):1498–1505

229. Moutinho C, Martinez-Cardus A, Santos C, Navarro-Perez V, Martinez-Balibrea E, Musulen E, Carmona FJ, Sartore-Bianchi A, Cassingena A, Siena S, Elez E, Tabernero J, Salazar R, Abad A, Esteller M (2014) Epigenetic inactivation of the BRCA1 interactor SRBC and resistance to oxaliplatin in colorectal cancer. J Natl Cancer Inst 106(1):djt322

Part II

Methods for the Measurement of DNA Methylation at CpG Islands

Chapter 5

Infinium DNA Methylation Microarrays on Formalin-Fixed, Paraffin-Embedded Samples

Sebastian Moran and Manel Esteller

Abstract

Formalin-fixed, paraffin-embedded (FFPE) samples usually yield fragmented DNA, which is incompatible with traditional Infinium Methylation beadchips. In this chapter, we aim to explain in detail all the processes carried out in order to obtain high-quality methylation profiles from FFPE samples through an FFPE restoration procedure. High-throughput methylation profiling platforms such as the Infinium Methylation beadchips have been extensively used by the scientific community to elucidate many hypotheses in multiple scenarios, now with the incorporation of FFPE samples to these platforms, a myriad of new opportunities is being opened.

Key words Epigenetics, DNA methylation, Formalin-fixed, paraffin-embedded, FFPE, Methylation, Microarray, Restoration, Degraded DNA

1 Introduction

DNA methylation, the most extensively studied type of epigenetic modification, consists in the covalent addition of a methyl group to cytosine bases, mainly in CpG context. It is extensively known that these modifications are inversely correlated with gene expression when occurring on gene promoter regions, and a growing number of studies have demonstrated its role in metabolic diseases, neurodegenerative diseases, aging, cancer, and drug response to chemotherapy, among others. Thus, the importance of DNA methylation signatures as a tool in many biomedical studies is becoming increasingly well established.

To interrogate CpG methylation sites at a genome-wide manner, apart from sequencing based methods, different platforms have been developed; from the GoldenGate Methylation panels [1], passing through Infinium Methylation27k [2], HumanMethylation450k [3], and the recently launched MethylationEPIC microarrays [4]. However, all these platforms suffers from the same limitation; they required high-quality (nonfragmented) DNA

Tanya Vavouri and Miguel A. Peinado (eds.), *CpG Islands: Methods and Protocols*, Methods in Molecular Biology, vol. 1766, https://doi.org/10.1007/978-1-4939-7768-0_5,

starting material, hence, restricting the analysisto only those DNA extracted from fresh-frozen (FF) tissues, blood samples or in vitro cultured cells. The reason behind these strict requirements (DNA fragments >1 kb) [5] are a consequence of the inefficiency of the whole-genome amplification step included in methylation microarrays procedures.

Conversely, millions of samples are routinely formalin-fixed and paraffin-embedded (FFPE) for histopathological diagnosis, samples holding a tremendous valuable clinical information. However, this fixation process frequently leads to nucleic acid degradation, preventing these valuable samples for use in methylation microarray methods. Here we detail a DNA restoration procedure to overcome this problem using a commercially available restoration kit, in order to process FFPE samples yielding comparable results to those obtained from fresh-frozen (FF) samples [6].

2 Materials

The following reagents and equipment are required in order to process 96 samples through the methylation microarrays. In case fewer samples will be processed, contact suppliers for adequate size kits.

2.1 Kits and Reagents

1. E.Z.N.A. FFPE DNA kit (contains FTL, BL, HBC, DNA wash, elution buffers, proteinase K, microelute DNA mini columns, and collection tubes) (Omega Bio-Tek).
2. QUBIT DSDNA BR assay kit (Thermofisher Scientific).
3. EZ-96 DNA methylation kit (deep-well format) (contains ct conversion reagent, m-dilution, m-wash, m-binding, m-desulphonation, m-elution buffers, zymo-spin plate, and collection plate) (Zymo).
4. ZR-96 DNA clean and concentrator-5 (contains DNA wash, DNA binding buffers, zymo-spin i-96 midi plate, and collection plates) (Zymo).
5. Illumina FFPE QC Kit (contains qcp, qct reagents) (Illumina).
6. Infinium HD FFPE DNA Restoration Kit (contains ppr, amr, erb, and cmm reagents) (Illumina).
7. Infinium Methylation EPIC kit (contains MA1, RPM, MSM, FMS, PM1, RA1, PB2, XC4, PB1, XC1, XC2, TEM, STM, ATM reagents, and beadchips) (Illumina).
8. 96-well 0.8 mL microtiter plate (MIDI) (VWR).
9. Adhesive foil seal (Microamp Clear) (Applied Biosystems).
10. Thermo-Applicable Film (Easypeel) (Fisher Scientific).
11. Formamide (Sigma-Aldrich).
12. EDTA (Sigma-Aldrich).

2.2 Equipment

1. Illumina iScan or HiScan scanner.
2. Illumina hybridization oven.
3. Alignment beadchip fixture.
4. Glass back plates.
5. Hybridization cambers (including gaskets, inserts, clamps, and spacers).
6. Water circulator and flow-through chamber rack.
7. Bench microcentrifuge capable of 14,000 × *g*.
8. Centrifuge for 96-well plates capable of more than 3000 × *g* and 4 °C (*see* **Note 1**).
9. Heat blocks capable of 37, 55, 70, and 90 °C.
10. Thermocycler.
11. Adjustable pipettes (2, 20, 200, 1000 μL).
12. 8-tip multichannel pipette.
13. Fluorimeter (Qubit).
14. NanoDrop (optional).
15. Microtome.
16. Vortex.
17. Scissors, self-locking tweezers.

3 Methods

The following sections will guide you through all the procedures needed to generate methylation profiles for Illumina Infinium Methylation beadchips. At each section a brief explanation of steps and the aim of the section will be given (*see* **Note 2**).

3.1 DNA Extraction

In our experience any FFPE DNA extraction method compatible with PCR (column based, phenol–chloroform, salting out, etc.), will result in high-quality methylation microarray data. However, we will explain in detail the method we have used (Omega Bio-Tek. E.Z.N.A.® FFPE DNA Kit) for thousands of samples during the last years, a selection made based on easy operation, the quality of DNA obtained, yield and costs of the DNA extraction method.

3.1.1 Preparation for DNA Extraction

1. Upon opening of the E.Z.N.A.® FFPE DNA Kit for the first time, **DNA Wash** buffer and **HBC** buffer must be diluted as follows:
   (a) Add 60 mL 100% ethanol to the **DNA Wash** buffer bottle. Mark the bottle so that you know ethanol has been added.

(b) Add 10 mL 100% isopropanol (2-propanol) to **HBC** buffer bottle. Mark the bottle so that you know isopropanol has been added.

2. Heat a heat block to 37 °C.
3. Heat a heat block to 55 °C.
4. Heat a heat block to 90 °C.
5. Heat **Elution** buffer to 70 °C for the elution step.

3.1.2 Procedure for DNA Extraction

Depending on the number of samples to be extracted, this protocol can be adapted to the use of 96-well plate DNA extraction kits (check suppliers for such kits); however, we will explain in detail the method used for smaller batches of samples, where the number of samples usually is limited by the number of ports in the centrifuge's rotor. Yield of extracted DNA generally meets the requirements of input DNA needed for methylation microarrays (250 ng gDNA); however, certain tissues (especially those containing high fat content) might not yield enough quantity of DNA, and thus multiple extractions should be performed to achieve the amount needed (*see* **Note 3**).

1. Cut 2–8 paraffin sections of 10 μm thick, in order to have a total area of approximately 0.5 cm^2 of tissue and 10 μm thick (*see* **Note 4**).
2. Place the curl sections on a 1.5 mL microcentrifuge tube.
3. Add 1 mL xylene to the 1.5 mL tube.
4. Invert the 1.5 mL tubes for about 1 min, or until all paraffin from the sections has been resuspended.
5. Centrifuge at maximum speed for 2 min. A pellet with the tissue should have been formed.
6. Aspirate and discard the supernatant without disturbing the pellet, extracting as much xylene as possible.
7. Add 1 mL of 100% ethanol to the tube.
8. Invert the 1.5 mL tube 20 times to resuspend the pellet.
9. Centrifuge at maximum speed for 2 min.
10. Aspirate and discard the supernatant without disturbing the pellet, extracting as much ethanol as possible.
11. Leaving the tube lid opened, incubate the pellet at 37 °C for 15 min (*see* **Note 5**).
12. Add 200 μL of **FTL** buffer, pipetting up-down to resuspend the pellet.
13. Add 20 μL Proteinase K solution and vortex to mix thoroughly.

14. Incubate at 55 °C for 3 h, or until no tissue particles are observable (*see* **Note 6**).
15. Inactivate Proteinase K, by incubating at 90 °C for 20 min.
16. Centrifuge the tube briefly to collect any liquid from the lid (*see* **Note 7**).
17. Add 220 μL **BL** buffer, and vortex to ensure proper mixing (*see* **Note 8**).
18. Place a MicroElute® DNA Mini Column in a 2 mL collection tube.
19. Transfer all the volume from **step 17**, including any precipitate that may have formed, to the MicroElute DNA Mini column.
20. Centrifuge at 10,000 × *g* for 1 min.
21. Transfer the column to a new 2 mL collection tube, discard the previous collection tube that contains the filtrate.
22. Add 500 μL **HBC** buffer to the column (*see* **Note 9**).
23. Centrifuge at 10,000 × *g* for 1 min.
24. Transfer the column to a new 2 mL collection tube, and discard the previous collection tube that contains the filtrate.
25. Add 700 μL **DNA Wash** buffer to the column (*see* **Note 10**).
26. Centrifuge at 10,000 × *g* for 1 min.
27. Discard the filtrate, reusing the collection tube.
28. Add 700 μL **DNA Wash** buffer to the column, for a second wash step.
29. Centrifuge at full speed for 2 min to completely dry the membrane of the column (*see* **Note 11**).
30. Place the column in a new 1.5 mL tube.
31. Add 50 μL **Elution** buffer preheated to 70 °C to the center of the column membrane (*see* **Note 12**).
32. Incubate the tube at room temperature for 3 min.
33. Centrifuge at maximum speed for 1 min.
34. Repeat **steps 31–33** for a second elution process.
35. Obtained DNA may be stored at −20 °C.

3.2 DNA Quantification

DNA concentration should be measured by fluorescent dye assay such as Picogreen. Avoid measuring DNA concentration by spectrometric methods such as NanoDrop, as it over estimates double stranded DNA (dsDNA) by a factor of 2–4 times, and this behavior is exacerbated when using degraded DNAs, such as the ones obtained from FFPE. NanoDrop should be used to detect contaminants in the DNA, as for example in case of DNA extraction by phenol–chloroform, to detect any possible phenol or organic contaminants, in which case 260/230 ratio will be lower than 1.8.

Likewise, protein contamination will be indicated by 260/280 ratio lower than 1.8, while RNA contamination will be indicated by 260/280 ratio around 2.0.

To facilitate the handling of multiple samples, it is recommended to normalize each sample, by making a working solution of DNA of 50 ng/μL (*see* **Note 13**).

3.3 DNA FFPE Quality Control

The following quality control procedure is used to determine if a sample can be subject to restoration, or if it is too fragmented to be restored, hence having to be discarded from the project. For such control, a quantitative real-time PCR (qPCR) is performed using Illumina's HD FFPE QC kit primers and standards. Depending on your available real-time instrument, reaction volumes might be adjusted (96-well plates: 20 μL; 384-well plates: 10 μL), and will be detailed accordingly.

3.3.1 Preparation of FFPE QC

1. From the working solution of DNA (50 ng/μL), perform a serial dilution of each DNA to obtain a final solution of 1 ng/μ L concentration.
2. Thaw to room temperature **QCP** and **QCT** reagents from the HD FFPE QC kit (*see* **Note 14**).
3. Thaw as much **2×** **qPCR Master Mix** tubes as needed based on the amount of samples to be processed (*see* **Note 15**).

3.3.2 Procedure for FFPE QC

1. Add 990 μL of DiH_2O to a fresh 10 μL aliquot of **QCT** reagent, in order to create a 100-fold dilution.
2. Briefly vortex and spin the tube at 280 × *g* to collect droplets.
3. Depending on your final reaction volume for the qPCR, follow the steps below accordingly:
   (a) For 10 μL reaction volumes:
      - Per plate of real time being processed, dispense 2 μL of the 100-fold diluted **QCT** (**step 1**) into three wells of the qPCR plate. These three wells will constitute the standard used to assess sample's quality for restoration.
      - Per plate of real time being processed, dispense 2μL of DiH_2O into three wells of the qPCR plate. These three wells will constitute the no-template control (NTC) used to assess sample's quality for restoration.
      - For each FFPE sample to be processed, dispense 2 μL of genomic DNA (1 ng/μL) into three wells of the qPCR plate.

   (b) For 20 μL reaction volumes:
      - Per plate of real time being processed, dispense 4 μL of the 100-fold diluted **QCT** (**step 1**) into three wells of

Table 1
FFPE quality control qPCR volumes used to prepare the premix, depending on the final volume used in the qPCR reaction (10 or 20 μL)

Volumes per reaction	10 μL reaction volumes, μL	20 μL reaction volumes, μL
2× qPCR Master Mix	5	10
QCP	1	2
DiH_2O	2	4
Total volume added per well	8	16

the qPCR plate. These three wells will constitute the standard used to assess sample's quality for restoration.

- Per plate of real time being processed, dispense 4 μL of DiH_2O into three wells of the qPCR plate. These three wells will constitute the no-template control (NTC) used to assess sample's quality for restoration.
- For each FFPE sample to be processed, dispense 4 μL of genomic DNA (1 ng/μL) into three wells of the qPCR plate.

4. Prepare as much qPCR premix as needed (*see* Table 1), depending on the final number of reactions, using the following table. Take into account that each sample, as well as each one of the controls (QCT and NTC) will run per triplicate on the qPCR.
5. Mix the qPCR container where you have prepared qPCR premix, by inverting it ten times. Tap the container on the lab bench to collect the droplets.
6. Using a eight-tip multichannel pipette dispense 8 μL (for 10 μL reaction volume) or 16 μL (for 20 μL reaction volume) of qPCR premix into each one of the qPCR plate wells used for samples and controls (QCT and NTC) (*see* **Note 16**).
7. Seal the plate with an optical film, according to the manufacturer's instructions and briefly centrifuge the plate at 280 × *g*.
8. Use the thermal profile described in Table 2 to run the qPCR instrument (*see* **Note 17**).

3.3.3 Analysis of qPCR

1. Check NTC wells, and ensure that near zero amplification has occurred (*see* **Note 18**).
2. Per sample, remove any replicate that deviates from more than half C_q unit.
3. Obtain C_q values for all qPCR wells used, and compute per FFPE sample, its average C_q.
4. Compute QCT average using the three wells intended for this purpose.

Table 2
Thermal profile used during qPCR FFPE quality control

	Temperature, °C	Time
	50	2 min (*see* **Note 17**)
	95	10 min (*see* **Note 17**)
× 40	95	30 s
	57	30 s
	72	30 s

5. Calculate the sample's ΔC_q, by subtracting the QCT average C_q from the sample average C_q value.

$$\Delta C_q^{\text{Sample i}} = \overline{C_q^{\text{Sample i}}} - \overline{C_q^{\text{QCT}}}.$$

where i denotes each one of the different FFPE gDNA samples being processed.

6. All resulting samples with $\Delta C_q < 5$ will pass this quality control, hence being suitable for restoration. Samples with $\Delta C_q \geq 5$ *will not be* suitable for restoration, and should be discarded from the project.

3.4 Bisulfite Conversion

The aim of this step is the basics of the methylation measurement by microarrays, as in this step we transform a methylation status into a nucleotide change, by the use of sodium bisulfite, a reaction called bisulfite conversion. When exposing DNA to sodium bisulfite, the methyl group located on the carbon 5 of methylated cytosines (5-methyl-cytosine) will act as a protecting agent against the action of sodium bisulfite. On the contrary, unmethylated cytosines will not be protected from the action of sodium bisulfite, thus being subject to a number of modifications that will lead to a nucleotide change, where the cytosine will become a uracil. Subsequent reaction steps of the microarray processing will transform those uracils coming from unmethylated cytosines into thymines.

3.4.1 Preparation for Bisulfite Conversion

1. Prepare **CT Conversion** reagent by adding 7.5 mL water and 2.1 mL of **M-Dilution** buffer. Mix at room temperature with frequent vortexing for 10 min (*see* **Note 19**).
2. Prepare **M-Wash** buffer by adding 144 mL of 100% ethanol to the 36 mL M-Wash buffer concentrate.

3. Make sure that your centrifuge equipment has buckets deep enough to hold the Zymo-Spin MIDI plate assembly. Some types of centrifuge equipment do not provide buckets deep enough or block access to two stacked plates.

3.4.2 Procedure for Bisulfite Conversion

1. For each sample, dispense the adequate volume of gDNA solution (5 μL if using 50 ng/μL working solution) in order to have 250 ng of gDNA in a 96-well plate (*see* **Note 20**).
2. Add as much DiH_2O as needed to adjust the volume for each well in the plate to a final volume of 45 μL (5μL gDNA +40 μL DiH_2O).
3. Dispense 5μL of **M-Dilution** buffer to each position of the plate. Mixing by pipetting up and down.
4. Incubate plate at 37 °C for 15 min.
5. Add 100μL of prepared **CT Conversion** reagent (*see* **step 1** of Subheading 3.4.1) to each well of the plate.
6. Use the thermal profile defined in Table 2 to incubate the plate in a thermal cycler:
7. Ensure that samples have been incubated for at least 10 min at 4 °C (*see* **Note 21**).
8. Place one Zymo-Spin TM I-96 Binding plate on top of a collection plate.
9. Add 400 μL of **M-Binding** buffer to each well of a Zymo-Spin TM I-96 Binding plate.
10. Transfer all the volume from the incubated plate (**step** 7) into the binding plate containing the **M-Binding** buffer (**step 6**), ensuring proper mixing by pipetting up and down ten times.
11. Centrifuge at ≥3000 × *g* for 5 min, after which you should discard the flow-through.
12. Dispense 400 μL of prepared **M-Wash** buffer (*see* **step 2** of Subheading 3.4.1), into each well of the plate.
13. Centrifuge at ≥3000 × *g* for 5 min, after which you should discard the flow-through.
14. Add 200 μL of **M-Desulphonation** buffer to each well and incubate at room temperature for 20 min.
15. Centrifuge at ≥3000 × *g* for 5 min, after which you should discard the flow-through.
16. Add 500 μL of **M-Wash** buffer to each well.
17. Centrifuge at ≥3000 × *g* for 5 min, after which you should discard the flow-through.
18. Add 500 μL of **M-Wash** buffer to each well for a second wash.

19. Centrifuge at $\geq 3000 \times g$ for 10 min, after which you should discard the flow-through and collection plate (*see* **Note 22**).
20. Place the Zymo-Spin TM I-96 Binding plate onto an elution plate.
21. Add 12 μL of **M-Elution** buffer into the center of each well.
22. Incubate for 2 min at room temperature.
23. Centrifuge at $\geq 3000 \times g$ for 3 min.
24. The elution plate will approximately have 8–10 μL of eluate.
25. Proceed immediately to restoration procedure.

3.5 FFPE Restoration

FFPE samples usually yields fragmented DNA, a handicap when wanting to apply high-throughput methylation platforms, as fragmented DNA is incompatible with those. With the FFPE restoration procedure, the aim is to pass from multiple fragments of DNA, to just a few but longer, by ligating the multiple fragments between them. To achieve this, a two enzymatic reactions will be performed. In the first reaction, enzymes will fix the extremes of the DNA, which usually are non-blunt ended, and transform them into blunt ends. Once all fragments of DNA are properly fixed, the second enzymatic reaction occurs, by which the different blunt fragments are ligated between them, creating longer fragments of DNA.

3.5.1 Preparation for Restoration

1. Prepare the **DNA Wash** buffer from the Zymo Purification kit (ZR-96 DNA Clean & Concentrator™-5), by adding 192 mL 100% ethanol to the 48 mL DNA Wash buffer bottle. Mark the bottle so that you would know ethanol has been added.
2. Preheat one heat block to 37 °C and another one to 95 °C, allowing the temperatures to equilibrate.
3. Thaw **PPR** and **AMR** reagents tubes (use 1 tube each per 24 FFPE DNA samples), from Illumina FFPE Restoration kit, to room temperature. Once thawed, invert them to mix the contents.
4. Prepare a fresh 1 mL 0.1 N NaOH solution (*see* **Note 23**).
5. Make sure that your centrifuge equipment has buckets deep enough to hold the Zymo-Spin I-96 MIDI plate assembly. Some types of centrifuge equipment do not provide buckets deep enough or block access to two stacked plates.

3.5.2 Procedure for Restoration

1. Transfer 8 μL of bisulfite-converted gDNA (**step 24** from Subheading 3.4.2) into a new 96-well 0.8 mL plate (MIDI).
2. Dispense 4 μL of 0.1 N NaOH to each well containing gDNA.
3. Incubate at room temperature for 10 min.
4. Dispense 34 μL of **PPR** to each sample well in the plate.
5. Dispense 38 μL of **AMR** to each sample well in the plate.

6. Seal the plate with an adhesive film (*see* **Note 24**).
7. Invert the plate ten times and spin it 1 min at 280 × *g*.
8. Incubate the plate for 1 h at 37 °C (*see* **Note 25**).
9. Centrifuge the plate 1 min at 280 × *g* to collect droplets.
10. Place a Zymo-Spin I-96 plate on top of a collection plate (both supplied on ZR-96 DNA Clean & Concentrator™-5 kit).
11. Add 560 μL of **Zymo DNA Binding** buffer to each well of the plate containing the gDNA (**step 9**), mixing the buffer and the DNA pipetting up and down five times (*see* **Note 26**).
12. Transfer all the volume of the sample mixtures (previous step) to the wells of the Zymo-Spin I-96 plate (**step 10**).
13. Centrifuge the Zymo-Spin I-96 plate at 2250 × *g* for 2 min. Discard the flow-through from the collection plate.
14. Dispense 600 μL of **Zymo DNA Wash** buffer (check ethanol has been added) to each well of the Zymo-Spin I-96 plate.
15. Centrifuge the Zymo-Spin I-96 plate at 2250 × *g* for 2 min. Discard the flow-through from the collection plate.
16. Transfer the Zymo-Spin I-96 plate on top of a new MIDI plate.
17. Dispense 13 μL of **ERB** reagent on the center of each column matrix, without disturbing the membrane.
18. Incubate the plate at room temperature for 5 min.
19. Centrifuge the Zymo-Spin I-96 plate at 2250 × *g* for 1 min to elute the DNA. Approximately 10 μL of eluate will be present on the collection plate.
20. Seal the collection plate containing 10 μL of eluted DNA with a foil adhesive seal.
21. Incubate at 95 °C for 2 min.
22. During the incubation, prepare an ice bucket with ice, capable of fitting a 96-well plate.
23. Immediately after the 2-min incubation, transfer the plate to the ice for 5 min. Press the plate into the ice to make sure that the bottom of all wells contacts the ice.
24. Without moving the plate from the ice, dispense 10 μL of **CMM** reagent to each well.
25. Seal the plate and vortex it for 1 min at 1600 rpm.
26. Centrifuge the plate for 1 min at 280 × *g*.
27. Incubate the plate for 1 h at 37 °C.
28. After incubation, briefly centrifuge the plate for 1 min at 280 × *g*.

29. Dispense 140 μL of **Zymo DNA Binding** buffer to each well of the plate, pipetting up/down five times to ensure proper mixing.
30. Place a new Zymo-Spin I-96 plate on top of a new collection plate.
31. Transfer all volume from **step 29** to the Zymo-Spin I-96 plate.
32. Centrifuge the Zymo-Spin I-96 plate at 2250 × *g* for 2 min. Discard the flow-through from the collection plate.
33. Dispense 600 μL of **Zymo Wash buffer** (check that ethanol has been added) to each well of the Zymo-Spin I-96 plate.
34. Centrifuge the Zymo-Spin I-96 plate at 2250 × *g* for 2 min. Discard the flow-through from the collection plate.
35. Place the Zymo-Spin I-96 plate on top of a new MIDI plate.
36. Dispense 10 μL of DiH_2O on the center of each column matrix, without disturbing the membrane.
37. Incubate the plate at room temperature for 5 min.
38. Centrifuge the Zymo-Spin I-96 plate at 2250 × *g* for 1 min to elute the DNA. Approximately 8 μL of eluate will be present on the MIDI plate.
39. Label the plate as MSA5 plate.
40. Either proceed with the Infinium Methylation Microarray procedure immediately, or store the purified DNA at −20 °C.

3.6 Infinium Methylation Microarrays

Once we have transformed the methylation status (methylated/nonmethylated) into a nucleotide base change (C to U, which under the amplification step will become a C to T SNP), a genotyping array, designed to interrogate specific cytosines is applied to measure the level of methylation by monitoring the allele frequencies.

3.6.1 Amplification

There are two aims for this procedure: On the one hand, to amplify the starting material (250 ng of DNA) to the levels of DNA needed for the hybridization of the array; and on the other, with this amplification procedure, we transform those uracils (an RNA base) coming from the unmethylated cytosines after bisulfite conversion into thymines (a DNA base).

1. Preheat Illumina Hybridization Oven to 37 °C, allowing the temperature to equilibrate.
2. Thaw **MA1**, **RPM**, and **MSM** tubes to room temperature. Once thawed, invert tubes 5 times to mix contents.
3. Dispense 20 μL **MA1** to all wells in the MSA5 plate (from **step 39** Subheading 3.5.2), containing bisulfite-converted FFPE-restored DNA.

4. Dispense 4 μL 0.1 N NaOH to each well of the previous plate.
5. Seal the plate with an adhesive film, ensuring that all wells are properly sealed to avoid cross-contamination.
6. Vortex the plate at 1600 rpm for 1 min.
7. Pulse-centrifuge to 280 × *g*, to collect droplets.
8. Incubate the MSA5 plate 10 min at room temperature.
9. Dispense 68 μL **RPM** into each well of the plate.
10. Dispense 75 μL **MSM** into each well of the plate.
11. Seal the plate with a thermo-adhesive film, ensuring that all wells are properly sealed to avoid cross-contamination (*see* **Note 27**).
12. Vortex the plate at 1600 rpm for 1 min.
13. Pulse-centrifuge to 280 × *g*, to collect droplets.
14. Incubate the plate at 37 °C for 20 h (*see* **Note 28**).
15. Once incubation has ended, proceed immediately to fragmentation procedure.

3.6.2 Fragmentation

Amplified restored bisulfite converted DNA has to be fragmented to a 350–500 bp size to properly be able to hybridize each fragment of DNA with the beads, each one containing a designed oligo to interrogate one specific cytosine. This fragmentation process, although not much detailed information is provided by the manufacturer, is specified by them as being an enzymatic end-point fragmentation process that avoids the overfragmentation of the DNA.

1. Preheat a heat block (compatible with MIDI plates) to 37 °C.
2. Thaw **FMS** tubes to room temperature. Once thawed, invert tubes five times to mix contents.
3. Pulse-centrifuge MSA5 plate to 280 × *g*, to collect droplets.
4. Dispense 50 μL **FMS** to each well of the MSA5 plate.
5. Seal the plate with an adhesive film, ensuring that all wells are properly sealed to avoid cross-contamination.
6. Vortex the plate at 1600 rpm for 1 min.
7. Pulse-centrifuge to 280 × *g*, to collect droplets.
8. Incubate the plate at 37 °C for 1 h (*see* **Note 29**).

3.6.3 Precipitation

In order to purify the DNA from the enzymes that have been previously used, and also to performed a buffer change, the DNA has to be precipitated using isopropanol.

1. Preheat a heat block (compatible with MIDI plates) to 37 °C.

2. Thaw **PM1** tubes to room temperature. Once thawed, invert tubes five times to mix contents.
3. Dispense 100 µL **PM1** to each well of the MSA5 plate.
4. Seal the plate with an adhesive film, ensuring that all wells are properly sealed to avoid cross-contamination.
5. Incubate the plate at 37 °C for 5 min.
6. Pulse-centrifuge MSA5 plate to 280 × *g*, to collect droplets.
7. While proceeding with following steps, set the centrifuge's temperature to 4 °C, allowing the temperature to equilibrate.
8. Dispense 300 µL 100% 2-propanol (isopropanol) to each well containing sample.
9. Seal the plate with a thermo-adhesive film, ensuring that all wells are properly sealed to avoid cross-contamination. Let the filmed plate cool for 1 min at room temperature.
10. Invert MSA5 plate ten times to mix contents thoroughly.
11. Incubate MSA4 plate at 4 °C for 30 min (*see* **Note 30**).
12. Ensure that the centrifuge temperature has reached 4 °C (from **step** 7).
13. Centrifuge MSA5 plate to 3000 × *g* at 4 °C for at least 20 min (*see* **Notes 31** and **32**).
14. Prepare an absorbent pad over the bench, which will be used on **step 16**.
15. Removing the film of the MSA5 plate, over a waste container, decant the supernatant by quickly inverting the MSA5 plate. Once the plate has been inverted, do not turn it side-up until you are done with the incubation on **step 17**.
16. Tap firmly the plate upside-down several times for 1 min over the absorbent pad in the bench, to ensure that wells are devoid of liquid (*see* **Note 33**).
17. Leave the uncovered, inverted plate on a tube rack for 1 h at room temperature.
18. Proceed immediately to Resuspension step.

3.6.4 Resuspension

Precipitated DNA is resupended in the proper buffer formulated by the manufacturer (RA1) that is optimal for the latter hybridization on the beadchips. Finally, resuspended DNA will be denatured, as the input DNA material for the hybridization is single stranded DNA.

1. Preheat Illumina Hybridization Oven to 48 °C.
2. Thaw **RA1** in a room temperature water bath. Once thawed, gently mix to dissolve any crystals that may be present.

3. Add 46 μL **RA1** to each well of the MSA5 plate containing a DNA pellet.
4. Seal the plate with a thermo-adhesive film, ensuring that all wells are properly sealed to avoid cross-contamination. Let the filmed plate cool for 1 min at room temperature.
5. Incubate the plate at 48 °C for 1 h.
6. Vortex MSA5 plate to 1800 rpm for 1 min in order to resuspend the pellet.
7. Pulse-centrifuge MSA5 plate to 280 × *g*, to collect droplets (*see* **Notes 34** and **35**).

3.6.5 Hybridization

Single stranded DNA will specifically bind by complementary to its corresponding bead. Each bead carries multiple molecules of a 50 bp oligonucleotide designed so the cytosine being interrogated is located just one base ahead of the complementary region between the oligonucleotide and the DNA template.

1. Preheat a heat block (compatible with MIDI plates) to 95 °C.
2. Preheat Illumina Hybridization Oven to 48 °C.
3. Set Illumina Hybridization Oven rotation speed to 5 (Press "F" button twice; press "S" button; use increment/decrement dial to select 5 on the SPd option; press "S" to finalize).
4. Incubate MSA5 plate at 95 °C for 20 min.
5. During incubation, remove beadchips from fridge, leaving the beadchips in their ziplock bags until you are ready to begin hybridization.
6. Also, during incubation prepare hybridization chamber, by placing the gasket (Fig. 1b) in the hybridization chamber (Fig. 1c). Dispense 400 μL **PB2** in each of the humidifying reservoirs (Fig. 1c) in the hybridization chamber. Then place the array inserts on their position (Fig. 1d), and close the hybridization chamber (Fig. 1a) until ready to load the beadchips.
7. After the 20-min incubation, remove the plate and place it over the benchtop at room temperature for 30 min.
8. Pulse-centrifuge MSA5 plate to 280 × *g*, to collect droplets.
9. Remove the beadchips from their bags, and place them on top of the array inserts of the hybridization chambers, making sure that barcodes are facing on the same orientation as the diagram present on the array inserts.
10. Dispense 26 μL of each sample onto the appropriate beadchip position, according to the chart (Fig. 2) (*see* **Note 36**).
11. Once all arrays have been loaded with the samples, close the hybridization chambers, place them on the hybridization oven with the clamps facing left and right side of the oven, and incubate them at 48 °C for a period of 16–24 h.

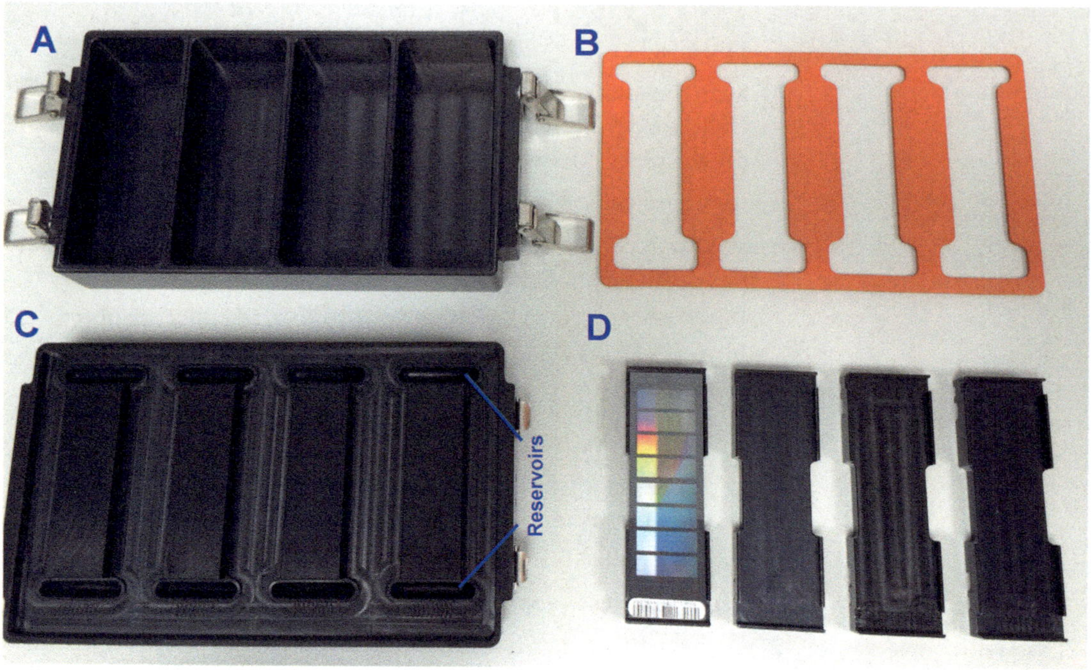

Fig. 1 Hybridization chamber assembly parts. Cover plate (**a**), gasket (**b**), bottom plate (**c**), and array inserts (**d**)

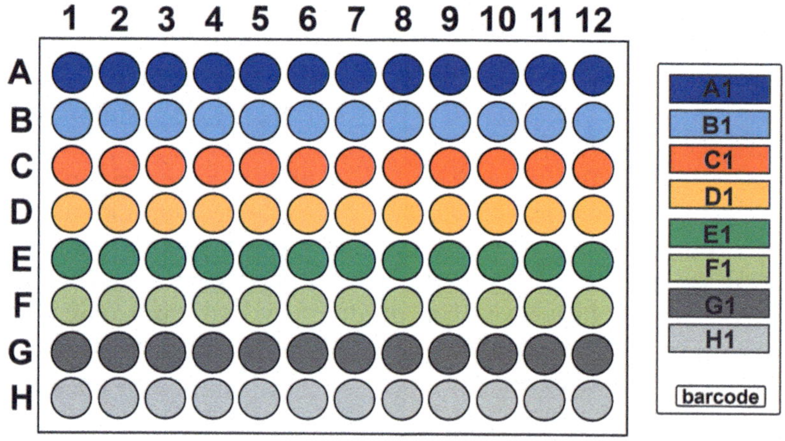

Fig. 2 Diagram of sample position in a 96-well plate and its corresponding array distribution. In each array, eight samples can be hybridized, resulting in each column of the 96-well plate corresponding to a different methylation microarray

12. Prepare **XC4** reagent for use in forthcoming steps, by adding 144 mL of 100% ethanol to a **XC4** bottle. Shake vigorously the bottle for 1 min and store it until next day at room temperature in a dark place.
13. Preparation for the forthcoming steps, glass back plates should be clean prior to their use (*see* **Note 37**).

3.6.6 Wash

Unspecific DNA bound to the beads present on the beadchip, as well as unbounded DNA will be washed away to avoid interferences or low-quality results.

1. Once incubation has finished, remove the hybridization chambers from the oven and let them cool down on the benchtop at room temperature for 30 min before opening.
2. During the 30-min cool down step, prepare two wash dishes with 200 mL **PB1** buffer and fill with 150 mL **PB1** the beadchip alignment fixture.
3. Attach the wire handle to the rack and submerge the wash rack in the wash dish containing 200 mL **PB1**.
4. Remove each beadchip from the hybridization chamber.
5. Using paper towels on the benchtop, remove the cover seal from each beadchip and immediately place the beadchip on the submerged rack at the wash dish previously prepared (*see* **Note 38**).
6. After al beadchips are placed on the wash rack, move the rack up and down for 1 min, making sure to break the surface of the **PB1** with slow agitation.
7. Move the rack to the other wash dish containing 200 mL **PB1** and move the rack up and down for another minute.
8. Let the beadchips submerged on the wash dish after the previous 1-min incubation, while you now prepare the beadchip alignment fixture.
9. For each beadchip being processed, place a black frame into the alignment fixture prefilled with **PB1**.
10. Place each beadchip to be processed into the black frame, aligning its barcode with the ridges stamped onto the alignment fixture (*see* **Note 39**).
11. For each beadchip, place a clear spacer on top of each beadchip.
12. Place the alignment bar onto the alignment fixture.
13. Place a clean glass back plate on top of the clear spacer covering each beadchip. T he reservoir space of the glass back plate should be placed facing inward at the barcode end of the beadchip, to create a reservoir against the beadchip surface.
14. Attach two metal clamps to the flow-through chambers, at approximately 5 mm from both ends.
15. Using a scissors, trim both ends of the spacers at the glass back plate level.
16. Place all assembled beadchips in horizontal position, and proceed immediately to XStain procedure.

3.6.7 XStain

Finally, single nucleotide extension using marked nucleotides (ddCTP and ddGTP are labeled with biotin, while ddATP and ddTTP are labeled with 2,4-dinitrophenol) is performed. Interrogation of which marked nucleotide has been incorporated on each bead is performed by using two specific antibodies, one against biotin (antibody labeled with a green fluorophore) and the other against 2,4-dinitrophenol (antibody labeled with a red fluorophore).

1. Thaw **RA1** in a room temperature water bath. Once thawed, gently mix to dissolve any crystals that may be present.
2. Thaw **XC1**, **XC2**, **TEM**, **STM**, and **ATM** tubes to room temperature. Once thawed, invert tubes five times to mix contents.
3. Prepare 15 mL solution of 95% formamide—1 mM EDTA.
4. Turn on the water circulator of the flow-through chamber rack, and set its temperature to 44 °C. Allow the temperature to equilibrate before starting the subsequent steps. Measure the temperature of the rack at different positions to ensure that the temperature is correct.
5. Once chamber rack reaches 44 °C (±0.5 °C), place the assembled beadchips into the chamber with the reservoir facing upwards.
6. Dispense 150 μL **RA1** to each beadchip and incubate 30 s.
7. Repeat **step 6** for a total of five times.
8. Dispense 450 μL **XC1** to each beadchip and incubate for 10 min.
9. Dispense 450 μL **XC2** to each beadchip and incubate for 10 min.
10. Dispense 200 μL **TEM** to each beadchip and incubate for 15 min.
11. Dispense 450 μL 95% formamide-1 mM EDTA to each beadchip and incubate for 1 min.
12. Repeat **step 11** one time.
13. Incubate beadchips for 5 min.
14. Set the temperature of the water circulator of the flow-through chamber rack to the temperature specified on the STM tubes (32 °C).
15. Once temperature has been set, dispense 450 μL **XC3** to each beadchip and incubate 1 min.
16. Dispense once again 450 μL **XC3** but let the beadchip incubate until rack has reached the temperature set on **step 14** (32 °C).
17. Dispense 250 μL **STM** to each beadchip and incubate 10 min.

18. Dispense 450 μL **XC3** to each beadchip and incubate 1 min.
19. Repeat previous step one time letting incubate 5 min.
20. Dispense 250 μL **ATM** to each beadchip and incubate 10 min.
21. Dispense 450 μL **XC3** to each beadchip and incubate 1 min.
22. Repeat previous step one time letting incubate 5 min.
23. Dispense 250 μL **STM** to each beadchip and incubate 10 min.
24. Dispense 450 μL **XC3** to each beadchip and incubate 1 min.
25. Repeat previous step one time letting incubate 5 min.
26. Dispense 250 μL **ATM** to each beadchip and incubate 10 min.
27. Dispense 450 μL **XC3** to each beadchip and incubate 1 min.
28. Repeat previous step one time letting incubate 5 min.
29. Dispense 250 μL **STM** to each beadchip and incubate 10 min.
30. Dispense 450 μL **XC3** to each beadchip and incubate 1 min.
31. Repeat previous step one time letting incubate 5 min.
32. Remove immediately the flow-through chambers form the chamber rack and place them horizontally on a lab bench at room temperature.
33. Fill a wash dish with 310 mL **PB1**, placing the staining rack inside the wash dish.
34. One beadchip at the time, using the dismantling tool, remove the two metal clamps. Then remove carefully the glass back plate and the spacers without disturbing the bead area. Place disassembled beadchip on the staining rack inside the wash dish containing **PB1**. Repeat the operation for all beadchips.
35. Slowly move the staining rack up and down ten times breaking the surface of the **PB1**.
36. Incubate beadchips for 5 min in **PB1**.
37. Shake vigorously the prepared **XC4** bottle, and dispense 310 mL into a wash dish.
38. Move the staining rack from the **PB1** into the **XC4** dish.
39. Slowly move the staining rack up and down ten times breaking the surface of the **XC4**.
40. Incubate beadchips for 5 min in **XC4**.
41. Lift the staining rack out of the solution and place it on a tube rack with the staining rack and beadchips horizontal, barcodes facing up.
42. Remove each beadchip, one at the time, with a locking tweezers, and rub the undersize surface of the chip (the one that does not contain any bead area) in a paper towel impregnated with 70% ethanol. This is done in order to remove **XC4** from

the undersize part of the beadchip, facilitating the proper laying of the array in the scanning trays.

43. Place each beadchip on a tube rack. Once all chips have been placed, move the rack with the beadchips to a vacuum desiccator for 50–60 min at 0.9 bar (*see* **Note 40**).

3.6.8 Scan

Measurements of the red and green fluorescence intensities are performed with a high-resolution scanner. Those fluorescence levels will be later transformed into methylation levels by a software package (GenomeStudio).

1. Turn on the scanner at least 1 h previous to scanning the arrays to let the lasers equilibrate.
2. Make sure that you have downloaded the DMAP files of the processed beadchips (*see* **Note 41**).
3. Place the beadchips on the scanning trays and follow the instructions on the scanner software using the Methylation NXT settings (*see* **Note 42**).

4 Notes

1. Ensure that your centrifuge is compatible with the height of the deep well plates which will have on top the Zymo-Spin TM I-96 Binding plate, used during the bisulfite conversion and FFPE Restoration steps.
2. All centrifugation steps unless otherwise specified should strictly be performed at room temperature.
3. Each MicroElute® DNA Mini Column can bind approximately 100 μg DNA. It is not recommended to use more than 30 mg FFPE tissue.
4. It is best to discard the first 2–3 sections from the block to obtain high-quality material. If you use 5 μm thick sections, a total of 1 cm^2 of tissue will be needed.
5. Ensure that no ethanol remains on the tube before proceeding to the next step. If so, carefully aspirate the residual ethanol with a pipette or incubate the pellet until it is dry.
6. Incubation can proceed overnight.
7. (Optional) if RNA-free gDNA is required, add 10 μL RNase A (20 mg/mL) and incubate for 5 min at room temperature.
8. Under cool ambient conditions, a precipitate may form in the BL Buffer. In case of such an event, heat the bottle at 37 °C to dissolve. Store BL Buffer at room temperature.
9. Ensure that isopropanol has been added to the HBC buffer bottle, when the kit was used for first time.

10. Ensure that ethanol has been added to the DNA Wash buffer bottle, when the kit was used for first time.
11. It is important to dry the column membrane before elution. Residual ethanol may interfere with downstream applications.
12. Using preheated 70 °C Elution buffer is optional, but if used, it will maximize the amount of DNA recovered from the column.
13. It is advisable to requantify the 50 ng/μL working solution, to ensure that a correct dilution of the stock DNA has been made. Big discrepancies between the desire and measured concentration might be attributable to improper resuspension of DNA, in which case incubating the DNA for 1 h at 37 °C with slight agitation, might help to resuspend homogenously the DNA.
14. To avoid multiple thawing–freezing cycles over the QCT reagent, create 10 μL aliquots of QCT and stored them at −20 °C. For each PCR run use a fresh QCT aliquot.
15. qPCR 2× Master Mix aliquots volumes will depend on your lab's procedures; however, it is highly advisable to aliquot qPCR Master Mix, to avoid multiple thaw–freeze cycles that will lead to a suboptimal performance of polymerase contained on the Master Mix.
16. Variations in the volume dispensation will impact the results, thus, maximum care should be taken when pipetting.
17. Depending on the qPCR Master Mix used, the activation steps (50 °C 2 min, 95 °C 10 min) will be required. In case the manufacturers do not recommend them, proceed directly to the 40 qPCR cycles.
18. It is acceptable if the amplification of NTC samples is occurring 10 cycles after the QCT standard.
19. CT Conversion reagent is light sensitive, so minimize its exposure to light as much as possible. For best results, use freshly prepared CT Conversion buffer.
20. Randomization of sample distribution among the 96-well plate, is always recommended, as it will avoid any bias in the results due to differences associated to certain positions of the arrays, or even differences among the different arrays used. In case of having samples with variable concentrations, dispense the volume needed for having 250 ng. If concentrations are variable among samples, so it will be the volumes dispensed. Once all samples are distributed over the plate, evaporate completely the plate on a vacuum spinning concentrator at a maximum of 45 °C. Add 45 μL of DiH_2O to each well of the plate, and incubate 37 °C for 15 min. Then proceed directly to **step 3**.

21. Optionally incubate plate on ice for 10 min once the last cycle has ended.
22. Ensure that the membrane is completely dry and free ofethanol, as it might interfere with downstream applications. In case presence of ethanol, discard flow-through and repeat centrifugation.
23. Preparation of 0.1 N NaOH solution shall be prepare fresh each time the procedure is applied. We recommend having a 10 M NaOH stock solution, and dilute it accordingly whenever necessary.
24. Thermo-adhesive films might also be used, if preferred. Make sure that plate well openings are not deformed when applying heat to the plate to seal the film.
25. At the end of the 1-h incubation, thaw ERB and CMM reagent tubes (use 1 tube each per 24 FFPE DNA samples) to room temperature, from the Illumina FFPE Restoration kit. Once thawed, invert them to mix the contents.
26. Care should be taken to avoid cross-contamination between wells.
27. We have extensive experience with multiple thermo-applicable films. We have found that multiple films, because of their thickness, require longer times under the heat sealer, and as a consequence deforming the overture of the wells of the plate. Although if performing the protocol manually, the deformities will not deter the operation, when performing an automated version of the protocol, liquid handler robots will have difficulties accessing the wells. In our experience, the best resulting thermo-applicable films have been Fisher-Scientific EasyPeel (ref. 11,540,284), as they seal really well the plates, while minimizing the deformities of the wells, as just 2–3 s heat sealing application is enough to seal the plate.
28. Incubation can extend up to 24 h, but should not exceed more than 24 h.
29. After the incubation, if you do not plan to proceed immediately to the precipitation procedure, you can store fragmented DNA up to 24 h at −20 °C.
30. Incubation time can extend up to 1 h without affecting the results, in order to accommodate timings to better fit working hours.
31. A blue precipitate will be formed on the bottom of the plate's wells after centrifugation step.
32. It is crucial to proceed to next step as soon as the centrifugation step has ended, to avoid any resuspension of the precipitate

formed during centrifugation. If for any reason you cannot proceed immediately, you can extend centrifugation time up to 10 extra min without affecting results. Although theoretically, extension time can be longer, we have not yet faced the need to extend centrifugation time more than 10 min from the indicated 20 min.

33. Although it is normal to proceed with extreme caution when tapping the plate over the absorbent pad on the bench (to not loosen the pellets from the bottom of the wells), make sure that you dry as much liquid from the wells as possible, as isopropanol traces might affect downstream procedures. Generous tapping can be applied without making the pellets to loosen from the bottom of the plate.
34. Make sure that the pellet is fully resuspended. Poorly resuspended samples will lead to poor data quality. If needed repeat the previous vortexing and centrifugation steps until the pellets are completely resuspended.
35. If you do not plan to proceed immediately to the hybridization procedure, you can store resuspended DNA up to 24 h at −20 °C.
36. With near-to-expiry beadchips, it is not rare to see the formation of air bubbles within the hybridization area of a certain sample. Wait until all samples have been load onto their corresponding beadchips, and then ensure that there are no air bubbles present. If so, if they are small, no interference with quality of data will happen, as each one of the 850,000 beads is present on each sample on an average of 16 times, plus they are randomly placed over the sample section of the array. In the contrary, if the section affected by the bubble is quite large, we have been successfully applied the following trick. With a small diameter intradermal needle, tap the cover film of the beadchip in order to make a small overture up to the center of the air bubble. You will immediately see how the air bubble will be replaced by respupended DNA solution, and the air will exit by the created opening.
37. It is very important to not clean the glasses on the same room as where the XStaining would be carried out, as bleach fumes will affect fluorescence signals. Clean the glass back plates with a 10% bleach solution in water. Dip the glass back plates into the solution and let them incubate for 1 h. Then individually, rinse the glasses with deionized water and place them on a rack. Submerge the rack up and down 20 times in deionized water. Rinse the glasses with deionized water and submerge the rack up and down 20 times. Let the glasses soak for 5 min. Prepare a

1% Alconox detergent powder solution with deionized water. Place the glasses on the rack on a container containing the Alconox 1% dilution. Submerge the rack up and down 20 times and let them incubate 1 h. After the 1 h, rinse carefully the glasses individually with deionized water. Let the glasses dry completely. Prior to use the glass back plates, use a paper towel (low lint) with 70% ethanol to clean the glasses individually.

38. To remove properly the seal without leaving any black adhesive on the array surface, the movement to remove the seal should be continuous starting from the corner on the barcode end and pull with a continuous upward motion away from you and towards the opposite corner on the top side of the BeadChip. Do not touch the exposed arrays. If any adhesive has been left on the array, do not try to remove it now, but proceed immediately to place the array on the PB1 wash rack.

39. In case there is any black adhesive left from the cover seal, now it is time to remove it carefully once arrays are lying horizontally and completely submerged with PB1 on the alignment fixture. With a 1000 μL pipette tip, and taking care to not touch the bead area, remove any residual adhesive.

40. Ensure that XC4 has completely dried out before continuing to next process. Placing the arrays at your eye level will facilitate checking if XC4 has completely dried out.

41. It is always advisable to download all DMAP files upon the reception of the kits, and not wait until processing them, as the download might take up to 25 min per beadchip (depending on your Internet connection).

42. It is not recommended to delay the scanning of the beadchips. If unviable to scan them right away from the XStaing procedure, store the beadchips in their enclosures with their Mylar bags, and store them at room temperature under dark conditions. Do not exceed more than 72 h to scan the beadchip as fluorescence will be degraded affecting quality of the obtained data.

References

1. Bibikova M, Lin Z, Zhou L, Chudin E (2006) High-throughput DNA methylation profiling using universal bead arrays. Genome Res 2006:383–393
2. Bibikova M, Le J, Barnes B, Saedinia-melnyk S, Zhou L, Shen R, Gunderson KL (2009) Genome-wide DNA methylation profiling using Infinium® assay. Epigenomics 1:177–200
3. Sandoval J, Heyn HA, Moran S, Serra-Musach J, Pujana MA, Bibikova M, Esteller M (2011) Validation of a DNA methylation microarray for 450,000 CpG sites in the human genome. Epigenetics 6:692–702
4. Moran S, Arribas C, Esteller M (2016) Validation of a DNA methylation microarray for 850,000 CpG sites of the human genome

enriched in enhancer sequences. Epigenomics 8:389–399

5. Thirlwell C, Eymard M, Feber A, Teschendorff A, Pearce K, Lechner M, Widschwendter M, Beck S (2010) Genome-wide DNA methylation analysis of archival formalin-fixed paraffin-embedded tissue using the Illumina Infinium HumanMethylation27 BeadChip. Methods 52:248–254
6. Moran S, Vizoso M, Martinez-Cardús A, Gomez A, Matías-Guiu X, Chiavenna SM, Fernandez AG, Esteller M (2014) Validation of DNA methylation profiling in formalin-fixed paraffin-embedded samples using the Infinium HumanMethylation450 Microarray. Epigenetics 9:829–833

Check for updates

Chapter 6

The Use of Methylation-Sensitive Multiplex Ligation-Dependent Probe Amplification for Quantification of Imprinted Methylation

Ana Monteagudo-Sánchez, Intza Garin, Guiomar Perez de Nanclares, and David Monk

Abstract

Imprinting disorders are a group of congenital diseases that can result from multiple mechanisms affecting imprinted gene dosage including cytogenetic aberration and epigenetic anomalies. Quantification of CpG methylation and correct copy-number calling is required for molecular diagnosis. Methylation-sensitive multiplex ligation-dependent probe amplification (MS-MLPA) is a multiplex method that accurately measures both parameters in a single assay. This technique relies upon the ligation of MLPA probe oligonucleotides and digestion of the genomic DNA–probe hybrid complexes with the HhaI methylation-sensitive restriction endonuclease prior to fluorescent PCR amplification with a single primer pair. Since each targeted probe contains stuffer sequence of varying length, each interrogated position is visualized as an amplicon of different size upon capillary electrophoresis.

Key words Imprinting, DNA methylation, Methylation-sensitive multiplex ligation-dependent probe amplification

1 Introduction

Imprinting disorders (IDs) are a group of congenital diseases caused by changes in allele-specific expression dosage of imprinted genes [1]. In contrast to the majority of biallelically expressed transcripts, imprinted genes are monoallelically expressed in a parent-of-origin specific manner [2]. Aberrations leading to IDs include epigenetic mutations, or changes in genomic sequence including point mutation or cytogenetic anomalies. Several molecular tests designed for single disease-specific loci have been developed for the characterization of various IDs, with the diagnostic workflow depending on the most frequent type of aberration. However very few techniques simultaneously determine copy-number variations and methylation changes within a single test.

Tanya Vavouri and Miguel A. Peinado (eds.), *CpG Islands: Methods and Protocols*, Methods in Molecular Biology, vol. 1766, https://doi.org/10.1007/978-1-4939-7768-0_6,

The methylation-sensitive multiplex ligation-dependent probe amplification (MS-MLPA) technique is a simple and reliable method for detecting copy-number and methylation changes in genomic DNA [3]. Unlike most commonly used methods to quantify methylation at imprinted loci this technique does not require prior bisulfite conversion. Genomic DNA is first denatured, followed by the addition of MS-MLPA probe mixes and an overnight hybridization step. Subsequently, the DNA–probe mix is split into two fractions. The first aliquot of the MS-MLPA reaction is processed as a normal MLPA reaction, providing information on copy number status of the target DNA, while the second is digested by the methylation-sensitive enzyme HhaI that recognizes GCGC sequences (Fig. 1). If the CpG site is methylated then there will be no digestion and a PCR amplicon will result which will be detected following capillary electrophoresis. However if the site is unmethylated the DNA–probe complex will be digested and no amplification product formed. In MLPA reactions it is not the underlying target sequence that is amplified, but the probe that hybridizes to the target sequence. Each MLPA probe consists of two separate oligonucleotides, each containing a hybridization sequence of ~55–80 base pairs and a unique PCR primer sequences allowing for single primer pair PCR amplification. One of the two

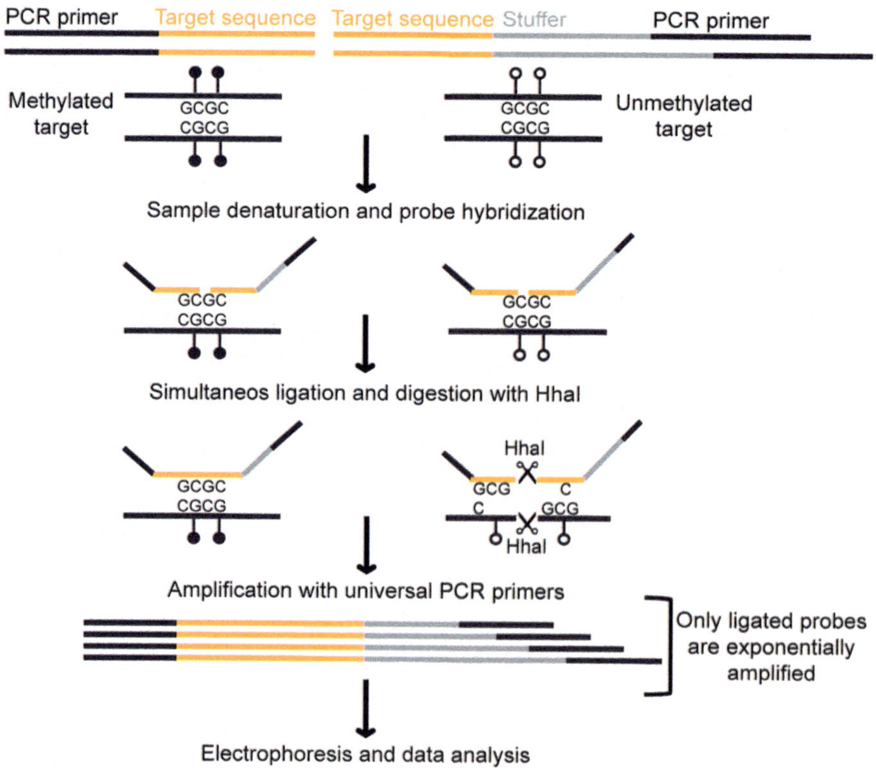

Fig. 1 Overview of probe design and the MS-MLPA assay

primers also contains a stuffer sequence that varies in length (typically 100 to >500 nucleotides) allowing for multiplex reactions to be separated using capillary electrophoresis. Importantly the two probe oligonucleotides hybridize to immediately adjacent target sequences and ligated. Because only ligated probes will be exponentially amplified during the PCR reaction, the number of probe ligation products is a measure for the number of target sequences in the sample.

Currently there are several commercially available MS-MLPA kits from MRC Holland (http://mrc-holland.com) targeting specific imprinted loci associated with IDs (Table 1), the majority of which are routinely used in diagnostic laboratories [4–11].

Table 1
Description of the available MS-MLPA kits from MRC Holland targeting imprinted loci indicating the position of Hha1 probes for methylation quantification

Probe ID	Imprinted DMR	Hha1 location (GRCh37/hg19)
ME033 transient neonatal diabetes mellitus		
18458-L25628	*PLAGL1*	chr6:144,329,309
18460-L26610	*PLAGL1*	chr6:144,328,934
15755-L25779	*PLAGL1*	chr6:144,329,107
ME032 UPD7/UPD14		
18460-L24410	*PLAGL1*	chr6:144,328,934
15755-L24417	*PLAGL1*	chr6:144,329,107
18458-L24538	*PLAGL1*	chr6:144,329,309
15744-L24405	*GRB10*	chr7:50,850,532
15742-L18941	*GRB10*	chr7:50,850,757
15756-L17775	*MEST*	chr7:130,131,403
15750-L17769	*MEST*	chr7:130,131,475
15749-L17768	*MEST*	chr7:130,132,161
15754-L18942	*MEG3*	chr14:101,292,098
SO912-L25282	*MEG3*	chr14:101,292,375
15759-L17778	*MEG3*	chr14:101,292,454
ME028 PraderWilli/Angelman syndrome		
nt250[a]	*SNURF*	chr15:25,200,055
nt178[a]	*SNURF*	chr15:25,200,133
nt190[a]	*SNURF*	chr15:25,200,420
nt 142[a]	*SNURF*	chr15:25,200,693

(continued)

Table 1
(continued)

Probe ID	Imprinted DMR	Hha1 location (GRCh37/hg19)
nt419[a]	*NDN*	chr15:23,932,370
ME030 Beckwith Wiedemann/Silver Russell syndrome		
06266-L05772	*H19/IGF2*	chr11:2,019,397
14792-L16503	*H19/IGF2*	chr11:2,019,566
14063-L08764	*H19/IGF2*	chr11:2,019,736
08743-L20532	*H19/IGF2*	chr11:2,020,030
1654-L19204	*KCNQ1OT1*	chr11:2,720,587
06276-L05782	*KCNQ1OT1*	chr11:2,720,649
07172-L06781	*KCNQ1OT1*	chr11:2,721,027
07173-L19191	*KCNQ1OT1*	chr11:2,721,437
ME031 GNAS		
04870-L23097	*NESP*	chr20:57,414,760
18126-SP0007-L22614	*NESP*	chr20:57,414,960
18194-SP0008-L21093xl	*NESP*	chr20:57,415,192
03877-L23093	*GNAS-AS1*	chr20:57,425,849
18127-SP0009-L25603	*GNAS-AS1*	chr20:57,425,950
06190-L22606	*GNAS-AS1*	chr20:57,426,057
03879-L23095	*GNAS-XL*	chr20:57,429,259
18125-L25598	*GNAS-XL*	chr20:57,429,312
07523-L17516	*GNAS-XL*	chr20:57,430,133
15646-L22676	*GNAS-XL*	chr20:57,430,214
12775-L21107	*GNAS-XL*	chr20:57,430,980
06191-L23094	*GNAS A/B*	chr20:57,464,129
03882-L22603	*GNAS A/B*	chr20:57,464,375
PO47 RB1		
S1065-L27111	*RB1*	chr13:48,892,668
15264-L25114	*RB1*	chr13:48,893,174
15265-L25148	*RB1*	chr13:48,893,598
19147-L17021	*RB1*	chr13:48,893,778

[a]No probe ID given, listed by size of PCR amplicon

2 Materials

2.1 Components of SALSA MLPA Kits

SALSA MLPA Buffer: KCl, Tris–HCl, EDTA, PEG-6000, oligonucleotides. pH 8.5.

SALSA Ligase-65: Glycerol, BRIJ (0.05%), EDTA, Beta-Mercaptoethanol (0.1%), KCl, Tris–HCl. pH 7.5, Ligase-65 enzyme.

Ligase Buffer A: NAD. pH 3.5.

Ligase Buffer B: Tris–HCl, nonionic detergents, $MgCl_2$. pH 8.5.

SALSA PCR Primer and polymerase: Synthetic oligonucleotides with fluorescent dye (FAM, Cy5, IRD800, ROX, or unlabeled), dNTPs, Tris–HCl, KCl, EDTA, BRIJ. pH 8 Glycerol, nonionic detergents, EDTA, DTT (0.1%), KCl, Tris–HCl, Polymerase enzyme (bacterial origin). pH 7.5.

2.2 Specific MLPA Probe Mixes

Probemix: Synthetic oligonucleotides, oligonucleotides purified from bacteria, Tris–HCl, EDTA. pH 8.0.

2.3 Additional Material

HhaI restriction endonuclease.

0.2 ml PCR tubes, strips or plates.

Formamide.

Thermocycler with heated lid.

Capillary Electrophoresis equipment, polymer and size standards.

Standard laboratory equipment.

3 Methods

3.1 Sample Preparation

1. The optimum amount of DNA for reach MLPA reaction is 50–100 ng at a concentration of ~20 ng/μl diluted in buffer TE 0.1 or water (final volume 5 μl) (*see* **Note 1**).
2. Always compare the MS-MLPA peak patterns observed after capillary electrophoresis of the patients' sample with reference samples to estimate reproducibility of the results. We recommend using at least 3–5 different control samples with normal copy-number and methylation profile per MS-MLPA run. We also strongly recommend to include samples with known underlying methylation defects or cytogenetic anomalies (deletions, duplications or uniparental disomies) to ensure correct detection.
3. As with standard PCR we recommend including a no template control (5 μl TE) to expose a possible contaminations of the MLPA reagents of electrophoresis capillaries.

3.2 MS-MLPA Reaction: Day 1

1. Set the programs for denaturation, hybridization, ligation and digestion and PCR into a thermocycler with a heated lid. Perform PCR with the following conditions: Initial DNA denaturation 98 °C min, 25 °C pause; hybridization 95 °C 1 min, 60 °C pause, 20 °C pause; ligation and ligation-digestion 48 °C pause, 48 °C 30 min, 98 °C 5 min, 20 °C pause; PCR reaction (35 cycles) 95 °C 30 s, 60 °C s, 72 °C 1 min; final extension 72 °C 20 min, 15 °C pause.
2. Add the 5 μl DNA sample into a labeled 0.2 ml PCR tube. Use TE for the no template control. Start the thermocycler program to denature the DNA samples for 5 min at 98 °C and subsequently cool the sample to 25 °C before continuing with the next step.
3. Vortex the MLPA buffer and MLPA probe mix before use. Pulse in a centrifuge to collect any liquid that may have collected in cap. Prepare the hybridization master mix which contains 1.5 μl MPLA buffer +1.5 μl probemix for each reaction. Mix well by repeated pipetting.
4. Add 3 μl of master mix to each denatured sample. Mix well by pipetting. Place the tubes back in the thermocycler and continue with the preset program: incubate for 1 min 95 °C followed by 16–20 h at 60 °C. At this stage the reaction can be left overnight.

3.3 Ligation and Ligation-Digestion Reactions: Day 2

1. Briefly vortex the two ligase buffers before use and pulse in a centrifuge to collect any liquid that may have collected in cap.
2. Prepare the Ligase-65 master mix. For each reaction add 8.25 μl water, 1.5 μl ligase buffer B and 0.25 μl ligase-65 enzyme. Mix well by pipetting (*see* **Note 2**).
3. Prepare the Ligase-digestion master mix. For each reaction add 7.75 μl water, 1.5 μl ligase buffer B. Then add 0.25 μl ligase-65 enzyme and 0.5 μl of HhaI enzyme (5 U). Mix well by pipetting.
4. Continue the thermocycler program, paused at 20 °C. Remove the tubes containing the DNA–probe complex and add 3 μl of ligase buffer A and 10 μl of water to each tube. Mix by gently pipetting and separate the mixture by transferring 10 μl of the whole mixture to a second tube.
5. Place the tubes back in the thermocycler, releasing the 20 °C pause step and continue the thermocycler program to 48 °C. When the thermocycler is at 48 °C, add 10 μl of the Ligase-65 master mix to the first tube (for copy number evaluation). Mix by gently pipetting.
6. Add 10 μl of the ligase-digestion mix to the second tube (for methylation quantification) and mix by gently pipetting.

7. Continue the thermocycler program with a 30 min incubation at 48 °C followed by a 5 min at 98 °C to inactivate enzymatic activity and return to 20 °C pause.

3.4 PCR Reaction: Day 2

1. Vortex the SALSA PCR primer mix. Prepare the PCR master mix, for each reaction add 3.75 μl water, 1 μl SALSA PCR primer mix, and 0.25 μl SALSA Polymerase. Mix well by pipetting up and down (*see* **Note 2**). Incubate on ice until use.
2. At room temperature add 5 μl PCR mix to each tube. Mix by gently pipetting. Continue thermocycler program for 35 cycles of 30 s 95 °C; 30 s 60 °C; 60 s 72 °C. End with 20 min incubation at 72 °C and pause the reaction at 15 °C. PCR product can be stored at 4 °C for 1 week or longer at −20 °C (*see* **Note 3**).

3.5 Capillary Electrophoresis: Day 2–3

1. Run the 0.7 μl of the PCR on electrophoresis machine. Size standard, run conditions, polymer and fluorescent dye depend on instrument type. *See* Table 2 for standard settings (*see* **Notes 4–6**).

Table 2
Conditions for capillary electrophoresis

Instrument	Primer dye	Capillaries	Injection mixture	Settings
Beckman CEQ-2000 CEQ-8000 CEQ-8800	Cy5	33 cm	0.7 μl PCR reaction 0.2 μl CEQ—size standard 600 32 μl HiDi formamide/ Beckman SLS Add one drop of high quality mineral oil	Run method: Frag Capillary temperature: 50 °C Denaturation: 90 °C for 120 s Injection voltage: 1.6 kV Injection time: 30 s Run time: 60 min Run voltage: 4.8 kV
ABI-Prism 3100 (Avant) ABI-3130 (XL) ABI-3500 ABI-3730 (XL)	FAM	36, 50 cm	0. 7 μl PCR reaction 0.3 μl ROX/0.2 μl LIZ Size standard 9 μl HiDi formamide Seal the injection plate, incubate for 3 min at 86 °C, cool for 2 min at 4 °C	Run module: Fragment Analysis (compatible with array length) Injection voltage: 1.6 kV Injection time: 15 s Run voltage: 15 kV run time: 1800 s Oven temperature: 60 °C
ABI-Prism 310	FAM	47 cm	0.75 μl PCR reaction 0.75 μl dH_2O 0.5 μl Size standard 13.5 μl HiDi formamide Incubate mix for 3 min at 86 °C, cool for 2 min at 4 °C	Injection voltage: 1.6 kV Injection time: 15 s Filter set: D-polymer: POP4

2. Visualize the result using raw data depending on instrument used:

   Raw data (.fsa, .scf, .cqf, .esd files prior to size-calling):

   (a) Beckman: Coffalyser.Net or Beckman GenomeLab software. (b) ABI: Coffalyser.Net, GeneMapper, GeneMarker, Peak Scanner, or Foundation Data Collection Software on CE instrument.

   Size-calling your data: (a) Beckman: use Coffalyser.Net or CEQ/GeXP GenomeLab software to size-call and visualize data. (b) ABI: use Coffalyser.Net, GeneMapper, GeneMarker, or Peak Scanner to size-call and then visualize data.

3.6 Assessing the Quality of the Data Using Internal Controls

1. Each MLPA probe set contains quality control (QC) fragment for diagnosing any potential problems. Only data fulfilling the QC is suitable for interpretation.

   A 92 nt benchmark probe is used for comparison with other QC fragments.

   Four Q-fragments (64, 70, 76 and 82 nt) determine whether sufficient DNA sample was used and if the ligation step was successful. The Q-fragments do not require hybridization or ligation for amplification and signal intensity is inversely related to sample DNA, i.e., the sample DNA competes with the Q-fragments, the more sample added the lower the Q-fragment signal. In the no template reaction, only the spikes Q-fragments should be visible following electrophoresis. This is a control for PCR contamination and nonspecific peaks.

   Two D-fragments (88, 96 nt) determine if denaturation was complete. Low peak heights correspond to denaturing problems often associated with CG-rich sequences.

3.7 Data Analysis

1. MRC Holland recommends the use of Coffalyser.Net for MS-MLPA data analysis, a specifically designed algorithm with QC checks; however many laboratories prefer custom excel file. The analysis of MS-MLPA consists of two parts corresponding to determining either copy-number using the undigested samples or for methylation that compares each undigested sample with its digested counterpart.
2. For determining copy number the absolute fluorescent detected by after electrophoresis requires careful intrasample and intersample normalization. For *intra-sample normalization* the peak heights for all probes are compared to those reference probes that quantify copy-number at normal loci and do not contain HhaI sites. This should be done by dividing the signal of each probe by the signal of every reference probe in that sample, thus creating as many ratios per probe as there

are reference probes. The normalization constant (NC) is represented as the median of all the ratios per probe.

For inter-sample normalization the peak height pattern of DNA sample of interest is compared to the control reference samples described in Subheading 3.1, **step 2**. The final probe ratio for each probe is calculated to give the dosage quotient (DQ):

$$\text{DQ probe} = \frac{[\text{NC for probe n in undigested test sample}]}{[\text{average NC for probe n in the undigested all reference samples}]}.$$

To facilitate the interpretation of results, the DQ probes can be arranged according to chromosomal location. This will reveal deletion or duplication of MLPA probe targets. For reliable results the standard deviation of all probes in the reference samples should be $\leq$0.10 and DQ between 0.85–1.15. Similar criteria should be observed for the reference probes in the test samples.

For reliably determine copy numbers, the exact cutoff values for the dosage quotients will differ for each probe as they depend on the standard deviation of that probe. For expected values for duplication and deletions *see* Table 3 (Fig. 2) (*see* **Note 7**).

3. The methylation values for each MS-MLPA probe can be determined by comparing the peak pattern of each digested DNA sample with the corresponding undigested sample. This give a relative methylation percentage per probe calculated as follows:

$$\%\text{Methylation} = \frac{[\text{NC for MS-MLPA probe in digested test sample}]}{[\text{NC for MS-MLPA probe n in undigested test sample}]} \times 100\%.$$

Table 3
Dosage quotients for deletions, duplications, triplications, and uniparental disomies

Copy number	Dosage quotient
Normal	0.85 < DQ < 1.15
Heterozygous duplication	1.35 < DQ < 1.55
Triplication	1.70 < DQ < 2.20
Heterozygous deletion	0.35 < DQ < 0.65
Homozygous deletion	0
Uniparental disomy	0.85 < DQ < 1.15
Ambiguous copy number	Any other values

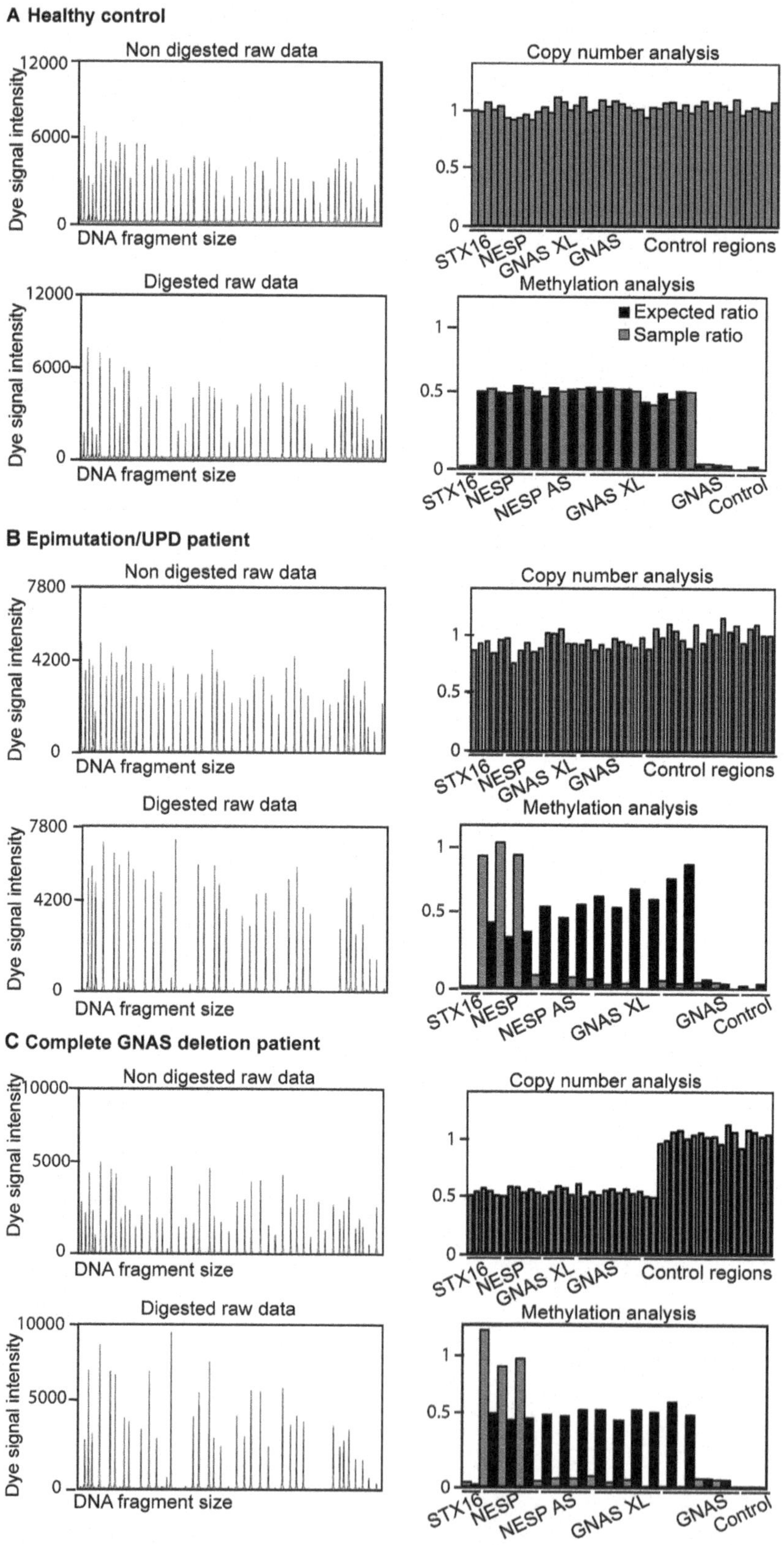

Fig. 2 Examples of electropherograms and the normalized resulting data for (**a**) control DNA sample, (**b**) epimutated/UPD case, and (**c**) a sample with complete maternal deletion of the GNAS locus

The methylation profile of a test sample is then assessed by comparing the probe methylation percentages obtained for the test sample to the percentages of the normal reference samples (*see* **Notes 8–10**). When patients with methylation defects and normal copy-number are identified it is important to perform follow-up genetic studies (SNP arrays or STR markers) to differentiate between epimutations and uniparental disomy.

For a reliable methylation analysis the probe signal of the "Digestion Control Probes" in all samples should be <4% of the corresponding probe signals in the undigested reactions. The digestion control probes map to regions that are devoid of methylation in all samples and map to non-imprinted loci. For all digested samples the DQ of the reference probes should be between 0.7 and 1.3.

4 Notes

1. The MLPA PCR step is very sensitive and all template DNA should be of high quality free of contaminants such as salt and phenol and ethanol. Always treat DNA with RNAse during the extraction protocol for optimal results. We recommend using the same DNA extract protocol for all samples and never to compare test samples and controls extracted using different methods.
2. Never vortex enzyme solutions, mixing by repeat pipetting is sufficient.
3. All fluorescent dyes are light-sensitive and undergo photobleaching. Therefore store PCR products should be placed in a dark box or wrapped in aluminum foil until use.
4. To avoid PCR contamination, do not open the PCR tubes in the area used for Subheadings 3.2 and 3.3. To avoid amplicon crossover, use different micropipettes for performing MLPA reactions and handling PCR products.
5. Change capillaries and polymer regularly to avoid polymer deterioration that can affect signal peak profiles.
6. If all PCR peak signals are low, do not add more amplicon to injection mixture as this will increases the salt concentration in the injection mixture, competing with the DNA for injection.
7. The standard deviation is affected by DNA sample quality, selection of reference DNA samples and quality of the MLPA reaction and capillary electrophoresis.
8. Reference samples should be derived from the same tissue source and of a similar age. They should also be expected to have a normal copy number and methylation status for the DMR of interest.

9. Note that only the CpG dinucleotides associated with the HhaI recognition sites are analyzed in each region. It is possible that discrete methylation aberrations mapping elsewhere within a DMR could be missed.

10. It is important to keep in mind that any SNP/mutation located within the hybridization region will create a mismatch interfering with probe ligation and amplification, leading to a false negative result (i.e., it could be interpreted as no methylation or aberrant dosage). Usually each MS-MLPA assay contains more than one probe pair for each imprinted DMR so if an isolated probe presents with an aberration, subsequent analysis using a different technique should be performed.

Acknowledgments

This work was supported by Spanish Ministry of Economy and Competitiveness (MINECO) (BFU2014-53093-R to D.M) cofunded with the European Union Regional Development Fund (FEDER); Institute of Health Carlos III of the Ministry of Economy and Competitiveness cofinanced with European Union ERDF funds (PI13/00467 to G.P.dN.); the Basque Department of Health (GV2014111017 to G.P.dN.) and the I3SNS Program of the Spanish Ministry of Health (CP03/0064; S.I.V.I. 1395/09 to G.P.dN.). AMS is a recipient of an F.P.I. Ph.D. studentship from MINECO. All authors are members of the "European Network for Human Congenital Imprinting Disorders" COST action (BM1208).

References

1. Soellner L, Begemann M, Mackay DJ et al (2016) Recent advances in imprinting disorders. Clin Genet. https://doi.org/10.1111/cge.12827
2. Monk D (2015) Germline-derived DNA methylation and early embryo epigenetic reprogramming: the selected survival of imprints. Int J Biochem Cell Biol 67:128–138. https://doi.org/10.1016/j.biocel.2015.04.014
3. Nygren AO, Ameziane N, Duarte HM et al (2005) Methylation-specific MLPA (MS-MLPA): simultaneous detection of CpG methylation and copy number changes of up to 40 sequences. Nucleic Acids Res 33:e128. https://doi.org/10.1093/nar/gni127
4. Procter M, Chou LS, Tang W et al (2006) Molecular diagnosis of Prader-Willi and Angelman syndromes by methylation-specific melting analysis and methylation-specific multiplex ligation-dependent probe amplification. Clin Chem 52:1276–1283. https://doi.org/10.1373/clinchem.2006.067603
5. Dawson AJ, Cox J, Hovanes K et al (2015) PWS/AS MS-MLPA confirms maternal origin of 15q11.2 microduplication. Case Rep Genet 2015:474097. https://doi.org/10.1155/2015/474097
6. Priolo M, Sparago A, Mammì C et al (2008) MS-MLPA is a specific and sensitive technique for detecting all chromosome 11p15.5 imprinting defects of BWS and SRS in a single-tube experiment. Eur J Hum Genet 16:565–571. https://doi.org/10.1038/sj.ejhg.5202001
7. Scott RH, Douglas J, Baskcomb L et al (2008) Methylation-specific multiplex ligation-dependent probe amplification (MS-MLPA) robustly detects and distinguishes 11p15 abnormalities associated with overgrowth and

growth retardation. J Med Genet 45:106–113. https://doi.org/10.1136/jmg.2007.053207
8. Garin I, Mantovani G, Aguirre U et al (2015) European guidance for the molecular diagnosis of pseudohypoparathyroidism not caused by point genetic variants at GNAS: an EQA study. Eur J Hum Genet 23:560. https://doi.org/10.1038/ejhg.2015.40
9. Yuno A, Usui T, Yambe Y et al (2013) Genetic and epigenetic states of the GNAS complex in pseudohypoparathyroidism type Ib using methylation-specific multiplex ligation-dependent probe amplification assay. Eur J Endocrinol 168:169–175. https://doi.org/10.1530/EJE-12-0548
10. Sachwitz J, Strobl-Wildemann G, Fekete G et al (2016) Examinations of maternal uniparental disomy and epimutations for chromosomes 6, 14, 16 and 20 in Silver-Russell syndrome-like phenotypes. BMC Med Genet 17:20. https://doi.org/10.1186/s12881-016-0280-8. PMID: 26969265
11. Price EA, Price K, Kolkiewicz K et al (2014) Spectrum of RB1 mutations identified in 403 retinoblastoma patients. J Med Genet 51:208–214. https://doi.org/10.1136/jmedgenet-2013-101821

URLs

http://www.mrc-holland.com
https://coffalyser.wordpress.com
https://www.beckmancoulter.com

Chapter 7

The Pancancer DNA Methylation Trackhub: A Window to The Cancer Genome Atlas Epigenomics Data

Izaskun Mallona, Alberto Sierco, and Miguel A. Peinado

Abstract

The Cancer Genome Atlas (TCGA) epigenome data includes the DNA methylation status of tumor and normal tissues of large cohorts for dozens of cancer types. Due to the moderately large data sizes, retrieving and analyzing them requires basic programming skills. Simple data browsing (e.g., candidate gene search) is hampered by the scarcity of easy-to-use data browsers addressed to the broad community of biomedical researchers. We propose a new visualization method depicting the overall DNA methylation status at each TCGA cohort while emphasizing its heterogeneity, thus facilitating the evaluation of the cohort variability and the normal versus tumor differences. Implemented as a trackhub integrated to the University of California Santa Cruz (UCSC) genome browser, it can be easily added to any genome-wide annotation layer.

To exemplify the trackhub usage we evaluate local DNA methylation boundaries, the aberrant DNA methylation of a CpG island located at the estrogen receptor 1 (ESR1) in breast and colon cancer, and the hypermethylation of the Homeobox HOXA gene cluster and the EN1 gene in multiple cancer types. The DNA methylation pancancer trackhub is freely available at http://maplab.cat/tcga_450k_trackhub.

Key words DNA methylation, Pancancer, Data visualization, TCGA, The Cancer Genome Atlas

1 Introduction

Cancer research is being benefited by the unprecedented availability of multilayered molecular data gathered by international consortia. Namely, The Cancer Genome Atlas (TCGA) [1] offers a wide assortment of genomics and epigenomics data for more than 30 cancer types, thus facilitating hypothesis generation and testing. The genome-wide quality of the data and hence its big size might however set back browsing individual candidates. This gap is largely due to the lack of specific tools to extract and explore parts of the data; noteworthy, this is the case for experimentalists or clinicians, who are interested in candidate-based queries rather than in accessing whole-genome datasets at once [2]. Although there are many refined methods of data mining that generate valuable information

Tanya Vavouri and Miguel A. Peinado (eds.), *CpG Islands: Methods and Protocols*, Methods in Molecular Biology, vol. 1766, https://doi.org/10.1007/978-1-4939-7768-0_7, © Springer Science+Business Media, LLC, part of Springer Nature 2018

from these big datasets [3, 4], interpretation by biological and clinical domain experts is still a major way to retrieve basic and applicable knowledge from them.

(Epi)genomics data are still lacking tools to be effectively displayed, integrating both the user-friendliness and the big biodatasets mining. For instance, a common gateway to genomic data is genome browsers, which offer chromosome views with some annotated layers over it. The University of California Santa Cruz (UCSC) genome browser is widely used in (epi)genomic data visualization: it renders a genomic region offering an overview of selected datasets. Among the data included there are transcript and genes models, regulation, expression, epigenetics, disease association, variation, and evolutionary pairwise genomic alignments [5]. Such availability makes the tool very insightful for visual correlation. The main drawback is that data are usually displayed in a single window, limiting the maximum number of tracks. Thus data tracks are mostly useful when they summarize models or generalizations, rather than individual samples.

This chapter describes a pancancer set of DNA methylation tracks for the UCSC Genome Browser to visualize datasets with dozens to thousands of patients in a compact manner while depicting their inner variability and differences between tumor versus normal tissue (Fig. 1a). To exemplify the trackhub usability we display examples of local DNA methylation boundaries between CpG islands and their flanking sequences, the aberrant DNA methylation of the estrogen receptor 1 (ESR1) in breast and colon cancer and the hypermethylation of the Homeobox HOXA cluster and the EN1 gene in multiple cancer types.

2 Trackhub Generation

We fetched the open access tier of highly processed DNA methylation data (TCGA level 3) of over 450,000 genomic locations as measured by the Illumina's Infinium array [6] (Table 1; *see* **Notes 1–3**). As the methylation metadata refers to biospecimens stratified by cancer cohort and sample type, we retrieved those corresponding to solid primary tumors and matched normals (sample codes 01 and 11, respectively). Data was downloaded using TCGA-Assembler v1.0.3 [7].

Theoretically, the methylation status of a given CpG is rather binary, being methylated or unmethylated; however, some cells might have an intermediate state, with the two epialleles present (imprinted). Strikingly, however, nonimprinted loci often follow a continuous bimodal distribution ranging from full unmethylation to complete methylation (beta values from 0 to 1) [6]. This is mainly due to the mixture of cell populations subjected to the measurement of the CpG methylation (except for imprinted

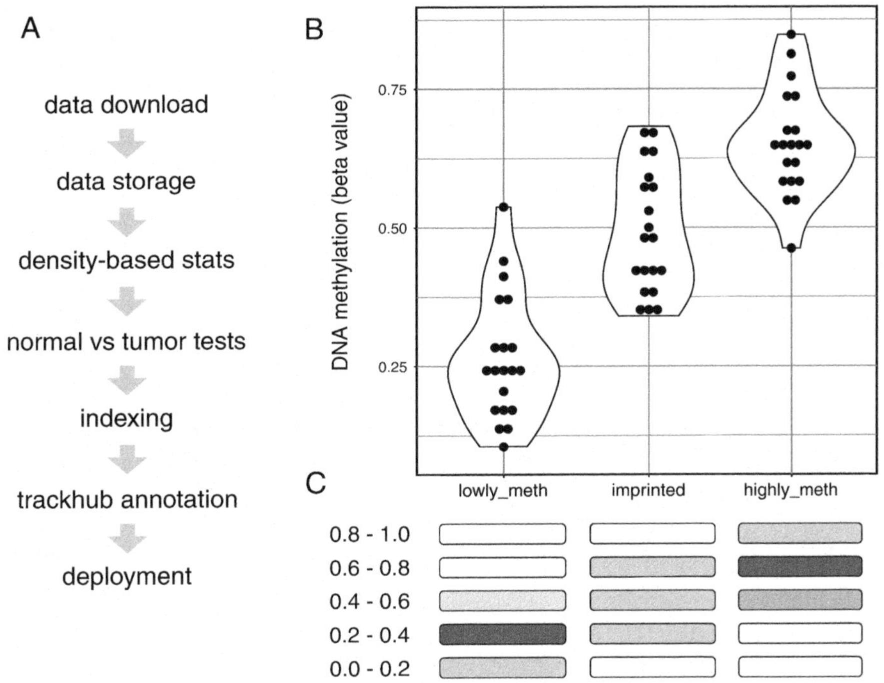

Fig. 1 Pancancer trackhub. (**a**) Data flow chart. (**b**) DNA methylation status for three loci (lowly methylated, imprinted and highly methylated) across a cohort (each point represents a sample) as depicted by violin plots. (**c**) Track-based representation of the same series summarizing cohort heterogeneity

sites). The DNA methylation status of numerous genomic regions, however, is conserved across samples belonging to the same tissue and physiological or pathological condition. Therefore, visualizing the DNA methylation barcode of a set of samples aid to interpret and/or infer their inner complexity. To do so we have developed a new visualization of the DNA methylation status in which the abundance of samples in a low, intermediate and high DNA methylation status is shown for each genomic position (Fig. 1b, c).

We represent a 5-row stacked track (namely, composite track) in which the methylation status is split and sorted into its layers from almost full methylation (beta value of 0.8–1) on top to minimal methylation (beta 0–0.2) on bottom, with three rows in between depicting intermediate methylation statuses (betas 0.8–0.6, 0.6–0.4 and 0.4–0.2). For each CpG, the color at each track reflects the proportion of samples which show that methylation status; such score ranges from 0 to 1000, with 1000 being the 100% of the data (Fig. 1c).

Apart from visualizing the methylation profiles, we generated direct comparisons between normal and tumor for cohorts in which both datasets are present. To statistically assess such differences, we performed independent nonparametric Wilcoxon rank sum tests on each CpG in which the beta value showed an absolute difference of at

Table 1
DNA methylation dataset sizes

Description	Code	Normal tissue	Tumor tissue
Adrenocortical carcinoma	ACC	0	80
Bladder urothelial carcinoma	BLCA	21	358
Brain lower grade glioma	LGG	0	511
Breast invasive carcinoma	BRCA	98	743
Cervical squamous cell carcinoma and endocervical adenocarcinoma	CESC	3	256
Colon adenocarcinoma	COAD	38	302
Esophageal carcinoma	ESCA	16	185
Glioblastoma multiforme	GBM	2	129
Head and neck squamous cell carcinoma	HNSC	50	528
Kidney chromophobe	KICH	0	66
Kidney renal clear cell carcinoma	KIRC	160	324
Kidney renal papillary cell carcinoma	KIRP	45	226
Liver hepatocellular carcinoma	LIHC	50	256
Lung adenocarcinoma	LUAD	32	463
Lung squamous cell carcinoma	LUSC	43	361
Lymphoid neoplasm diffuse large B + AC0-cell lymphoma	DLBC	0	48
Mesothelioma	MESO	0	37
Ovarian serous cystadenocarcinoma	OV	0	10
Pancreatic adenocarcinoma	PAAD	10	146
Pheochromocytoma and paraganglioma	PCPG	3	179
Prostate adenocarcinoma	PRAD	49	340
Rectum adenocarcinoma	READ	7	98
Sarcoma	SARC	4	242
Skin cutaneous melanoma	SKCM	2	92
Stomach adenocarcinoma	STAD	2	339
Thyroid carcinoma	THCA	56	507
Uterine carcinosarcoma	UCS	0	57
Uterine corpus endometrial carcinoma	UCEC	46	438
Uveal melanoma	UVM	0	80

Tumors depict the number of primary tumors samples scrutinized. Normal tissue refers to adjacent to the primary tumor; note that some cohorts lack the adjacent normal counterpart and therefore will lack the statistics track least 0.2 [6]. Tests did not assume normal-tumor pairing and compared the whole distribution of normal against that of tumor tissues (two-tailed, null hypothesis: there are no differences between normal and tumor samples). Significance p-value cutoff was set to 0.001 ($p < 0.001$ there are detectable differences between normal and tumor samples). To ease visualization we depicted no differences ($p >= 0.001$) in gray, tumor hypermethylations in red, and tumor hypomethylations in blue (*see* **Notes 4** and **5**).

3 Querying and Cohort Selection

UCSC browser can be launched by entering http://maplab.cat/tcga_450k_trackhub from any Web browser. This will render a default genomic location, which can be readily changed to another region of interest by typing a position (i.e., chr6:152,107,634–152,148,574), a gene symbol (e.g., GLDC) or Illumina Infinium probename (e.g., cg16029534) (*see* **Notes 3–9**). After pressing the "go" button the browser will point to the new location (zooming in and out might be necessary).

The setup panel named "Pan-cancer DNA methylation profiling" located below the track window allows to select the cohort to focus on. Due to the abundance of data, most cohorts are hidden by default, and manual selection of a target cohort is required. To start exploring the Bladder Urothelial Carcinoma (BLCA) dataset just click the second drop down menu (named "blca") and select "show." Then, click to the "refresh" button at the upper right corner of the setup panel (*see* **Note 4**). This will render a composite track describing the DNA methylation distribution across tumor samples, another equivalent for the matching normal; and a simple track for the normal vs. tumor statistical comparison.

4 Data Visualization and Interpretation

Both the normal and the tumor composite tracksets are composed of a stack of five tracks depicting the abundance of samples belonging to five strata of DNA methylation, from nearly fully methylated (top track, named "a_blca_1.0", with beta values between 0.8 and 1) to almost totally unmethylated (bottom track, named "e_blca_0.2", with beta values between 0 and 0.2), with three tracks of intermediate methylation between them. The shade of gray at each track depicts the abundance of samples at that level of DNA methylation: black represent that all samples belong to the DNA methylation interval; and white, none (Fig. 1).

Differences in DNA methylation between tumor and normal samples are depicted for those cohorts where both datasets exist

(Table 1). The statistics track simply depicts in grey the scrutinized CpGs; and tumor hypomethylation in blue and tumor hypermethylation in red.

The tracks depicting the DNA methylation distribution are designed to be visualized as dense tracks by default, so they arrange all the DNA methylation data inline. This display mode can be changed in order to expand the methylation data to multiple rows. Taking the BLCA normal dataset as an example, the display mode can be updated by right-clicking to the gray vertical bar on their left at the browsing panel ("configure Methylation in blca normal track set"). The settings page present a drop-down named "display mode" which permits bulk updates (i.e., expanding or collapsing many datasets at once). If desired, each track can be fine tuned (e.g., hidden) independently. Clicking to "submit" after selecting the proper visualization method will save the changes and reload the browser window with the updated display mode (*see* **Notes 4** and **5**).

4.1 Pancancer Tumor vs. Normal Evaluation

The stacked DNA methylation density tracks allow to visually correlate hypermethylation and hypomethylation. Independently, the statistics track simplifies the visualization as it summarizes in a single line the significant changes. To perform a pancancer analysis evaluating the statistical differences for a given region across cohorts (i.e., in bulk), the configure button (below the browsing window) launches a form for the "Pan-cancer DNA methylation profiling" in which all the statistics tracks display options must be set to "dense" (they appear as "hidden" by default). Then, clicking the "submit" button at the top will save the changes and draw an updated browsing window in which shared and divergent patterns can be easily detected (as in Figs. 2, 3, 4, and 5).

4.2 Candidate CpGs Numeric Data Acquisition

The pancancer trackhub renders the DNA methylation distribution with a grayscale color palette; and, for cohorts with both normal and tumor data, with an extra track testing hypermetylations and hypomethylations. The user, however, can extract the numeric data of both types of representation (for instance, the *p*-value of the statistical test).

To do so, if starting from the statistical test track, a two-step procedure must be followed. First, clicking to any CpG from a default densely stacked track will force the browser to render a "full" display version of the track, representing each individual CpG separately. Then, if clicking at a given CpG, a new Web page depicting the mean DNA methylation difference between normal and tumor will be generated. This page includes extra information, such as the Wilcoxon (Mann–Whitney) test *p*-value and statistic (two tailed, unpaired) and details about the CpG, such as the cytoband or genomic coordinate. The DNA sequence of the region (which can be extended up and downstream according to the user

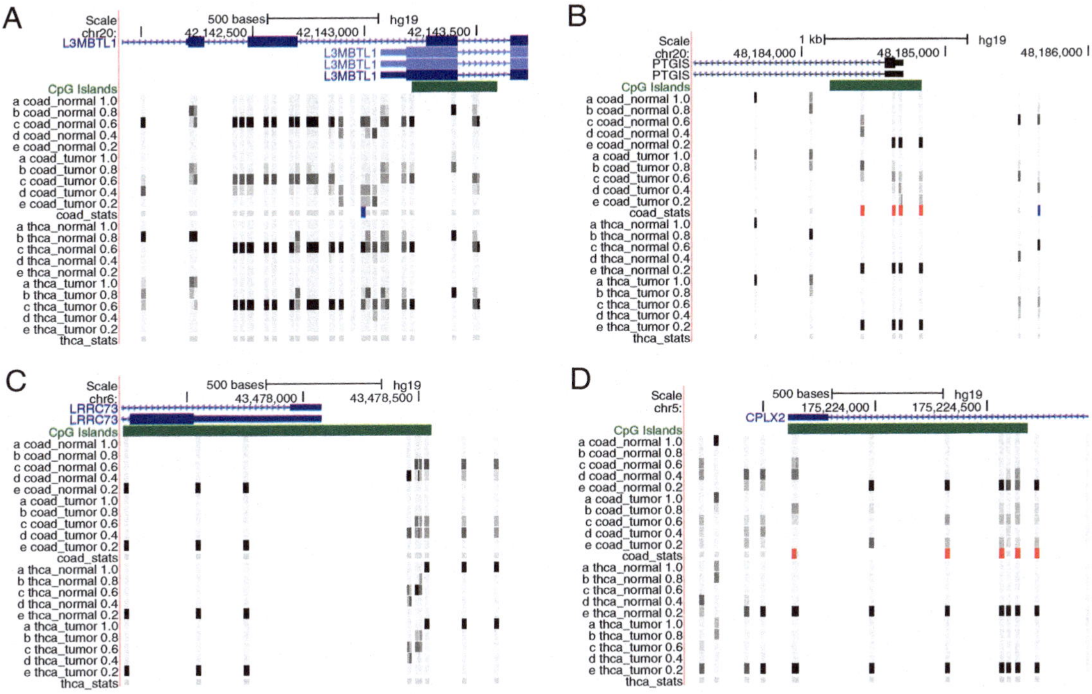

Fig. 2 Four DNA methylation landscapes at the local level. (**a**) Paternally imprinted region in the L3MBTL1 gene, in which beta values are at the 0.4–0.6 range (middle track). (**b–d**) Different types of DNA methylation boundaries in CpG islands and their flanking regions in thyroid and colon tissues. Track descriptions have been omitted

request) can be readily downloaded by clicking "view DNA for this feature."

The same two-step procedure can be applied to the DNA methylation profile composite track. In this case, a full description of the CpG including its score, represented as the percentile of samples at the CpG range multiplied by a factor of 10, will be rendered. That is, if 50% of the samples are located at that given DNA methylation range the score of the CpG will be 500.

4.3 Data Sharing

A raster image of the browsing window can be exported at any time by right-clicking at any point of the visualization window and selecting "view image". To save a high-resolution, vector image the user must click to the "View" menu at the blue navigation bar at the top of the screen ("PDF/PS" link).

As opposed to taking just a static image of the browsing window, the sessions tool allows to save a snapshot of the location and a full set of annotation tracks, including the overall aspect (display settings, track order, and zoom level, among others). This snapshot can be launched and browsed interactively. To do so, the sessions tool https://genome.ucsc.edu/cgi-bin/hgSession allows to download a file with the session details ("Save Settings" section).

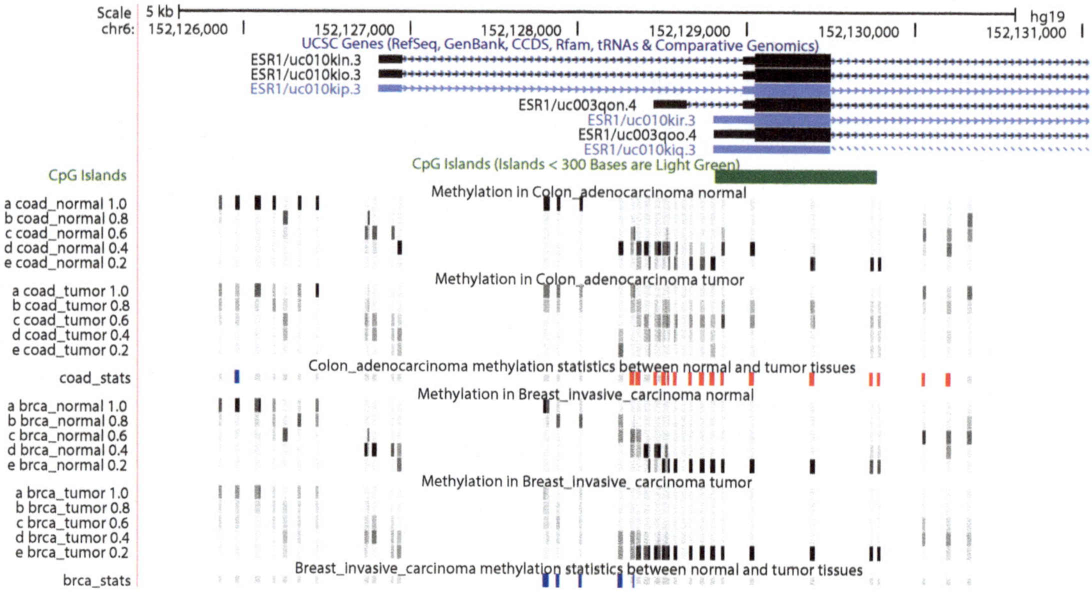

Fig. 3 DNA methylation landscape of the ESR1 gene illustrating hypermethylation in colon tumor (red ticks in coad_stats) and hypomethylation of the 5′ region upstream of the CpG island in breast cancer (blue ticks at brca_stats)

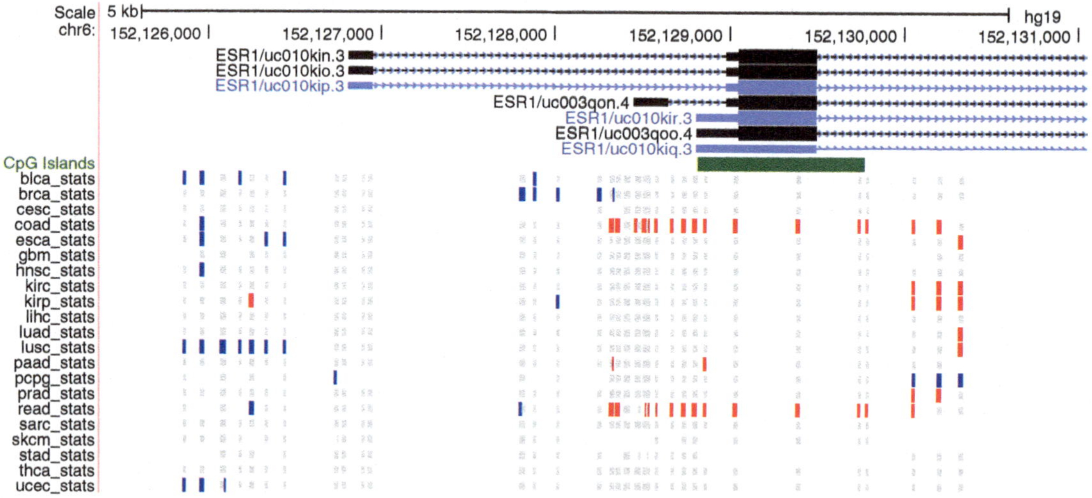

Fig. 4 Pancancer DNA methylation landscape of differential methylation in the ESR1 gene illustrating hypermethylation and hypomethylation of the 5′ region upstream of the CpG island in different tumor types. Track descriptions have been omitted

The upload of such a file will result in genome browser configured exactly as it was when the session was saved. Users can avoid the need to download a configuration file by registering to the UCSC

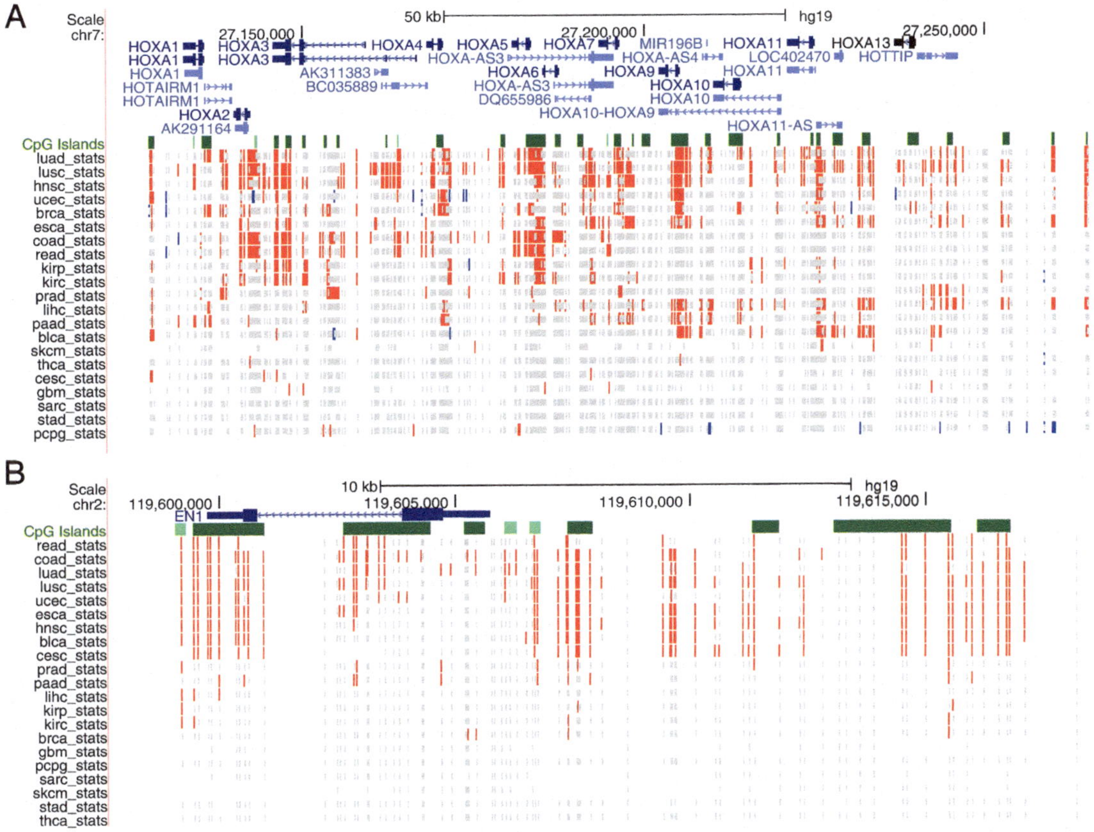

Fig. 5 (**a**) Pancancer visualization of the local aggregation of significantly hypermethylated CpGs (red ticks) at the HOXA cluster. Despite of the fact that differential methylation between normal and tumor samples are plotted independently, long-range effects can be easily detected. Track descriptions have been omitted. (**b**) Pancancer visualization of the local aggregation of significantly hypermethylated CpGs (red ticks) at the CpG island cluster neighboring the EN1 gene. Track descriptions have been omitted

(free, requires a contact email, username and password). After registering, the UCSC browser permits to store and browse sessions for private use, as well to share them with other investigators.

5 Usage Examples

The pancancer DNA methylation trackhub depicts DNA methylation variability for discrete CpG loci. Independently, it also provides a statistically assessed straightforward tumor vs. normal comparison. Despite of representing genomic positions, the trackhub can be used at different scales, from the fine-grained nucleotide level to regions of hundreds of genes. To help the local use case, the trackhub adaptively resizes the width of each CpG to effectively display highly packed areas; and, when zooming out, it increases the thickness of the bars depicting the statistically significant changes between normal and tumor samples to emphasize them.

To exemplify the trackhub usage versatility we briefly describe three use cases, from the closest to the nucleotide level to gene clusters.

5.1 Fine-Grained CpG Evaluation

Despite of the preponderant bimodal distribution of DNA methylation beta values, with CpGs being either methylated or unmethylated, DNA methylation landscapes are zonally heterogeneous in both mean values (e.g., the methylation status) and variances (e.g., the heterogeneity of methylation levels within the cohort). Therefore, different signatures can be detected by visual inspection of candidate regions, which might be interpreted in terms of regulatory potential. For instance, most samples display intermediate methylation values (0.4–0.6 beta values range, central density track) at the L3MBTL1 locus, a Polycomb family member located in a region of chromosome 20 and imprinted in human, indicating hemimethylation and low variability (Fig. 2a).

On the other hand, extreme methylation statuses are predominant. Namely, CpG islands are often fully unmethylated whereas other compartments, such as repetitive elements, are methylated, in normal tissue. Nonetheless, transitions between unmethylated CpG islands and their methylated flanked sequences can produce both sharp and sloping boundaries. For instance, in the downstream region of the PTGIS gene associated CpG island thyroid tissues display a blunt boundary with most samples in either the 0–0.2 and 0.8–1 beta values range, while colon samples show a gradual DNA methylation profile (Fig. 2b). Another interesting example is the upstream region of the LRRC73 gene CpG island that is fully methylated in thyroid samples but shows partial methylation in colon tissues, in which the boundary is not so precisely set, indicating variability at the cohort level (Fig. 2c). Finally, the CPLX2 gene promoter CpG island also displays a tight profile in thyroid tissues but a broad variability in colon with a hypermethylation trend in tumor samples (Fig. 2d). As a whole, the differential DNA methylation profiles pinpoint potential diversity in the genomic regulation of the examined genes.

5.2 Pancancer Close-Up Comparison

The aberrant methylation of the estrogen receptor 1 (ESR1) gene drives hormone insensitivity in breast cancer and has been reported for other malignancies [8, 9]. According to the UCSC gene annotation ("knownGene" table), ESR1 encodes for 27 isoforms, including uc003qon.4 (NM_001122740; transcript variant 2 translated to P03372), whose transcription start site lies near a CpG island which is extensively scrutinized by the Infinium 450k chip.

In breast cancer and matched normal samples (BRCA cohort) the CpG island is almost completely unmethylated, with all the samples concentrated at the 0–0.2 beta values range (the density track at the bottom of each trackset; Fig. 3). The upstream flanking region, however, points to a significant tumor hypomethylation, as the profile of proximal CpGs to the island displace the samples located at almost full methylation in normal tissues (0.8–1 range, density track at the top of the trackset) to an almost continuous representation at intermediate methylation statuses (i.e., all the tracks from the trackset with a uniform shade of gray). This tumor hypomethylation reached statistical significance according to the Mann–Whitney test (blue ticks at the stats track; Fig. 3).

In colon cancer, the CpG island and its flanks are reshaped in a different manner (COAD cohort; Figs. 3 and 4). In this case, the CpG island gets significantly hypermethylated (red ticks at the stats track), while non detectable changes occur at its 5′, where most breast tumor did hypomethylated. Noteworthy, different DNA methylation changes at the ESR1 CpG island and its flanking region can be found across the different cohorts (Fig. 4), pointing to the locus propensity to dysregulation and its putative functional relevance. On the other hand, colon and rectum carcinomas show a highly consistent island hypermethylation pattern, in line with their clinicopathological similarities.

5.3 Pancancer Regional Landscapes

Discrete genomic loci repression by means of de novo methylation of CpG islands (epigenetic silencing) is a hallmark of human cancer. Remarkably, local CpG island hypermethylation can spread to megabase-sized blocks, altering the epigenetic landscape of large genomic regions whose gene expression gets disrupted (long-range epigenetic gene silencing [10]). Among the long range coordinated events in multiple cancer types, hypermethylations of the Homeobox clusters at chromosomes 2, 7, 12, and 17 are widespread [11].

Even though the trackhub is built upon the methylation status of discrete CpGs, the statistics tracks effectively display long range phenomena when zooming out. Namely, the cluster HOXA depicts a consistent tumor vs normal hypermethylation in an almost 100 KB range for most of cancer types, with the higher abundance of differentially methylated positions for lung adenocarcinoma (LUAD) and lung squamous cell carcinoma (LUSC) to the lower for prostate adenocarcinoma (PRAD) and thyroid carcinoma (THCA) (Fig. 5a, red ticks indicate tumor hypermethylation).

Another interesting example illustrating the usefulness of this tool refers to the hypermethylation of the EN1 CpG island, that has been postulated as a biomarker with diagnostic and prognostic applications in colorectal cancer [12]. A quick view of the EN1 CpG island cluster differential DNA methylation panorama in other tumor types (Fig. 5b) may aid the selection of the most appropriate loci to be screened in each specific tumor type.

6 Notes

1. The trackhub corresponds to data from human assembly GRCh37 (hg19).
2. TCGA produces DNA methylation readouts with a standard pipeline which includes quality checking and normalization, making final results comparable across datasets at their highest processing level, which corresponds to level 3. Data processing includes the removal of ambiguous readouts or data matching to regions with low mappability, such as most of the genomic repetitive elements.
3. The Infinium chip evaluates the DNA methylation status of individual cytosines (Cs) in CpG contexts, which are scattered along the genome, though particularly abundant in CpG islands [6]. Given the difficulty of visualizing 2-bp long features immersed in the roughly 3000 million basepairs-long human genome, we expanded each CpG 10 bp up and downstream, thus increasing its effective size. When CpG lied close together, we adaptively resized the intervals to accommodate them (*see* **Note 8**).
4. Unsorted or messed DNA methylation tracks (i.e., after manual drag and dropping, or showing different cohorts interleaved) can be recovered to the default sorting by pressing the "Default track order" link located at the "View" button on the blue bar on top of the page; or by clicking to the "default order" below the browsing window.
5. To simplify the display of pancancer analysis and increase its readability, track descriptions (the text over each trackset) and guides (the fade blue vertical lines) can be disabled by pressing the "Configure" button below the browsing window and then unchecking both "Display description above each track" and "Show light blue vertical guidelines."
6. An useful track summarizing all the human CpG islands is available at the UCSC's "Regulation trackset." This depicts all CpG Islands over 300 bp in green and the rest in light green.
7. The UCSC Genome Browser described here corresponds to version 341; button names/locations can vary between different releases.
8. The procedure to build the trackhub can be generalized to any collection of samples evaluated by the Illumina's Infinium array. The source code to generate the trackhub (starting from a populated postgres database) can be accessed at https://bitbucket.org/imallona/pancancer_dnameth_trackhub.

9. The pancancer trackhub can be disabled by clicking "disconnect" at the top-right corner of its configuration box, just below the browsing window.

Acknowledgments

We thank Iñaki Martinez de Ilarduya for his excellent technical support. The trackhub published here is based upon data generated by the TCGA Research Network: http://cancergenome.nih.gov/. This work was supported by the Spanish Ministry of Economy and Competitiveness [SAF2011/23638 and SAF2015-64521-R to M.A.P.]. CERCA Program/Generalitat de Catalunya.

References

1. Zhang J, Baran J, Cros A, Guberman JM, Haider S, Hsu J, Liang Y, Rivkin E, Wang J, Whitty B, Wong-Erasmus M, Yao L, Kasprzyk A (2011) International Cancer Genome Consortium Data Portal—a one-stop shop for cancer genomics data. Database (Oxford) 2011: bar026
2. Schroeder MP, Gonzalez-Perez A, Lopez-Bigas N (2013) Visualizing multidimensional cancer genomics data. Genome Med 5(1):9
3. Cerami E, Gao J, Dogrusoz U, Gross BE, Sumer SO, Aksoy BA, Jacobsen A, Byrne CJ, Heuer ML, Larsson E, Antipin Y, Reva B, Goldberg AP, Sander C, Schultz N (2012) The cBio cancer genomics portal: an open platform for exploring multidimensional cancer genomics data. Cancer Discov 2(5):401–404
4. Cline MS, Craft B, Swatloski T, Goldman M, Ma S, Haussler D, Zhu J (2013) Exploring TCGA Pan-Cancer data at the UCSC Cancer Genomics Browser. Sci Rep 3:2652
5. Speir ML, Zweig AS, Rosenbloom KR, Raney BJ, Paten B, Nejad P, Lee BT, Learned K, Karolchik D, Hinrichs AS, Heitner S, Harte RA, Haeussler M, Guruvadoo L, Fujita PA, Eisenhart C, Diekhans M, Clawson H, Casper J, Barber GP, Haussler D, Kuhn RM, Kent WJ (2016) The UCSC Genome Browser database: 2016 update. Nucleic Acids Res 44 (D1):D717–D725
6. Bibikova M, Barnes B, Tsan C, Ho V, Klotzle B, Le JM, Delano D, Zhang L, Schroth GP, Gunderson KL, Fan JB, Shen R (2011) High density DNA methylation array with single CpG site resolution. Genomics 98 (4):288–295
7. Zhu Y, Qiu P, Ji Y (2014) TCGA-assembler: open-source software for retrieving and processing TCGA data. Nat Methods 11(6):599–600
8. Esteller M (2007) Cancer epigenomics: DNA methylomes and histone-modification maps. Nat Rev Genet 8(4):286–298
9. Muller HM, Widschwendter A, Fiegl H, Ivarsson L, Goebel G, Perkmann E, Marth C, Widschwendter M (2003) DNA methylation in serum of breast cancer patients: an independent prognostic marker. Cancer Res 63 (22):7641–7645
10. Frigola J, Song J, Stirzaker C, Hinshelwood RA, Peinado MA, Clark S (2006) Epigenetic remodeling in colorectal cancer results in coordinate gene suppression across an entire chromosome band. Nat Genet 38(5):540–549
11. Rauch T, Wang Z, Zhang X, Zhong X, Wu X, Lau SK, Kernstine KH, Riggs AD, Pfeifer GP (2007) Homeobox gene methylation in lung cancer studied by genome-wide analysis with a microarray-based methylated CpG island recovery assay. Proc Natl Acad Sci U S A 104 (13):5527–5532
12. Mayor R, Casadome L, Azuara D, Moreno V, Clark SJ, Capella G, Peinado MA (2009) Long-range epigenetic silencing at 2q14.2 affects most human colorectal cancers and may have application as a non-invasive biomarker of disease. Br J Cancer 100 (10):1534–1539

Chapter 8

Methylation-Sensitive Amplification Length Polymorphism (MS-AFLP) Microarrays for Epigenetic Analysis of Human Genomes

Sergio Alonso, Koichi Suzuki, Fumiichiro Yamamoto, and Manuel Perucho

Abstract

Somatic, and in a minor scale also germ line, epigenetic aberrations are fundamental to carcinogenesis, cancer progression, and tumor phenotype. DNA methylation is the most extensively studied and arguably the best understood epigenetic mechanisms that become altered in cancer. Both somatic loss of methylation (hypomethylation) and gain of methylation (hypermethylation) are found in the genome of malignant cells. In general, the cancer cell epigenome is globally hypomethylated, while some regions—typically gene-associated CpG islands—become hypermethylated. Given the profound impact that DNA methylation exerts on the transcriptional profile and genomic stability of cancer cells, its characterization is essential to fully understand the complexity of cancer biology, improve tumor classification, and ultimately advance cancer patient management and treatment.

A plethora of methods have been devised to analyze and quantify DNA methylation alterations. Several of the early-developed methods relied on the use of methylation-sensitive restriction enzymes, whose activity depends on the methylation status of their recognition sequences. Among these techniques, methylation-sensitive amplification length polymorphism (MS-AFLP) was developed in the early 2000s, and successfully adapted from its original gel electrophoresis fingerprinting format to a microarray format that notably increased its throughput and allowed the quantification of the methylation changes. This array-based platform interrogates over 9500 independent *loci* putatively amplified by the MS-AFLP technique, corresponding to the *Not*I sites mapped throughout the human genome.

Key words DNA methylation, Methylation-specific amplified fragment length polymorphism, Methylation microarray, Epigenomics

Abbreviations

CGI	CpG island
MSRE	Methylation specific restriction enzyme

Electronic supplementary material: The online version of this chapter (https://doi.org/10.1007/978-1-4939-7768-0_8) contains supplementary material, which is available to authorized users.

Tanya Vavouri and Miguel A. Peinado (eds.), *CpG Islands: Methods and Protocols*, Methods in Molecular Biology, vol. 1766, https://doi.org/10.1007/978-1-4939-7768-0_8, © Springer Science+Business Media, LLC, part of Springer Nature 2018

1 Introduction

In this chapter we describe the MS-AFLP array [1, 2], a microarray-based methodology to analyze changes in DNA methylation patterns in human samples. This platform is based on the methylation-sensitive amplified length polymorphism technique (MS-AFLP) [3, 4], previously employed to study somatic DNA methylation alterations in human cancers [5, 6].

DNA methylation is a crucial epigenetic mechanism modulating the structure and transcriptional activity of the genome. The genomic DNA methylation profile is cell-type specific, conferring the very large plasticity required for tissue functional specialization. In humans, DNA methylation occurs almost exclusively in the carbon 5 of cytosines within CpG dinucleotides. These dinucleotides are underrepresented [7] (about 23% of the expected frequency), accounting for approximately 28 million sites in the latest version of the human genome sequence. In somatic cells, the vast majority of CpG sites are methylated (70–90%, depending on the tissue). The minority of the CpG sites that remain unmethylated are typically in CpG islands associated with gene promoters [8, 9]. In many cases, the methylation status of these CpG islands dictates whether the adjacent genes are accessible to the transcriptional machinery. Methylation of gene promoter associated CpG islands generally leads to a denser chromatin conformation and transcriptional silencing [8, 10].

DNA methylation alterations are ubiquitous in many, if not most, human cancers [11, 12]. Cancer cells exhibit global reduction of their 5-methylcytosine genomic content, mostly as a consequence of the demethylation of repetitive elements and also in gene bodies [13–15]. This genome-wide DNA hypomethylation associates with genomic instability, reactivation of transposable elements and alteration of the transcriptional profile. DNA hypermethylation, on the other hand, seems to preferentially take place inside or neighboring CpG islands located in gene promoters and transcriptional enhancers, potentially leading to transcriptional silencing of the genes that they control. While most of the DNA methylation alterations found in tumors are most likely inconsequential, some of them are known to play crucial roles in cancerogenesis, tumor progression, and/or tumor phenotype [12].

Many different methods to explore and quantify DNA methylation alterations have been developed in the last decades [16, 17]. Currently, analytical methods relying on bisulfite transformation of DNA have become widespread [18, 19], some of them reaching an impressive resolution. For example, the newest generation of Illumina Infinium methylation arrays, one the most frequently used methods, can interrogate over 850,000 CpG sites [20–23]. Even deeper resolution can be achieved by whole genome bisulfite sequencing, but at a much larger cost both economical and in terms of computational resources [24, 25].

In addition to the successful implementation of the genome-wide bisulfite-based platforms, alternative technologies relying on methylation-sensitive restriction enzymes (MSREs) have been devised and widely used, especially when the former were not yet fully developed [4, 16]. These methods were based on the methylation-specific activity of bacterial restriction enzymes, which are components of the methylation-restriction defense mechanisms that protect bacteria against foreign DNA. These mechanisms allow bacteria to differentiate their own genome by methylating it in specific sequences right after DNA synthesis, and expressing a restriction enzyme that will digest the foreign unmethylated DNA but not the endogenous methylated sequences. Consequently, in most cases the activity of MSREs is blocked by the presence of methylated bases in their DNA recognition sequence, although there are exceptions such as *Dpn*I, which exclusively cuts when its recognition sequence is methylated. To investigate methylation alterations in the human genome, MSRE-based techniques typically employed enzymes whose recognition sequence included one or more CpG sites. Depending on the technique, MSREs with recognition sequences of 4, 6 or even 8 bp in length have been applied. Many different MSREs techniques have been designed and successfully employed to study methylation alterations in various research fields, prominently in the study of human cancers [4, 16, 26, 27].

Methylation-sensitive amplification fragment length polymorphism (MS-AFLP) is a fingerprinting technique that, in its original implementation, relied in the use of the *Not*I methylation-sensitive restriction enzyme to detect differences in DNA methylation between two (or more) samples [3]. *Not*I recognizes the 8 bp-length sequence 5′GC$^{\vee}$GGCCGC3′ (cutting site indicated by the symbol $^{\vee}$). This sequence is relatively infrequent in the human genome, with approximately 9500 occurrences. *Not*I activity is blocked by methylation in either one of the two CpG sites within its recognition sequence.

The distribution of *Not*I recognition sites (as well as many other sequences) along the human genome is not random, since the human genome is itself not randomly organized. This is particularly evident for sequences containing CpG dinucleotides, due to the depletion of CpG dinucleotides in the human genome [7]. The distance between consecutive *Not*I sites (distance to nearest neighbor, DNN) follows an extremely clear bimodal distribution that significantly deviates from the expected geometric distribution (Fig. 1a). This bimodal distribution indicates that about 40% of the *Not*I sites tend to be closer than expected by pure random distribution. About 75% of the *Not*I sites are clustered within or in the vicinity (±500 bp) of CpG islands. This is not surprising because,

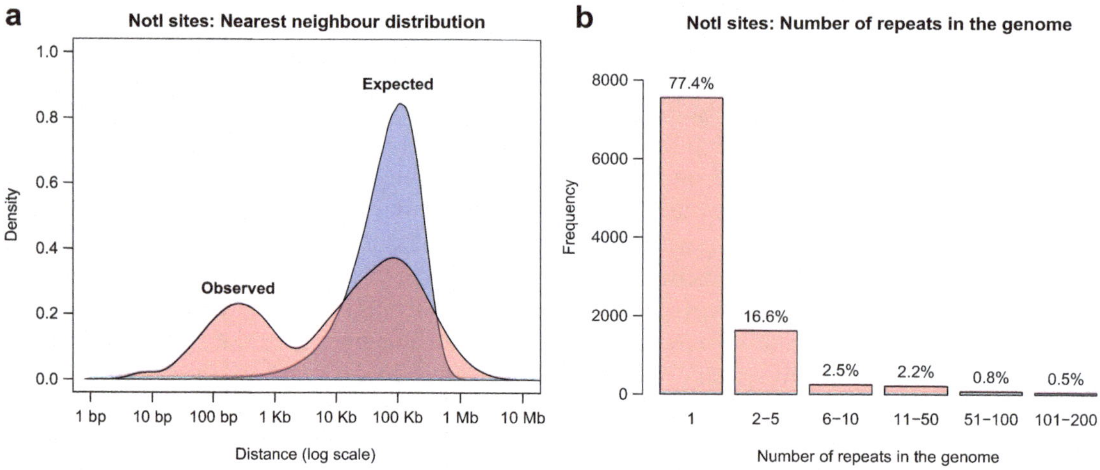

Fig. 1 (**a**) Distance to the nearest-neighbor (DNN) of the *NotI* site distribution in the human genome. The observed distribution (in pink) follows a clear bimodal distribution that greatly deviates from the geometric distribution (in blue) expected if the location of NotI sites were random along the genome. This deviation indicates that *NotI* sites tend to cluster, with nearest-neighbor distances below 1 Kb. These regions correspond to high GC content sequences that are enriched in CpG islands. (**b**) Repeats identified by Blast-search using the sequences surrounding every *NotI* site located in the human genome sequence. Any hit with more than 90% identity over 50 bp was considered positive. According to this criterion, most of the *NotI* sites are in unique sequences or in low-repetitive sequences

due to the very high GC content in its recognition sequence and the fact that it contains two CpG sites within 8 bp, these sequences are underrepresented in the human genome, clustering at GC-rich regions. Other 8 bp GC-rich sequences behave in a similar way, with DNN deviating from the expected geometric distribution (Supplementary Fig. 1). Several commercially available methylation-sensitive restriction enzymes targeting CpG-containing sequences also exhibit bimodal DNN distributions, and could be employed to generate MS-AFLP profiles that interrogate other different genomic locations (Supplementary Fig. 2).

In addition, most of the *Not*I sites are located in unique sequences (77%), or in sequences that are repeated less than 10 times in the genome (19%). Only a very small proportion of *Not*I sites (<4%) are in sequences repeated more than ten times in the human genome (Fig. 1b). Hence, most of the data generated by MS-AFLP arrays reflect changes in methylation taking place at unique locations of the genome. The technique, however, is not restricted to unique sequences and can also interrogate other low-to-moderately repetitive genomic sequences.

Therefore, the selection of *Not*I to generate methylation MS-AFLP fingerprints interrogating CpG island methylation without neglecting methylation alterations in other genomic locations provided a relatively unbiased and panoramic view of the cancer genome methylation status. Studying colorectal cancer with

MS-AFLP we showed the lack of evidence for a methylator phenotype when focusing on hypermethylation alterations [5], and the association between DNA hypomethylation changes with aging and genomic damage [6].

1.1 Methodological Aspects of MSFLP DNA Fingerprinting

MS-AFLP starts with the generation of a genomic restriction fragment library by double digestion of the DNA with the enzymes *Not*I and another common 4 bp cutter, like *Mse*I (cutting in 5′T˅TAA3′). *Mse*I is blocked by methylation in either one of the adenosines within its recognition sequence, but since adenosine methylation is virtually absent in the genome of somatic cells, this enzyme is generally considered to be methylation insensitive for the analysis of human samples. The double digestion generates a DNA fragment library containing three types of restriction fragments, i.e., *Mse*I-*Mse*I fragments, *Not*I-*Mse*I fragments, and *Not*I-*Not*I fragments. *Mse*I sites are extremely frequent, with more than 19 million mapped in the human genome sequence, and almost 2000 times more frequent that *Not*I sites. Consequently, over 99.9% of the restriction fragments correspond to methylation-independent *Mse*I-*Mse*I fragments (Fig. 2). The *Not*I-*Mse*I and *Not*I-*Not*I fragments, however, would be generated only if the associated *Not*I sites are unmethylated. Hence, changes in the DNA methylation pattern alter the relative frequency of *Not*I-*Mse*I and *Not*I-*Not*I fragments in the final library. Overall, the methylation-sensitive fragments are larger than the methylation-insensitive, because *Not*I sites are mostly located in GC rich regions where the AT-rich target sequence of *Mse*I is less frequent. In any case, since *Mse*I is a very frequent cutter, 99.7% of the generated restriction fragments in the library are shorter than 1.5 kb, therefore not posing an important hurdle for the subsequent PCR amplification.

After digestion, synthetic adaptors compatible with the cohesive ends left by *Mse*I and *Not*I are ligated. Then, the whole DNA library is amplified by PCR using two primers that recognize the sequence of these adaptors. This is a critical and distinctive characteristic of the MS-AFLP: since the DNA is not subjected to bisulfite treatment—a harsh procedure that degrades between 84–96% of the DNA [28]—and the restriction fragment library is amplified by PCR, minute amounts of starting material can be employed. The original technique was proven to generate reproducible results with as low as 2 ng of adaptor-ligated DNA (approximately 300 cell equivalents) [3]. This is of special interest for studies where DNA quantity might be limited. This is a significant advantage over the subsequent developments in the technologies to analyze DNA methylation aimed to increase the resolution and throughput. When this high resolution is not essential, the MS-AFLP DNA fingerprinting still offers a sound choice for analysis.

Differences in methylation in the original DNA sequences result in differences in the relative abundance of PCR-amplified

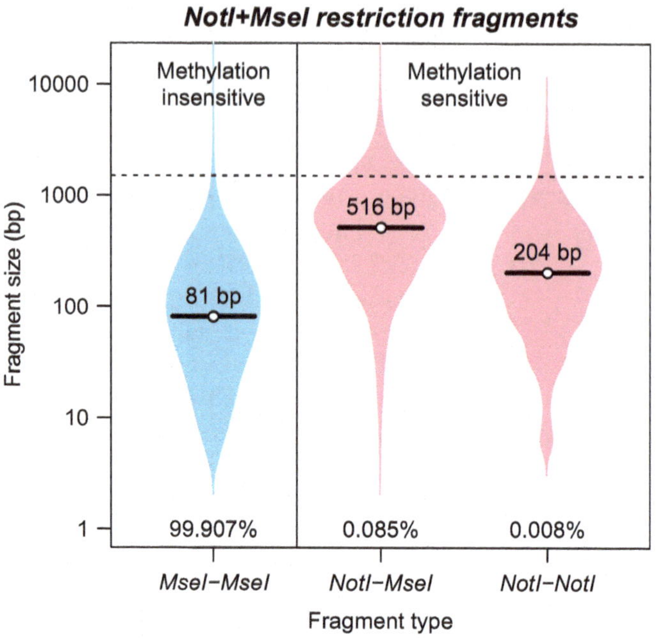

Fig. 2 Computationally derived restriction fragment size distributions after digestion of the human genome with *Not*I and *Mse*I enzymes, considering all *Not*I sites to be unmethylated. In this virtual library, the methylation-insensitive fragments with two *Mse*I ends (*Mse*I-*Mse*I) are vastly more abundant than the methylation-sensitive *Not*I-*Mse*I and *Not*I-*Not*I fragments. The majority of the fragments are shorter than 1500 bp (horizontal dashed line). *Mse*I-*Mse*I fragments are shorter (median 81 bp) than the *Not*I-*Mse*I (median 516 bp) and the *Not*I-*Not*I (median 204 bp)

*Not*I-*Mse*I and *Not*I-*Not*I fragments. As mentioned above, the *Mse*I-*Mse*I fragments are vastly more abundant but provide no information about DNA methylation. Theoretically, less than 0.1% of the amplified fragments would be methylation informative. Empirical results, however, demonstrated that the adaptor ligation and PCR amplification process favored the amplification of fragments with *Not*I-ends. Hence, the *Not*I-*Mse*I and *Not*I-*Not*I PCR-amplified fragments, albeit less abundant than the *Mse*I-*Mse*I, represent much more than 1% in the final PCR-amplified library.

To generate methylation-sensitive profiles, in the original version of MS-AFLP the *Not*I-*Mse*I and *Not*I-*Not*I amplification fragments were radioactively labeled by using a ^{32}P-labeled primer recognizing the *Not*I-compatible adaptor, resolved in vertical electrophoresis sequencing gels and then exposed to autoradiography films. The radioactive labeling guaranteed that only the methylation-informative *Not*I-*Mse*I and *Not*I-*Not*I amplicons were visible in the films. In further refinements of this technique, the radioactive labeling was substituted by fluorescent labeling that facilitated the amplified fragments profiling in automated DNA sequencers [3, 29].

In contrast with some coetaneous analytical techniques, *Not*I-*Mse*I MS-AFLP did not focus exclusively on hypermethylation events. Since this technique compared fingerprints from tumor tissues with their matching normal samples, both somatic hypermethylation and hypomethylation events could be detected. Somatic hypermethylation of a particular *Not*I site resulted in the decrease of the corresponding MS-AFLP band in the tumor fingerprint. Conversely, somatic hypomethylation produced a higher intensity band in the tumor fingerprint [3].

Thanks to its ability to detect both hypermethylation and hypomethylation alterations, MS-AFLP was instrumental to discover that these two types of alterations correlated between them, accumulated in a gradual fashion [5, 6], and correlated with patient age in gastrointestinal cancers [6]. These findings strongly suggested that the many of the DNA methylation alterations detected in tumors actually took place in the cancer cell-of-origin during normal aging and before, or shortly after its clonal expansion [6]. Notably, both hypomethylation and hypermethylation correlated with genomic damage, i.e., genomic copy number changes, but the multivariate analysis demonstrated that hypomethylation had a stronger association with genomic instability, and, moreover, it was a stronger predictor of patient survival than hypermethylation [6]. As additional examples, MS-AFLP in its fingerprinting platform was instrumental for the discovery of the frequent hypermethylation of *ADAMTS19* in gastrointestinal cancers [30], the demethylation of *RAPGEF1/C3G* in gastrointestinal and gynecological cancers [31] and the demethylation of the SST1 repetitive elements in a subset of gastrointestinal cancers characterized by a higher frequency of *TP53* mutations [32].

1.2 MS-AFLP in Microarray Format

The original fingerprint-format of the MS-AFLP technology was constrained by the resolution of the electrophoretic gels employed to resolve the different PCR amplicons. Also, the quantification of the alterations was performed by visual inspection and was difficult to automatize. To overcome these drawbacks, MS-AFLP was soon applied to a microarray platform [33]. This first approach to adapt the MS-AFLP to a microarray platform, however, was laborious since it involved the cloning and sequencing of MS-AFLP amplicons before printing them onto the arrays.

When the microarray technology improved, allowing a much larger number of probes to be directly synthesized onto the slides at a much lower cost, a second generation of MS-AFLP arrays was developed [1]. In these arrays, probes targeting the genomic sequences immediately adjacent to every single *Not*I site mapped in the genome were designed and synthesized onto the arrays, regardless of whether those *Not*I sites generated a detectable amplicon during the MS-AFLP PCR amplification. Hence, the second generation MS-AFLP array could interrogate the 9654 *Not*I sites mapped in the Build 36 version of the human genome sequence, the latest at that time. Every *Not*I site was potentially interrogated by two probes, located immediately upstream and downstream of the *Not*I, respectively.

MS-AFLP array employed the two-color technology of Agilent, widely used to analyze chromosomal copy number differences between two samples by CGH array [34]. Essentially, the normal and tumor DNA samples were separately subjected to the regular MS-AFLP protocol and subsequently labeled with two different fluorophores. Then, both reactions were cohybridized onto the arrays. Differences in the methylation of a particular *Not*I site between normal and tumor DNA resulted in deviations in the light emission intensity ratio between the two fluorophores. One of the advantages of this technology was that image capture and processing (intra-array probe averaging, background subtraction, normalization, etc.) were essentially identical to the standard protocols already optimized for CGH arrays.

Similarly to CGH arrays, MS-AFLP array values were represented as the log2 ratio of the normalized intensity signal in the tumor sample vs. the normalized intensity signal in the normal sample. Hence, unlike the now more common Illumina methylation arrays, MS-AFLP arrays did not provide the *absolute level* of methylation at the interrogated *loci*. Instead, they provided an estimation of the difference (as a log2 ratio) in unmethylated molecules between the two compared samples. Hypomethylation events resulted in higher signal intensity in the tumor, and therefore a log2 ratio above 0, while hypermethylation events resulted in lower signal intensity in the tumor, and therefore a log2 ratio below 0. Since methylation differences were represented as log2 ratios between two samples, standard two-color arrays analysis software could be employed to process and interpret the data. The analysis could be made by comparing tumors vs. their matching normal samples to investigate individually somatic changes in methylation, or by including a common reference sample in all arrays and comparing tumors vs. that reference sample.

These new-generation MS-AFLP arrays have been employed in the analysis of DNA methylation alterations in the colonic epithelium of patients with ulcerative colitis [1], revealing epigenetic alterations similar to those already present in colorectal tumors. They have been also employed to investigate tumors from patients with colorectal cancer, finding a novel DNA methylation signature that differentiates tumors of the serrated pathway [2].

The use of MS-AFLP arrays is not restricted to cancer. Any other application that would require the direct comparison of the DNA methylation profiles between different types of samples could benefit from this platform. Moreover, MS-AFLP arrays can be custom designed for other species as long as the complete or partial sequence of their genomes is available. In this chapter we describe the MS-AFLP array and provide the list of probes for the human genome. Custom arrays can be ordered from Agilent through their SureDesign web platform (https://earray.chem.agilent.com/suredesign/).

2 Materials

Prepare all the reagents using ultrapure, distilled, and sterilized water.

High-quality DNA is essential to obtain good results. The purification protocol depends on the type of available sample, i.e., fresh, frozen, paraffin embedded, etc. We have obtained excellent results with DNA purified with phenol-chloroform and with Qiagen commercial silica columns based kits.

2.1 Enzymes

- *Restriction enzymes*

  *Not*I (Promega).

  *Mse*I (New England Biolabs).
- *DNA ligase*

  T4 DNA ligase (Promega, Madinson).
- *DNA polymerases*

  PCR amplification: AmpliTaq® DNA Polymerase with Buffer I (1000 units/tube) from ThermoFisher Scientific (*see* **Note 1**).

2.2 Oligonucleotides

Oligonucleotides can be purchased from the most convenient supplier. The MS-AFLP oligonucleotides do not require any particular modification at their 5′ or 3′ ends. We suggest using HPLC-purified oligonucleotides (*see* **Note 2**).

- *Not*I adaptor:

  *Not*I-adaptor-A: 5′-CTCGTAGACTGCGTAGG-3′.

  *Not*I-adaptor-B: 5′-GGCCCCTACGCAGTCTAC-3′.
- MseI adaptor:

  *Mse*I-adaptor-A: 5′-GACGATGAGTCCTGAG-3′.

  *Mse*I-adaptor-B: 5′-TACTCAGGACTCAT-3′.
- PCR amplification:

  *Not*I-primer: 5′-GACTGCGTAGGGGCCGCN-3′.

  *Mse*I-primer: 5′-GATGAGTCCTGAGTAA-3′.

2.3 MS-AFLP Library Fluorescent Labeling Reagents

Bioprime labeling system (Invitrogen).

Both the CY5 and CY3 mix solution are available from Amersham-Pharmacia.

CY5 mix solution: 1.56 mM each of dGTP, dATP, and dTTP, 0.22 mM dCTP, and 0.11 mM Fluorolink CY5-dCTP.

CY3 mix solution: 1.56 mM each of dGTP, dATP, and dTTP, 0.22 mM dCTP, and 0.11 mM Fluorolink CY3-dCTP.

2.4 Computational Requirements

The MS-AFLP array data analysis does not require very large computational resources. A computer capable of running MeV or R/Rstudio is sufficient. However, the participation of an expert bioinformatician/statistician is highly advisable. All the suggested analytical software is freeware and multiplatform, running in UNIX/Linux, Mac, and Windows computers:

- Multiple experiment viewer (MeV). Freeware. Multiplatform. Available at http://mev.tm4.org.
- R. Freeware. Multiplatform. Available at https://www.r-project.org.
- Rstudio. Freeware. Multiplatform. Available at https://www.rstudio.com.
- Limma. Freeware. Multiplatform. Avilable at http://bioconductor.org.

3 Methods

The MS-AFLP platform requires two samples per array: the sample to be studied and the reference sample. These samples are cohybridized onto the MS-AFLP arrays. To minimize experimental bias, we strongly recommend performing all the steps previous to the hybridization (i.e., DNA purification, digestion, ligation, PCR amplification and labeling) of the studied sample(s) and associated reference(s) in parallel. In case that the number of samples or conditions is very large and more than one batch is required we suggest randomizing different conditions across the batches, but it is imperative to perform the prehybridization steps in parallel for every sample–reference pair.

3.1 DNA Digestion

For every sample and reference, separately: mix 500 ng of DNA, 2 Uof *Mse*I, 5 U of *Not*I, and 2.5 μL of 10× digestion buffer in a final volume of 25 μL in a 1.5 mL Eppendorf tube. Incubate overnight at 37 °C.

3.2 Adaptor Preparation

Mix the two oligonucleotides for the *Not*I adaptor (*Not*I-adaptor-A and *Not*I-adaptor-B, 5 μM each) and, in a separate tube, the two oligonucleotides for the *Mse*I adaptor (*Mse*I-adaptor-A and *Mse*I-adaptor-B, 50 μM each). Heat at 65 °C for 15 min, and then incubate overnight at 37 °C.

3.3 Adaptor Ligation

To every one of the digestion reaction tubes, add 17 μL of 1–1 TE buffer, 5 μL of 10× T4 DNA ligase buffer, 1.25 μL of 5 μM *Not*I adaptor, 1.25 μL of 50 μM *Mse*I adaptor, and 1 unit of T4 DNA ligase. Incubate overnight at 16 °C (*see* **Note 3**).

Incubate the ligated DNA at 37 °C for 2–6 h (*see* **Note 4**), and then inactivate the reaction by incubation at 70 °C for 20 min. Place the inactivated reactions on ice.

Dilute the reaction to 1 ng/μL with H_2O (i.e., add H_2O up to 500 μL, for the suggested 500 ng of original genomic DNA).

3.4 PCR Amplification

2 ng of template (2 μL of the 1 ng/μL solution).

6 ng of *Not*I-primer.

30 ng of *Mse*I-primer.

0.4 mM dNTPs.

1 U of AmpliTaq DNA polymerase.

Final volume of 20 μL.

The PCR program (Fig. 3) includes a first step of 72 °C for 30 s to repair the nick that it is left after the ligation of the adaptors (*see* Fig. 4 and **Note 5**). Then it continues with standard PCR amplification steps.

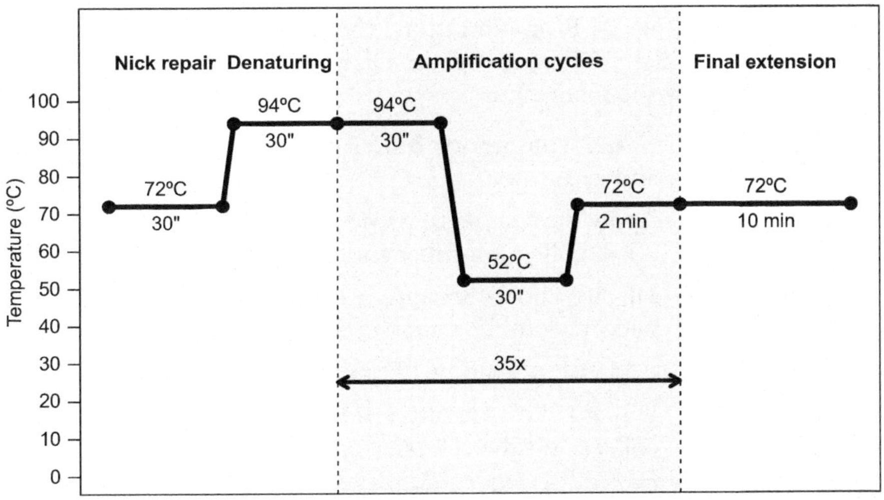

Fig. 3 PCR program for MS-AFLP. The program consists of six steps. In step 1 (72 °C, 30 s), the nick between the adaptors and the restriction fragments is repaired. Step 2 (94 °C, 30 s) is an initial denaturing phase. Steps 3 (94 °C, 30 s), 4 (52 °C, 30 s), and 5 (72 °C, 2 min) are the typical denaturing, annealing and extension steps of PCR, and are repeated 35 times. Step 6 (72 °C, 10 min) is a final elongation step to complete the synthesis of all unfinished amplification products

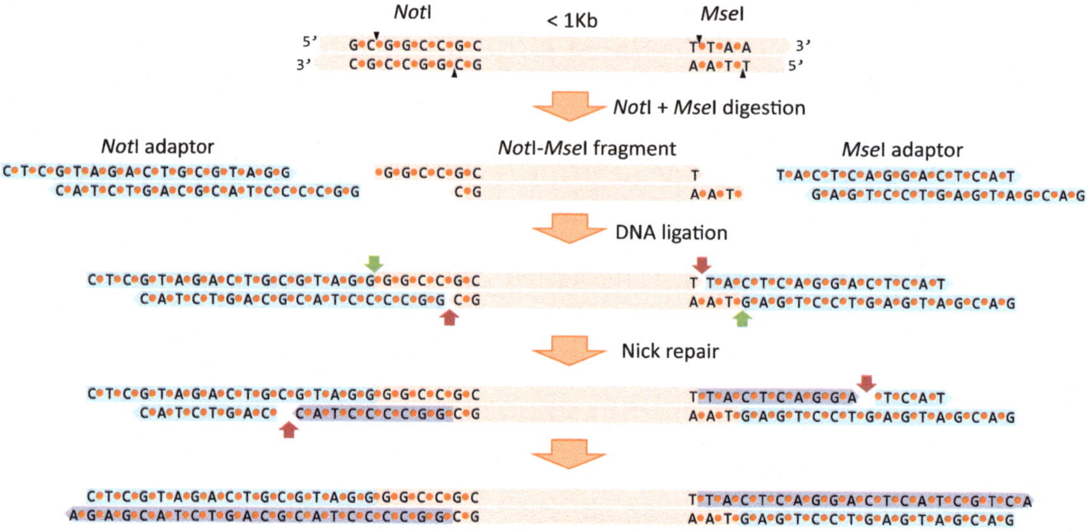

Fig. 4 Scheme of the restriction, ligation, and nick translation steps of the MS-AFLP. From top to bottom: the genomic DNA (in orange) is digested with *Not*I and *Mse*I, cutting at unmethylated GCGGCCG and TTAA sites, respectively. Orange dots represent the internucleoside phosphate groups. Black triangles indicate the cutting position. Due the high frequency of *Mse*I sites, the vast majority of the fragments are smaller than 1 kb. After digestion, two adaptors (in blue) complementary to the *Not*I and *Mse*I cohesive ends are ligated. Note that, due to the lack of phosphate groups in the 5′ of the adaptors, in each end of the fragment only one of the DNA strands can be covalently ligated (green arrows) while a DNA nick remains in the other strand (red arrows). During the first step of the MS-AFLP PCR protocol, the Taq polymerase extends the 3′ end of the DNA strands (in purple) while its 5′–3′ exonuclease activity degrades the 5′ end of the adaptors, resulting in the nick translation (red arrows) and finally in the complete repair of the DNA fragments ends. A Taq polymeras lacking the proofreading activity (3′–5′ exonuclease) is employed, thus the majority of the fragments exhibit a 3′-protuberant adenine residue, but it does not interfere with the subsequent steps

3.5 PCR Products Purification

After PCR amplification, the amplicons are purified using QIAquick PCR clean-up kit (Qiagen), following the protocol provided by the manufacturer:

- Add 5 volumes of Buffer PB to 1 volume of the PCR sample and mix.
- Apply the sample to a QIAquick column, previously placed on a 2 mL collection tube, and centrifuge for 20–60 s.
- Discard flow-through and place the column in the same collection tube.
- Add 750 μL Buffer PE, and centrifuge for 30–60 s.
- Discard flow-through and place the column in the same collection tube.
- Centrifuge for 1 min.
- Place QIAquick column in a clean 1.5 mL microcentrifuge tube.

- Add volume of 50 μL of elution buffer (10 mM Tris·Cl, pH 8.5) to the center of the membrane, let the column stand for 2–3 min at room temperature (this step improves the recovery of DNA).
- Centrifuge for 30 s.

3.6 DNA Labeling

To fluorescently label the MS-AFLP library, we suggest to employ the Bioprime labeling system from Invitrogen), following the manufacturer's protocol:

- Mix 2.5 μL of PCR-amplified DNA with 5 μL of water and 5 μL of random primer mix solution.
- Heat the mixture at 100 °C for 2 min. Then quickly transfer the sample to ice for 1 min, and briefly centrifuge for 10 s.
- Add 1 μL of CY5 or CY3 mix solution (*see* **Note 6**).
- Add *E. coli* DNA polymerase Klenow fragment to a final concentration of 0.8 U/μL. Incubate the mixture at 37 °C for 1 h, then add 2 μL of 0.5 M EDTA to terminate the reaction. Alternatively, samples can be heated to 65 °C for 10 min to terminate the reactions.
- Place the samples in ice.
- Samples can be stored up to a month at −20 °C in the dark.

3.7 Purification of Labeled DNA

- Spin the labeled DNA samples for 1 min.
- Add 430 μL of TE 1× pH 8.0 to each reaction tube.
- Load the labeled DNA onto a purificaction column, previously placed onto a 2 mL collection tubes.
- Cover the column with a cap and spin for 10 min at full speed in a microcentrifuge at room temperature. Discard the flow-through and place the column back in the 2 mL collection tube.
- Add 480 μL of 1× TE pH 8.0 to each column. Spin for 10 min at full speed in a microcentrifuge at room temperature. Discard the flow-through.
- Invert the column into a fresh 2 mL collection tube. Spin for 1 min at 1000 × *g* in a microcentrifuge at room temperature to collect purified sample.
- Add 1× TE pH 8.0 up to 50 μL.
- Samples can be stored up to a month at −20 °C in the dark.

3.8 Probe Synthesis and Array Printing

Probe synthesis onto the arrays is performed by Agilent. Custom arrays can be designed and ordered using the SureDesign online platform from Agilent (requires registration, https://earray.chem.agilent.com/suredesign/).

To generate MS-AFLP arrays with the predesigned probes interrogating the *Not*I sites of the Build 36 of the human genome, use the list of probes provided in Supplementary File 1. To design new sets of probes, targeting other restriction sites or other genomes, *see* Subheading 3.12.

3.9 DNA Hybridization and Scanning

We recommend the MS-AFLP array hybridization to be performed at a facility or service with ample experience in Agilent two-color arrays processing. This is a complex and error-prone procedure that requires very specific equipment and expertise not generally available in a regular molecular biology laboratory.

All the steps for the microarray hybridization, washing, and scanning are detailed in the Subheading 4 of the Agilent Oligonucleotide Array-Based CGH for Genomic DNA Analysis technical manual (Publication Part Number: G4410-90020) available at http://www.agilent.com/cs/library/usermanuals/public/G4410-90020_CGH_ULS_3.5.pdf.

3.10 Initial Data Processing

The initial array data processing is performed with the proprietary Agilent Scan Control and Feature Extraction software, following the exact same procedure used for CGH arrays.

3.11 Final Data Analysis

Many different tools are available to analyze the MS-AFLP arrays generated data, including proprietary and free software. In general, any software designed to analyze the output of two-color arrays (expression and/or CGH) would be applicable to MS-AFLP arrays. In our original publications, we employed *ad-hoc* R scripts that would be too specific for other projects. Hence, for the analysis of MS-AFLP we suggest two general-purpose options:

1. Multiple Experiment Viewer (MeV, https://sourceforge.net/projects/mev-tm4/files/mev-tm4). This software provides an intuitive and user-friendly graphical interface for the analysis of single and two-color arrays. It is an optimal start point for many microarray data analyses, providing the majority of required analytical tools.
2. For a more powerful and personalized alternative, albeit less user-friendly since it requires knowledge of R language, we suggest Limma [35], a R/Bioconductor [36] package (available at http://bioconductor.org/packages/release/bioc/html/limma.html).

As previously mentioned, MS-AFLP methylation data is represented as log2 ratios of the normalized signal intensity between the sample and the reference. *Important consideration*: since the signal intensity is proportional to the number of digested, i.e., unmethylated, *Not*I sites, the data should be interpreted as the ratio of unmethylated molecules at the *loci* under study between the sample and the reference. Please note that the data does not indicate absolute methylation levels, nor these can be directly inferred. This has to be carefully taken into consideration for data interpretation (*see* **Note 7**). Also, it has to be taken into consideration that some of the probes do not represent unique sequences (Fig. 1b) due to the location of their interrogated *Not*I sites in or adjacent to low to moderately repetitive elements. The information about the number of genomic hits of every probe is included in the Supplementary File 1.

3.12 Probe Design

We provide a list of 19,308 probes to interrogate the human genome (Supplementary File 1). This list was generated by extracting the ±90 bp sequences around the 9654 *Not*I sites located on the Build 36, to generate two probes per *Not*I site, one upstream and the other downstream, using the first version of Oligoarray [37], a freeware tool that provided excellent results with very little computational requirements. Chromosomal coordinates have been updated to the hg19 using liftover (http://genome.ucsc.edu/cgi-bin/hgLiftOver). Given that the currently the number of probes that can be synthesized onto an array largely exceeds the 9654 *Not*I sites, it is advisable to synthesize every probe two or more times, at separate locations of the array, thus facilitating the subsequent evaluation of the fluorescent signal robustness.

4 Notes

1. HotStart polymerases should be avoided. The nick-translation step in the first step of the MS-AFLP PCR program requires the polymerase to be immediately active, at 72 °C and before denaturing the DNA.
2. In the original MS-AFLP protocol, the sequences of the oligos for the *Not*I adaptor and PCR amplification were slightly different [3]. While these oligos generated very similar MS-AFLP profiles in the fingerprinting platform, they have not been tested in the microarray platform. The *Not*I primer might have a different base at the 3′ end (indicated by a *N*) to selectively amplify a subset of the restriction fragment library.
3. The restriction enzymes might be still active during the ligase incubation. They do not interfere with the adaptors already ligated to the MS-AFLP fragments ends, because the adaptors sequence does not regenerate the restriction sites after ligation (*see* Fig. 4). Note that the adaptors lack phosphate groups in their 5′ ends. Hence, even though they can self-anneal, the ligase is not able to covalently repair the nicks, joining the DNA strands.

4. This extra incubation at 37 °C facilitates the digestion of religated restriction fragments that could generate spurious PCR products in the subsequent steps. As mentioned in **Note 1**, the adaptors remain unaffected by the restriction enzymes because their ligation does not regenerate the restriction sites.
5. A DNA single strand nick is present after adaptor ligation because the oligonucleotides employed to generate the adaptors lack the phosphate group in their 5′ end, and hence only one of the strands of the adaptor is covalently ligated to the restriction fragments.
6. There is no preference about the sample or reference being labeled with Cy5 or Cy3. However, depending on the labeling and hybridization conditions, these dyes might have slightly different signal intensity range (Fig. 5a, b). If several samples are to be compared, it is advisable to use the same dye for all the sample reactions, and the other dye for all the reference reactions. In our experiments, we employed Cy5 to label the samples, and Cy3 to label the references. The signal intensity range was generally larger in the Cy5-labeled libraries (Fig. 5a, b). This may lead to the overestimation of hypomethylation events (values above 0) at the high-signal probes. This undesirable artifact is readily avoided by performing LOWESS normalization on the data (Fig. 5c, d).
7. The data returned by the MS-AFLP arrays does not represent the level of methylation, but the log2 transformed ratio of amplified, i.e., *unmethylated*, molecules between the sample and the reference at every interrogated *Not*I site. Considering that at every *loci* there might be a certain proportion of methylated (M) and unmethylated (U) molecules, such as $U + M = 1$, the MS-AFLP values (v) would follow the Eq. 1.

$$v \approx \log_2\left(\frac{U_{\text{sample}}}{U_{\text{reference}}}\right) = \log_2\left(\frac{1 - M_{\text{sample}}}{1 - M_{\text{reference}}}\right) \tag{1}$$

From the equation above, it becomes clear that the MS-AFLP array values do not follow a linear dependence with the change in methylation between the sample and the reference (Fig. 6a). This has several important consequences for data interpretation:

First, *positive* values indicate *lower* levels of methylation while *negative* values indicate *higher* levels of methylation in the sample vs. reference. Therefore, hypermethylation events generate negative values and hypomethylation events generate positive values.

Second, the sample methylation level cannot be inferred from the MS-AFLP value, unless the reference methylation level is known (Eq. 2).

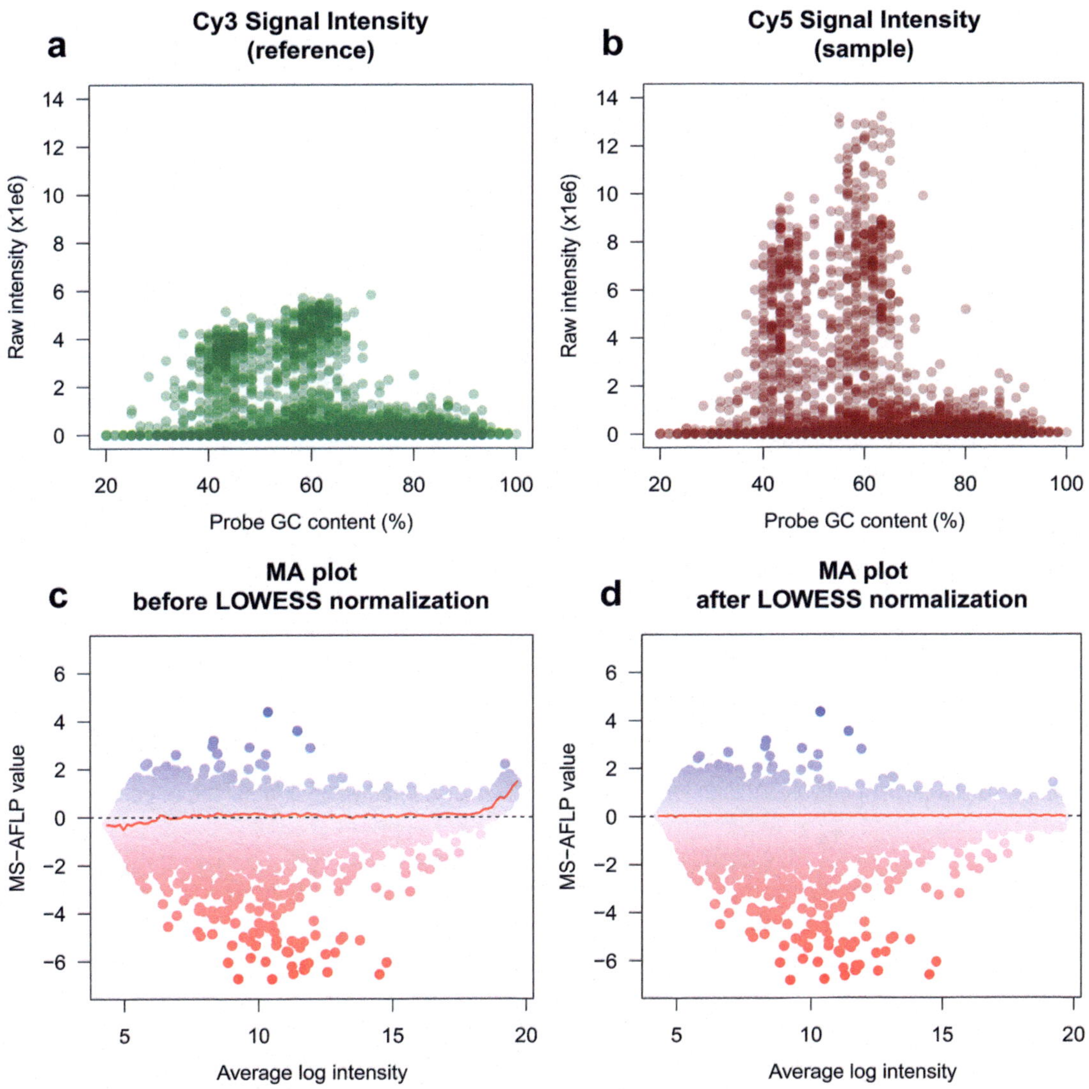

Fig. 5 Background-corrected signal intensity of the reference DNA, labeled with Cy3 (*y*-axis, panel **a**), and the sample DNA, labeled with Cy5 (*y*-axis, panel **b**), vs the GC content of the probes (*x*-axis). Cy5-labeled DNA exhibited a larger intensity range, resulting in a deviation above the 0 in the MS-AFLP log2 ratio of the sample vs the reference (panel **c**). Blue dots denote positive log2 ratios, i.e., hypomethylation, and red points denote negative values, i.e., hypermethylation (*see* **Note 7**). The red line indicates the moving average of the values, with a clear deviation toward hypomethylation in the high average intensity probes. This deviation is readily corrected by LOWESS normalization of the data (panel **d**)

$$M_{\text{sample}} = 1 - 2^{v} \times (1 - M_{\text{reference}}) \tag{2}$$

Identical values can be obtained from very different situations. As an example, a value of 1 would indicate that the proportion of unmethylated molecules in the sample is two times higher than in the reference. This value can be generated by a relatively small change in methylation from 95% to 90% (a twofold change in

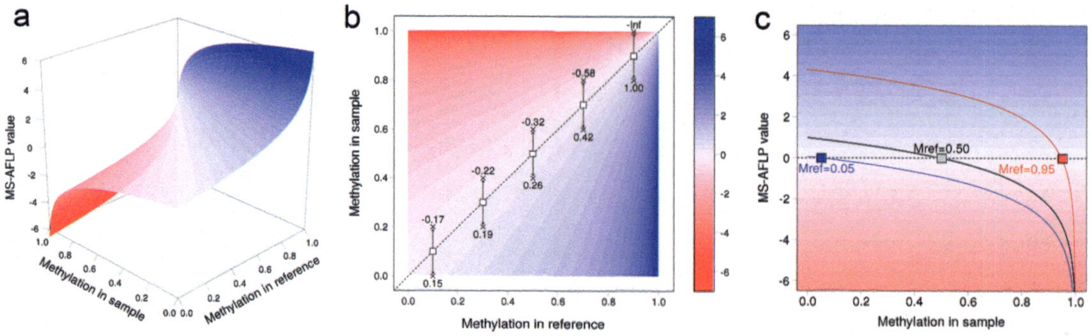

Fig. 6 (**a**) 3D representation of the theoretical MS-AFLP values as a function of the methylation level in the sample and reference DNA. Values are calculated according to Eq. 1. In red, values that deviate from 0 toward negative values, i.e., hypermethylation. In blue, values that deviate toward positive values, i.e., hypomethylation. Notice that MS-AFLP does not follow a linear dependence with the methylation. (**b**) 2D representation of the theoretical MS-AFLP values as a function of the methylation levels in the reference (*x*-axis) and sample (*y*-axis) DNAs. As in panel **a**, blue indicates positive values (hypomethylation) and red indicates negative values (hypermethylation). Note that similar changes in methylation (dashed arrows indicate a ±10% in methylation) have a very different impact in the MS-AFLP value depending on whether they occur at the low methylation *loci* or at the high methylation *loci*, and also there is a slight difference whether the change is toward hypermethylation or toward hypomethylation. (**c**) Theoretical MS-AFLP values as a function of the methylation level in the sample DNA and with three different levels of methylation in the reference DNA, i.e., 5% (blue line), 50% (black line), and 95% (red line). The range of change is much larger for the high-methylation *locus* (red line) than for the intermediate (black line) or the low-methylation (blue line) *loci*, and also larger for the negative values (hypermethylation) than for the positive values (hypomethylation) for all the three types of *loci*

demethylated molecules from 5% to 10%) or by a much larger change from 50% to 0% (equally a twofold change in demethylated molecules, from 50% to 100%).

Third, MS-AFLP technique is theoretically more sensitive to variations taking place at highly methylated *Not*I sites, were small variations in methylation may potentially generate large variations in the proportion of unmethylated molecules. It is also slightly more sensitive to hypermethylation events than to hypomethylation events, especially at these highly methylated sites (Fig. 6b, c). These highly methylated sites, in most cases, will also generate low signal intensities, albeit signal intensity is not exclusively affected by the methylation status of the interrogated *locus*, but also by the PCR amplification efficiency of that sequence and the probe affinity. Therefore, to avoid overestimating methylation changes, it is advisable to apply a relatively high minimun signal intensity cutoff before the final analysis.

Acknowledgments

We are grateful to our colleagues at the Jichi Medical School, Saitama, for the essential collaboration during the development, testing, validation, and application of the MS-AFLP arrays. We also are thankful to Dr. Lauro Sumoy for productive discussion during the writing of this chapter. This work was supported in part by the Spanish Ministry of Health Plan Nacional de I + D + I, ISCIII, FEDER, FIS PI09/02444, PI12/00511, and PI15/01763 grants, and 2014-SGR-1269 from the Agència de Gestió d'Ajuts Universitaris i de Recerca (AGAUR).

References

1. Koizumi K, Alonso S, Miyaki Y et al (2012) Array-based identification of common DNA methylation alterations in ulcerative colitis. Int J Oncol 40(4):983–994
2. Muto Y, Maeda T, Suzuki K et al (2014) DNA methylation alterations of AXIN2 in serrated adenomas and colon carcinomas with microsatellite instability. BMC Cancer 14:466
3. Yamamoto F, Yamamoto M, Soto JL et al (2001) Notl-Msell methylation-sensitive amplied fragment length polymorhism for DNA methylation analysis of human cancers. Electrophoresis 22(10):1946–1956
4. Samuelsson JK, Alonso S, Yamamoto F et al (2010) DNA fingerprinting techniques for the analysis of genetic and epigenetic alterations in colorectal cancer. Mutat Res 693 (1–2):61–76
5. Yamashita K, Dai T, Dai Y et al (2003) Genetics supersedes epigenetics in colon cancer phenotype. Cancer Cell 4(2):121–131
6. Suzuki K, Suzuki I, Leodolter A et al (2006) Global DNA demethylation in gastrointestinal cancer is age dependent and precedes genomic damage. Cancer Cell 9(3):199–207
7. Bird AP (1980) DNA methylation and the frequency of CpG in animal DNA. Nucleic Acids Res 8(7):1499–1504
8. Deaton AM, Bird A (2011) CpG islands and the regulation of transcription. Genes Dev 25 (10):1010–1022
9. Caiafa P, Zampieri M (2005) DNA methylation and chromatin structure: the puzzling CpG islands. J Cell Biochem 94(2):257–265
10. Illingworth RS, Bird AP (2009) CpG islands—a rough guide. FEBS Lett 583(11):1713–1720
11. Hanahan D, Weinberg RA (2011) Hallmarks of cancer: the next generation. Cell 144 (5):646–674
12. Baylin SB, Jones PA (2011) A decade of exploring the cancer epigenome – biological and translational implications. Nat Rev Cancer 11 (10):726–734
13. Gama-Sosa MA, Slagel VA, Trewyn RW et al (1983) The 5-methylcytosine content of DNA from human tumors. Nucleic Acids Res 11 (19):6883–6894
14. Feinberg AP, Gehrke CW, Kuo KC et al (1988) Reduced genomic 5-methylcytosine content in human colonic neoplasia. Cancer Res 48 (5):1159–1161
15. Ehrlich M (2002) DNA methylation in cancer: too much, but also too little. Oncogene 21 (35):5400–5413
16. Fraga MF, Esteller M (2002) DNA methylation: a profile of methods and applications. Biotechniques 33(3):632, 4, 6–49
17. Jorda M, Peinado MA (2010) Methods for DNA methylation analysis and applications in colon cancer. Mutat Res 693(1–2):84–93
18. Frommer M, McDonald LE, Millar DS et al (1992) A genomic sequencing protocol that yields a positive display of 5-methylcytosine residues in individual DNA strands. Proc Natl Acad Sci U S A 89(5):1827–1831
19. Clark SJ, Harrison J, Paul CL et al (1994) High sensitivity mapping of methylated cytosines. Nucleic Acids Res 22(15):2990–2997
20. Bibikova M, Le J, Barnes B et al (2009) Genome-wide DNA methylation profiling using Infinium(R) assay. Epigenomics 1 (1):177–200
21. Bibikova M, Barnes B, Tsan C et al (2011) High density DNA methylation array with single CpG site resolution. Genomics 98 (4):288–295
22. Sandoval J, Heyn H, Moran S et al (2011) Validation of a DNA methylation microarray for 450,000 CpG sites in the human genome. Epigenetics 6(6):692–702
23. Moran S, Arribas C, Esteller M (2016) Validation of a DNA methylation microarray for 850,000 CpG sites of the human genome enriched in enhancer sequences. Epigenomics 8(3):389–399
24. Lister R, Pelizzola M, Dowen RH et al (2009) Human DNA methylomes at base resolution show widespread epigenomic differences. Nature 462(7271):315–322

25. Stirzaker C, Taberlay PC, Statham AL et al (2014) Mining cancer methylomes: prospects and challenges. Trends Genet 30(2):75–84
26. Frigola J, Ribas M, Risques RA et al (2002) Methylome profiling of cancer cells by amplification of inter-methylated sites (AIMS). Nucleic Acids Res 30(7):e28
27. Jorda M, Rodriguez J, Frigola J et al (2009) Analysis of DNA methylation by amplification of intermethylated sites (AIMS). Methods Mol Biol 507:107–116
28. Grunau C, Clark SJ, Rosenthal A (2001) Bisulfite genomic sequencing: systematic investigation of critical experimental parameters. Nucleic Acids Res 29(13):E65–E65
29. Kageyama S, Shinmura K, Yamamoto H et al (2008) Fluorescence-labeled methylation-sensitive amplified fragment length polymorphism (FL-MS-AFLP) analysis for quantitative determination of DNA methylation and demethylation status. Jpn J Clin Oncol 38(4):317–322
30. Alonso S, Gonzalez B, Ruiz-Larroya T et al (2015) Epigenetic inactivation of the extracellular matrix metallopeptidase ADAMTS19 gene and the metastatic spread in colorectal cancer. Clin Epigenetics 7:124
31. Samuelsson J, Alonso S, Ruiz-Larroya T et al (2011) Frequent somatic demethylation of RAPGEF1/C3G intronic sequences in gastrointestinal and gynecological cancer. Int J Oncol 38(6):1575–1577
32. Samuelsson JK, Dumbovic G, Polo C et al (2016) Helicase lymphoid-specific enzyme contributes to the maintenance of methylation of SST1 pericentromeric repeats that are frequently demethylated in colon cancer and associate with genomic damage. Epigenomes 1 (1):2–18
33. Yamamoto F, Yamamoto M (2004) A DNA microarray-based methylation-sensitive (MS)-AFLP hybridization method for genetic and epigenetic analyses. Mol Gen Genomics 271 (6):678–686
34. Wolber PK, Collins PJ, Lucas AB et al (2006) The agilent in situ-synthesized microarray platform. Methods Enzymol 410:28–57
35. Ritchie ME, Phipson B, Wu D et al (2015) limma powers differential expression analyses for RNA-sequencing and microarray studies. Nucleic Acids Res 43(7):e47
36. Huber W, Carey VJ, Gentleman R et al (2015) Orchestrating high-throughput genomic analysis with Bioconductor. Nat Methods 12 (2):115–121
37. Rouillard JM, Zuker M, Gulari E (2003) OligoArray 2.0: design of oligonucleotide probes for DNA microarrays using a thermodynamic approach. Nucleic Acids Res 31 (12):3057–3062

Check for updates

Chapter 9

Genome-Wide Profiling of DNA Methyltransferases in Mammalian Cells

Massimiliano Manzo, Christina Ambrosi, and Tuncay Baubec

Abstract

Chromatin immunoprecipitation followed by high-throughput sequencing (ChIP-seq) is currently the method of choice to determine binding sites of chromatin-associated factors in a genome-wide manner. Here, we describe a method to investigate the binding preferences of mammalian DNA methyltransferases (DNMT) based on ChIP-seq using biotin-tagging. Stringent ChIP of DNMT proteins based on the strong interaction between biotin and avidin circumvents limitations arising from low antibody specificity and ensures reproducible enrichment. DNMT-bound DNA fragments are ligated to sequencing adaptors, amplified and sequenced on a high-throughput sequencing instrument. Bioinformatic analysis gives valuable information about the binding preferences of DNMTs genome-wide and around promoter regions. This method is unconventional due to the use of genetically engineered cells; however, it allows specific and reliable determination of DNMT binding.

Key words ChIP-seq, Immunoprecipitation, In vivo biotinylation, Next-generation sequencing, DNA methyltransferases, CpG islands

1 Introduction

Methylation of cytosine bases is one of the best mechanistically understood epigenetic modifications and plays various roles in genome regulation. Three conserved enzymes are responsible for the deposition of methyl groups to cytosine bases in mammals—the de novo DNA methyltransferases 3A (DNMT3A) and DNMT3B, as well as the maintenance DNA methyltransferase 1 (DNMT1) [1, 2]. In mammals, DNA methylation occurs at the majority of CpG dinucleotides throughout the entire genome, only CpG islands remain largely protected from methylation [3]. Based on biochemical studies, the mechanism of DNA methylation has been largely elucidated, but how DNA methylation patterns are precisely set along the genome and to what extent these cause further regulation remains to be understood in full detail.

Tanya Vavouri and Miguel A. Peinado (eds.), *CpG Islands: Methods and Protocols*, Methods in Molecular Biology, vol. 1766, https://doi.org/10.1007/978-1-4939-7768-0_9, © Springer Science+Business Media, LLC, part of Springer Nature 2018

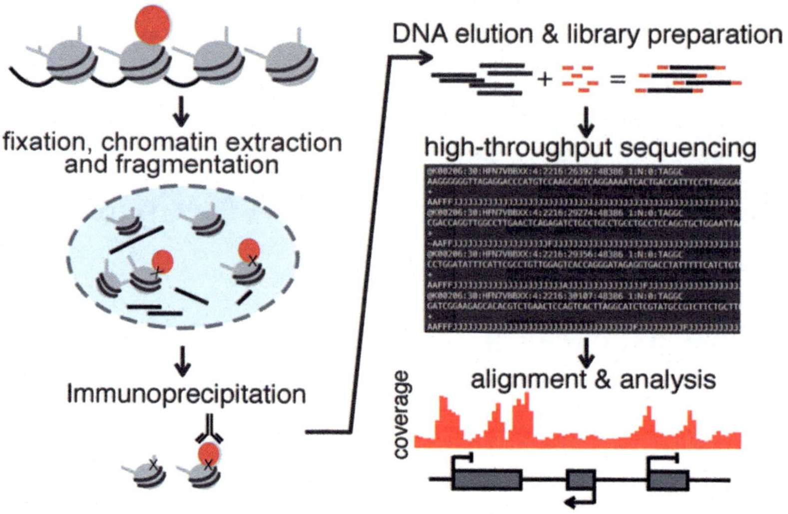

Fig. 1 Standard ChIP-seq workflow. Proteins of interest are first cross-linked to chromatin (fixation). Cells are lysed and chromatin is extracted. Fragmentation of chromatin is achieved by ultrasonication or enzymatic digest. Specific antibodies are used to enrich genomic regions bound by the protein of interest. Eluted DNA is sequenced and obtained reads are aligned to a reference genome for subsequent analysis

Genome-wide localization studies of epigenetic marks and their regulatory factors are essential for elucidating many biological processes and disease states. Recent technological advances in high-throughput sequencing have enabled a genomics revolution, making studies of chromatin-associated factors in a genome-wide manner affordable, faster and more precise. Among others, chromatin immunoprecipitation followed by high-throughput sequencing (ChIP-seq) can be utilized to determine how and where chromatin-associated proteins, such as DNMTs, bind to the genome (Fig. 1). In this method, proteins of interest are directly cross-linked to their site of interaction on chromatin, followed by selective enrichment using antibodies and high-throughput sequencing of bound DNA. Obtained sequencing reads are aligned to the genome and high-frequency interaction sites are determined based on local coverage (Fig. 1). However, this technique is often limited by various problems that result in unspecific or absent binding signals. In particular, the performance of antibodies can greatly vary and thus limit the applicability of ChIP for a wide range of proteins [4]. This problem of inadequate enrichment is in particular relevant for antibodies directed against DNA methyltransferases, of which only a few ChIP-grade antibodies exist, therefore limiting genome-wide profiling of DNMTs.

Epitope tags have been proven useful for ChIP-seq applications where suitable antibodies are not available [5–8]. Among these, we

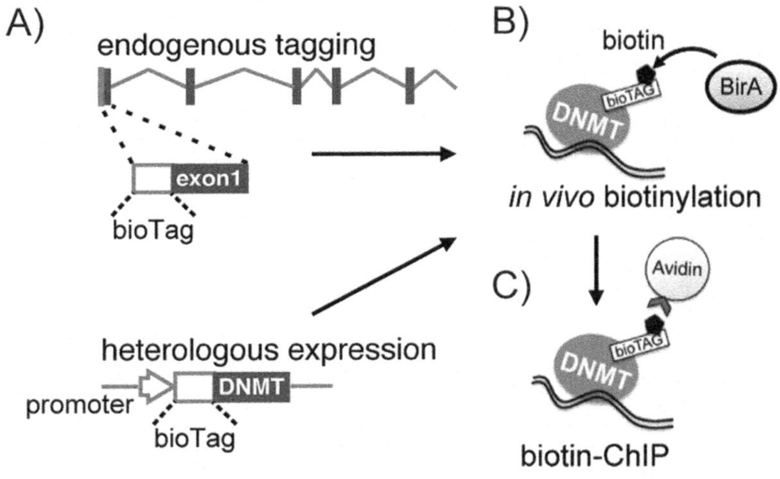

Fig. 2 (**a**) Expression of DNMT proteins fused to the biotin acceptor peptide can be either achieved from endogenously modified *Dnmt* genes or from heterologous integration sites. (**b**) The biotin acceptor peptide is covalently biotinylated in vivo by the BirA ligase. (**c**) This allows stringent and reproducible immunoprecipitation based on strong biotin–avidin interactions

recommend biotin tagging as an attractive technique for the purification of proteins of choice based on strong avidin–biotin interaction that can resist a large range of salt and detergent concentrations, temperatures and pH levels [9, 10]. This is especially beneficial for DNMTs that, unlike transcription factors, display a broad binding preference to the genome and where sequencing analysis requires robust signal-to-noise ratios [11]. Additionally, biotinylated proteins are rare in nature, dramatically minimizing the chances of cross-reactions that could distort the result of a performed precipitation assay [12].

Here, we first present a step-by-step protocol for genome-wide profiling of DNA-methyltransferases by biotin-ChIP followed by high-throughput sequencing. As a requirement, a short biotin acceptor site (16–23aa) needs to be added to the N- or C-terminus of the DNMT protein of interest, and the bacterial biotin ligase BirA has to be stably expressed in the cell line or tissue of interest (Fig. 2). Upon translation of the tagged DNMT protein, the acceptor site is specifically recognized and biotinylated by BirA in vivo [13, 14]. The biotin acceptor sequence can be either introduced to the endogenous *Dnmt* locus (via genome editing—*see* Flemr and Buhler (2015) for details [15]), or expressed as a tagged variant from a heterologous site [11]. Both approaches allow reliable measurements of DNMT–genome interactions, and should be used depending on the biological question.

For the subsequent ChIP, cells are fixed with formaldehyde, and their nuclei extracted and lysed. Following sonication of chromatin, streptavidin-coupled magnetic beads are used to enrich biotinylated-DNMT proteins. Stringent washing steps ensure removal of all unspecific interactions prior to DNA elution and high-throughput sequencing. Finally, the obtained sequencing reads from DNMT ChIP and input samples are aligned to the genome and genome-wide binding properties of DNMTs are extracted and visualized. We furthermore indicate potential strategies to analyze binding of DNMTs genome-wide, at promoters and at CpG islands, with particular interest in different genomic locations, DNA methylation, and CpG densities (Fig. 4).

2 Materials

Prepare all solutions using ultrapure water and analytical grade reagents. Prepare and store all reagents at room temperature, unless indicated otherwise. Diligently follow all waste disposal regulations when disposing of chemicals. *See* **Notes 1** and **2** for further general instructions.

2.1 Chromatin Cross-Linking

1. Cell culture medium.
2. Fixation Buffer: 50 mM HEPES (pH 8), 1 mM EDTA (pH 8), 0.5 mM EGTA (pH 8) (*see* **Note 3**), 100 mM NaCl, add H_2O to 500 mL.
3. 37.5% formaldehyde, mix with Fixation buffer to obtain 11% formaldehyde-solution (*see* **Note 4**).
4. 2.5 M glycine (sterile filtered).
5. Phosphate-buffered saline (PBS) at 4 °C.

2.2 Chromatin Extraction

1. PBS at 4 °C.
2. EDTA-free protease-inhibitor cocktail (PIC, make 25× stock in water, Sigma #11836170001).
3. Cell scraper.
4. Refrigerated swing-out, benchtop centrifuge for 15 mL falcons.
5. Chromatin Extraction Buffer 1: 10 mM EDTA (pH 8); 10 mM Tris–HCl (pH 8) (*see* **Note 3**); 0.5 mM EGTA (pH 8); 0.25% Triton X-100; ad H_2O to 500 mL.
6. Chromatin Extraction Buffer 2: 1 mM EDTA (pH 8); 10 mM TRIS (pH 8); 0.5 mM EGTA (pH 8); 200 mM NaCl; add H_2O to 500 mL.
7. ChIP Dilution Buffer: 50 mM HEPES (pH 7.5), 1 mM EDTA (pH 8); 1% Triton X-100; 0.1% sodium deoxycholate.
8. ChIP Lysis Buffer: ChIP Dilution Buffer with 0.2% sodium dodecyl sulfate (SDS) (*see* **Note 5**); 300 mM NaCl.

9. ChIP Buffer: ChIP Dilution Buffer with 0.1% sodium dodecyl sulfate (SDS) (*see* **Note 5**); 150 mM NaCl.
10. Optional: 1 mL syringe (Braun Injekt®-F #9166017V) and hypodermic needle (Braun Sterican® 25G× 5/8″, # 4658302).

2.3 Sonication

1. Bioruptor® Pico sonication device (Diagenode B01060001).
2. 1.5 mL Bioruptor® Pico Microtubes with Caps (Diagenode #C30010016).
3. Thermomixer (Eppendorf Thermomixer C).
4. PCR MinElute Kit (Qiagen #28004).
5. NanoDrop (ThermoFisher #ND-2000).
6. Agarose, 1× TBE, RedSafe Stain (20,000×, ChemBio #21141).
7. 6× gel loading dye (NEB #B7022).
8. Gel electrophoresis chamber (Bio-Rad #1704406), power supply (Bio-Rad #1645070).

2.4 Bead Preparation

1. Dynabeads® M-280 Streptavidin (ThermoFisher #11206D).
2. IP Buffer: ChIP Dilution Buffer with 0.1% SDS, 150 mM NaCl.
3. EDTA-free Protease-inhibitor cocktail (PIC, Sigma #11836170001).
4. Blocking buffer: 1 mL ChIP Buffer, 1% cold fish skin gelatin (Sigma #G7765), 100 μL *S. cerevisiae* tRNA (10 mg/mL, Sigma R5636-5X; *see* **Note 6**), 1× PIC.
5. Overhead rotator (ThermoFisher # 15920D).
6. Magnetic rack (ThermoFisher #12321D).

2.5 Chromatin Immunoprecipitation (IP)

1. 2% SDS in Tris–EDTA buffer (pH 8).
2. DOC Buffer: 250 mM LiCl; 0.5% 4-Nonylphenyl-polyethylene glycol (NP-40); 0.5% sodium deoxycholate; 1 mM EDTA (pH 8); 10 mM Tris (stock pH 8).
3. High salt buffer (HSB): ChIP Dilution Buffer with 0.2% sodium dodecyl sulfate (SDS); 500 mM NaCl.
4. TE Buffer: Tris–EDTA buffer (stock pH 8).

2.6 DNA Elution

1. Elution buffer: 1% SDS; 100 mM sodium bicarbonate (*see* **Note 7**).
2. RNase A (10 mg/mL, Sigma #1010969001).
3. Proteinase K (10 mg/mL, Sigma #03115828001).

2.7 DNA Purification

1. Option 1: PCR MinElute Kit (Qiagen #28004).
2. Option 2: Phenol (Sigma #P4557) (*see* **Note 1**).
3. Chloroform (Sigma # C0549), mix with isoamyl alcohol at 24:1 ratio.
4. Isoamyl alcohol.
5. Glycogen (20 μg/μL, Sigma #10901393001).
6. 3 M sodium acetate, ice-cold ethanol (70% and 100%).
7. Eppendorf® LoBind DNA/RNA microcentrifuge 1.5 mL tubes (Sigma #Z666548).

2.8 Library Preparation and Sequencing

1. NEBNext ULTRA DNA library prep kit (NEB #E7370).
2. NEBNext multiplex plugs for Illumina (NEB #E7335).
3. MinElute PCR Purification Kit (Qiagen #28004).
4. Nuclease-free water.
5. Ampure® XP Beads (Beckman #A63880).
6. 80% ethanol (*see* **Note 8**).
7. Magnetic stand.
8. DNA LoBind Tubes (Eppendorf #022431021).
9. PCR cycler.
10. Agilent DNA 1000 Kit (Agilent #5067-1504), including Ladder and Sample Buffer; D1000 ScreenTape (Agilent #5067-5582); 2200 TapeStation Instrument (Agilent #G2964AA).
11. Scalpel, UV-table.

2.9 Required Software and R Libraries

1. R package https://www.r-project.org.
2. BOWTIE http://bowtie-bio.sourceforge.net/index.shtml.
3. FASTX tools http://hannonlab.cshl.edu/fastx_toolkit/.
4. QuasR https://bioconductor.org/packages/release/bioc/html/QuasR.html.
5. GenomicRanges https://bioconductor.org/packages/release/bioc/html/GenomicRanges.html.
6. GenomicFeatures https://bioconductor.org/packages/release/bioc/html/GenomicFeatures.html.
7. Biostrings http://www.bioconductor.org/packages/release/bioc/html/Biostrings.html.
8. DESeq2 http://www.bioconductor.org/packages/release/bioc/html/DESeq2.html.
9. edgeR http://www.bioconductor.org/packages/release/bioc/html/edgeR.html.

10. genefilter http://www.bioconductor.org/packages/release/bioc/html/genefilter.html.
11. genomation https://www.bioconductor.org/packages/release/bioc/html/genomation.html.

3 Methods

3.1 Chromatin Cross-Linking

1. Culture cells under chosen conditions on standard cell culture dishes—e.g., one 10 cm dish of mouse embryonic stem cells is sufficient for one ChIP for which around 100 μg chromatin is needed (*see* **Notes 9** and **10**).
2. Add 1/10 vol. of 11% (v/v) formaldehyde solution to medium and ensure even distribution by shaking plate slightly. For 8 mL medium add 800 μL 11% formaldehyde solution to obtain 1%. Incubate for 8 min at room temperature.
3. Add 440 μL 2.5 M glycine (0.125 mM final concentration) to 10 cm dish and shake. Incubate for 10 min on ice to quench formaldehyde. When incubating avoid drying of cells.
4. Rinse cells twice with 10 mL ice-cold PBS. Keep dishes on ice and avoid drying. When handling several plates rinse plates one after the other.

3.2 Chromatin Extraction

1. Add 1 mL PBS + 1× PIC and scrape cells off. Resuspend and collect by flushing plate and transfer in appropriate tube depending on number of plates used.
2. Spin cells at 600 × *g* for 5 min at 4 °C and remove supernatant carefully. At this stage cells can be stored at −80 °C.
3. Resuspend in 5 mL/dish Chromatin Extraction Buffer 1, incubate for 10 min on ice.
4. Spin cells at 600 × *g* for 5 min at 4 °C and remove supernatant carefully.
5. Resuspend in 5 mL/dish Chromatin Extraction Buffer 2 (*see* **Note 11**), incubate for 10 min on ice.
6. Spin cells at 600 × *g* for 5 min at 4 °C and remove supernatant carefully.
7. Resuspend in 900 μL ChIP Lysis Buffer +1× PIC per ChIP and incubate for 1–2 h on ice by inverting tube sporadically to resuspend cell nuclei.

3.3 Sonication

1. Aliquot to prechilled tubes indicated by the manufacturer—a maximum of 300 μL per tube for 1.5 mL Diagenode reaction tubes.
2. Sonicate 300 μL aliquots in sonicator for 30 s ON and 45 s OFF for as many cycles as needed (*see* **Note 12**).

3. Pellet cell debris for 10 min at 12,000 × *g* and 4 °C, pool supernatants to a new, prechilled 1.5 mL-reaction tube and keep on ice.
4. Take out 40 μL of each sonicated chromatin for sonication test (*see* **Note 13**). Store remaining chromatin at 4 °C overnight.
5. Reverse cross-link all collected 40 μL aliquots overnight (procedure as described in Subheading 3.6—DNA Elution—Option 1).
6. Purify de-cross-linked chromatin with Minelute columns according to the manufacturer's protocol. Elute in 2× 20 μL EB and measure concentration on NanoDrop.
7. Pour a 1.2% agarose gel with 1× TBE and 1× RedSafe.
8. Add 6× Loading Dye to the DNA samples, load 500 and 1200 ng on a 1.2% agarose gel to test sonication. Add a DNA standard in another lane. Start electrophoresis at 60 V until the sample has entered the gel and then continue at 100 V till the dye front reaches the middle of the gel.
9. If sonicated chromatin ranges around 200–500 bp, continue with ChIP (Fig. 3). Otherwise sonicate as many more cycles as necessary to reach the recommended size (*see* **Note 14**).

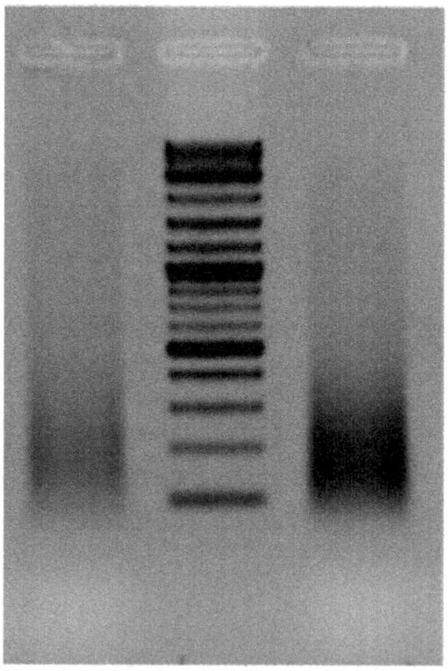

Fig. 3 DNA of sonicated chromatin from mouse embryonic stem cells loaded on 1.2% agarose gel. 500 and 1200 ng of DNA loaded on lane 1 and 3, respectively. Lower amounts of DNA allow better estimation of fragmentation efficiency. Furthermore gel was restained with 1× RedSafe in water for 1 h after electrophoresis to ensure homogenous detection of higher molecular weight DNA molecules

3.4 Bead Preparation

1. Wash 30 μL streptavidin–magnetic beads per ChIP 2× 5 min in 1 mL ChIP Buffer. Bind magnetic beads on magnetic rack for 2 min each time and remove the buffer.
2. Resuspend beads in 1 mL blocking buffer and incubate for 1–2 h at 4 °C with overhead rotation.
3. Wash blocked beads 3× with 1 mL ChIP buffer and take up in 30 μL ChIP buffer per ChIP after final washing step (*see* **Note 15**).

3.5 Chromatin Immunoprecipitation

1. Dilute NaCl- and SDS-concentration of lysed and sonicated chromatin by adding 1× volume of ChIP dilution buffer including 1× PIC. Aliquot ~100 μg chromatin based on the measured concentration obtained from the 5% de-cross-linked input elute (Subheading 3.3, **step 6**) and adjust sample volumes with ChIP buffer, supplemented with 1× PIC to 1 mL.
2. Store 5% of chromatin at −20 °C as an input sample for each cell line/tissue (*see* **Note 16**).
3. Add 30 μL preblocked streptavidin–magnetic beads for up to 2 mL volume of chromatin sample and incubate overnight at 4 °C on an overhead rotator (*see* **Note 17**).
4. Wash 2× 8 min with 2% SDS in 1× TE.
5. Pulse-spin and collect beads on magnetic rack for 2 min in order to remove supernatant between each step.
6. Wash 1× 8 min with High salt buffer.
7. Wash 1× 8 min with DOC buffer.
8. Wash 2× 8 min with 1 mL 1× TE. Change tube to 2 mL reaction tube at second wash to avoid carryover of unspecific material from the walls of the tube.

3.6 DNA Elution

1. Resuspend ChIP beads and saved input samples (from Subheading 3.3) in a final volume of 300 μL in fresh elution buffer.
2. Add 6 μL RNaseA and mix by inverting the tube.
3. Incubate for 30 min at 37 °C and shaking at 750 rpm to avoid settling of magnetic beads to the bottom of the tube.
4. Adjust elution buffer by adding 6 μL 0.5 M EDTA, 12 μL 1 M Tris–HCl (pH 8), and 6 μL Proteinase K.
5. Incubate for 3 h at 55 °C and de-crosslink overnight at 65 °C and mixing with 750 rpm to avoid settling of beads at the bottom of the tube.

3.7 DNA Purification

3.7.1 Option 1

1. Purify DNA from ChIP and input elutes with MinElute Kit following the manufacturer's instructions. Use 10 μL EB, and elute twice. Collect both elutes in DNA Low-bind tubes. Alternatively, use phenol-chloroform extraction as indicated below (*see* **Note 18**).

3.7.2 Option 2

1. Add 300 μL Phenol and mix by vortexing (*see* **Note 1**). The solution should become white throughout, although successive rapid phase separation can be observed.
2. Centrifuge for 3 min at 12,000 × *g* at room temperature. Transfer upper phase to new tube and add 300 μL of chloroform–isoamyl alcohol solution and mix by vortexing.
3. Centrifuge again for 3 min at 12000 × *g* and transfer upper phase to new 1.5 mL LoBind-tube.
4. Add 1 μL (20 μg) glycogen, 30 μL 3 M NaOAc, and 700 μL 100% ethanol cooled to −20 °C. Do not prepare master mix and add in given order, otherwise glycogen precipitates in ethanol.
5. Mix well and spin for about 2 h at 12,000 × *g* at 4 °C to ensure maximum recovery of precipitated DNA.
6. Remove supernatant and add 1 mL ice-cold 70% EtOH, vortex shortly and centrifuge for 30 min at 12,000 × *g* at 4 °C.
7. Remove supernatant completely through pipetting (*see* **Note 19**) and air-dry pellet for 10 min.
8. Take up pellet in 20 μL nuclease-free H_2O.
9. Measure DNA concentration using Qubit dsDNA HighSensitivity for ChIP material and NanoDrop for input material.
10. Samples can be stored at −20 °C for months or directly used for library preparation.

3.8 Library Preparation

1. End Repair: Add 6.5 μL 10× NEBNext End Repair Reaction Buffer and 3 μL NEBNext End Prep Enzyme Mix to 10–20 ng of ChIP or input DNA (diluted in water to final volume of 55.5 μL).
2. Incubate for 30 minutes at 20 °C, followed by 30 min at 65 °C.
3. Purify DNA with MinElute PCR Purification Kit according to manufacturer's instructions including 250 μL of PB buffer. Elute in 2× 22 μL EB.
4. Adaptor Ligation: Add 15 μL of Ligase Master Mix, 2.5 μL diluted adapter oligo mix (1:10 in water) and 1 μL Ligation enhancer to 65 μL sample obtained in **step 2**.
5. Incubate for 15 min at 20 °C.
6. Add 3 μL of USER enzyme, pipet up/down and incubate for 15 min at 37 °C.
7. Purify ligated DNA via Ampure® XP Beads in a beads-to-DNA ratio of 1.2 to 1 according to manufacturer's instructions and take up DNA in 20 μL water.
8. PCR-Amplification: Add 2.5 μL of Universal primer to 25 μL of NEBNext Q5 Hot Start HiFi PCR 2× MasterMix. Mix with

eluted DNA and add 2.5 μL of an Index primer of choice to gain a total volume of 50 μL (*see* **Note 20**).

9. Amplify library using the following cycling conditions: Initial denaturation: 30 s at 98 °C, 15 cycles amplification with 10 s at 98 °C and 75 s at 65 °C followed by a final extension of 5 min at 65 °C.
10. Purify samples again with Ampure® XP Beads as above.
11. Quality Control: Run 1 μL of each sample on a 2200 TapeStation with D1000 ScreenTape to check library size and concentration, and identify presence of primer dimers or self-ligated adapters.
12. Sample Pooling: The sample concentration is calculated according to the TapeStation result for fragments within a size of 150–400 bp. Samples with different indices can now be pooled at equimolar ratios.
13. Adapter removal: If required, the library can be purified from self-ligated adapters through a 2% agarose gel. Electrophorese at 60 V until the sample has entered the gel and then continue at 100 V until the dye front reaches the middle of the gel (*see* **Note 21**).
14. Check DNA size on a UV table and cut out fragments with a size of 150–400 bp. Extract DNA with the help of MinElute Gel Extraction Kit using 2× 11 μL EB. A total molarity of approximately 10 nM is sufficient for further processing and Illumina sequencing.
15. For this application, sequencing on an Illumina HiSeq2500 instrument on 125-bp-reads in a single-end manner is performed (*see* **Note 22**).

3.9 Sequencing Depth and Read Alignments

Due to the pervasive, genome-wide binding preference of DNMT proteins, we recommend to sequence at least 50 million reads per sample to guarantee sufficient coverage for downstream analysis. This is easily reached with current Illumina sequencing platforms. We also recommend including the corresponding input sample to the same sequencing reaction. This allows to identify and normalize potential biases in chromatin fragmentation or library preparation [17], and to calculate DNMT enrichments (see later steps). Once sequencing reads are obtained, the .fastq-files should be first filtered for low quality reads and PCR duplicates, and adapter sequences should be removed. This can be achieved using the FASTX-Toolkit [18] or similar tools. After preprocessing, the reads are aligned to the reference genome using BOWTIE [19] or similar aligners, allowing two mismatches and excluding reads that map to multiple locations in the genome (*see* **Note 23**). QuasR in R provides a simple, R-based interface for alignment of sequencing reads, count-based summary extraction in the GenomicRanges format and simple data visualization of genomic alignments [20].

3.10 Detection of DNMT-Enriched Sites

Due to the broad binding preference of DNMTs to the genome, standard peak calling algorithms fail to produce meaningful results and enriched sites have to be defined using alternative approaches. We suggest to bin the genome into 1-kb sized intervals and to calculate the $\log_2$-fold enrichments (LFE) of DNMT signals over the corresponding input sample (*see* **Note 24**). Binning and selection of genomic intervals can be achieved with the GenomicRanges package in R [21]. This package is especially useful for manipulation of genomic annotations, and for storing sequencing data along the defined annotations. Once the genomic intervals and the aligned sequencing reads are stored as GenomicRanges objects, the number of sequencing reads overlapping with the 1 kb-sized genomic intervals can be directly calculated for the ChIP and input samples and stored as a variable. LFE can be calculated using the following formula:

$$\text{LFE} = \log 2(n\_\text{IP} + p) - \log 2(n\_\text{inp} \times N\_\text{IP}/N\_\text{inp} + p),$$

whereas n_IP and n_inp represents the number of overlapping ChIP or input reads per 1 kb interval, respectively. N_IP and N_inp the library size of ChIP and input samples, and $p = 8$ pseudocounts to stabilize sampling noise. The obtained LFE can be utilized to rank and identify the enriched regions in the genome. Alternatively, and if replicates are available, the calculated reads per interval can be further used for the detection of significantly enriched tiles using edgeR or DESeq2 packages in R [22, 23]. For this we suggest to remove all genomic intervals with an LFE of <0.5 and > -0.5. This filtering step removes tiles that have no chance in passing significance tests, and therefore results in increased detection power by ameliorating multiple-testing normalization. Alternatively, the genefilter package in R provides functions that can be used to determine the most suitable cutoff for filtering [24].

3.11 Analysis of DNMT Binding at Promoters and CpG Islands

In order to analyze DNMT binding at promoters or CpG islands, first the correct genomic coordinates have to be defined. In case of promoters this can be easily achieved by using the GenomicFeatures package in R to obtain the regions of interest surrounding transcriptional start sites (TSS) in the appropriate GenomicRanges format [21]. Alternatively, promoter and TSS regions can be directly downloaded from the UCSC table browser [25]. The same is true for CpG island definitions which can be directly downloaded from UCSC. However, the UCSC definitions are based on Gardiner-Garden et al. [26], and more accurate CpG island detection algorithms exist [27, 28]. Again the decision on which definitions to use should be defined based on the biological question in mind. We suggest converting the obtained genomic coordinates to the GenomicRanges format in R [21]. This allows straightforward calculation and storage of CpG densities, GC percentages, CpG observed/expected ratios as well as calculations of DNA methylation percentages [29], DNA methylation densities [30], or DNMT protein enrichments and other chromatin features such as bivalency of H3K27me3 and H3K4me3 marks. For the detection and calculation of DNA sequence features (including occurrences of transcription factor binding motifs) we recommend the biostrings package in R [31]. This allows extraction of DNA sequence information and performing calculations in a straightforward manner. In case of calculating LFE for DNMTs or other proteins/histone modifications, the different sizes of the analyzed features have to be taken into account. This is especially important for CpG island intervals that strongly vary in size. Therefore we recommend performing a reads per kilobase (RPK) normalization prior to calculating the LFE (*see* **Note 25**). Once these measurements are obtained for the promoters and CpG islands of interest, binning or selection of regions of interest can be performed according to the research question. For example, CpG islands can be binned into islands of CpG high, intermediate, and low densities ([16], Fig. 4a), high or low DNA methylation [30], or association with annotated promoters or orphan CpG islands [32]. Total DNMT protein enrichment at promoters and CpG islands can now be visualized using various methods, including box plots, density plots, or violin plots (Fig. 4b) and compared to other genomic and epigenomic features extracted at the same promoter regions using scatterplots.

3.12 Visualization of DNMT Protein Localization Along Promoters and CpG Islands

However, these calculated DNMT enrichments do not necessarily allow to obtain information about the distribution of DNMT proteins along the analyzed genomic features. This is often required to understand the position of DNMT proteins along promoters and CpG islands, and to identify local binding preferences, requiring the user to inspect these sites one by one in genomic browsers. However, this is not feasible for large datasets and the results for individual intervals need to be visualized in bulk. For this we recommend the genomation package in R [33]. Heat map profiles allow visualizing the coverage of DNMT proteins around all promoters and CpG islands of interest simultaneously, and furthermore to arrange or cluster the intervals of interest based on coverage or binding patterns, respectively (Fig. 4c). Clustering is a straightforward method to separate promoters and CpG islands based on similar DNMT binding or other epigenetic properties. Additionally, the entire set of intervals, or selected intervals (e.g., by CpG density, genomic location or clustering), can be visualized as average metaprofiles that summarize general binding properties of DNMTs (Fig. 4d). Both heat map and metaprofile visualization of DNMT coverage require that the genomic intervals of interest are aligned to each other in order to obtain meaningful results. For

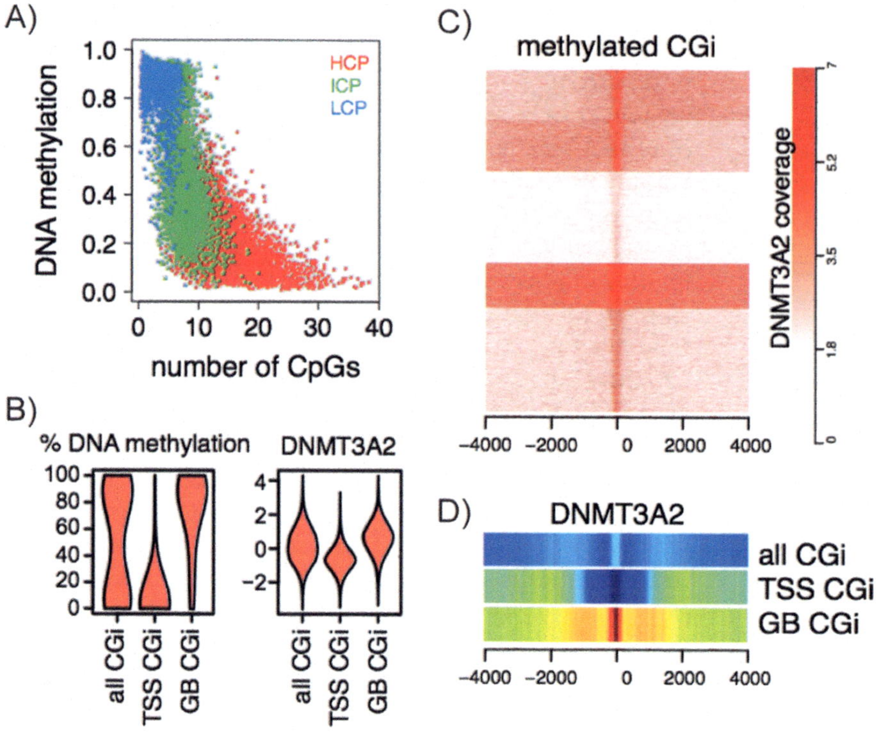

Fig. 4 Examples of data analysis to identify DNMT localization at promoters and CpG islands. (**a**) Scatter plot indicating variation in CpG density and CpG methylation at mouse gene promoters. Promoters are binned based on CpG densities into high, intermediate, and low CpG promoters, according to Weber et al. [16] (HCP, ICP, LCP). (**b**) Violin plots indicating the distribution of DNA methylation and DNMT3A2 binding at all CpG islands (all CGi), at CpG islands overlapping with promoters (TSS CGi) and CpG islands overlapping with gene bodies (GB CGi). (**c**) Heat map indicating coverage of DNMT3A2 reads around methylated (>80% m-CpG) CpG islands. Each row represents one ±4 kb interval surrounding individual CpG islands. *k*-means clustering identifies CpG islands with similar DNMT3A2 binding properties. (**d**) Average density profiles of DNMT3A2 binding at all CpG islands (all CGi), at CpG islands overlapping with promoters (TSS CGi) and CpG islands overlapping with gene bodies (GB CGi)

promoters, this requires that all intervals are centered and oriented based on a fixed position, such as the TSS. CpG island intervals extracted based on DNA sequence properties usually do not contain strand information and can be centered either on the midpoint or the border of the interval. For CpG islands that overlap with promoters, the orientation can be defined based on the directionality of the underlying gene, as performed in Manzo et al. [35]. Similar strategies can be also applied when analyzing DNMT coverage at exon–intron borders, TF binding sites or repetitive elements.

4 Notes

1. Do not handle chemicals until all safety precautions have been read and understood. Obtain special instructions before use. For instance, wear protective gloves, clothes and eye protection

when handling sodium dodecyl sulfate (SDS), phenol, chloroform, and sodium azide. In order to avoid formation of dust and/or vapors, work under a fume hood only is highly recommended. Be especially careful when handling phenol, always use SafeLock reaction tubes at all times.

2. All buffers can be stored at 4 °C unless stated otherwise.
3. pH adjustments: EDTA and EGTA are adjusted with NaOH; Tris solutions are adjusted with HCl; Hepes solution is adjusted with KOH or NaOH, as indicated.
4. Formaldehyde solution should be fresh and methanol-free.
5. SDS precipitates at 4 °C. Buffers containing more than 0.1% SDS should be stored at room temperature. These are stable for 1 year.
6. *S. cerevisiae* tRNA has to be denatured for 5 min at 95 °C prior to use. Keep on ice after denaturation.
7. We find that it is best to prepare the elution buffer fresh prior to de-crosslinking the chromatin.
8. Solutions containing ethanol should be prepared fresh each time. Provide good ventilation in process area to prevent formation of aerosols.
9. Depending on the study purpose, a wild-type or mock control should always be included as a separate sample for later evaluation.
10. Protocol performance varies from cell to cell and tissue type, since protein levels can vary and not all cell types are equally amenable to genetic engineering. Thus, the amount of cells/material and sonication cycles required for an optimal result is highly dependent on the cell line/tissue sample itself as well as the protein of interest in the study, and need to be tested beforehand.
11. Each step can be followed under the microscope, especially when performed for the first time (stain nuclei with trypan blue). Optionally, the extract can be passed three times through prechilled G26 syringe to dissociate cell clumps. In the end no clumps should be visible.
12. A sonication test run using different numbers of cycles is greatly recommended prior to starting due to cell type variability and properties of the sonicator in use.
13. Testing a successful sonication of the chromatin is an important step that should be skipped on no account. Fragmenting the chromatin to a correct size range is crucial for an optimal precipitation and sonication efficiency can vary greatly depending on used cell concentrations. Be aware of these fluctuations and ensure good chromatin ranges prior to immunoprecipitation.

If sonication is sufficient, DNA fragment should show highest signal in a small range of suitable sizes. For longer sonications, mix chromatin every ten cycles by inverting the tubes. Allow the machine and samples to cool down every 30 min of continuous usage, if not indicated by manufacturer otherwise.

14. Extended sonication should be treated with caution, since overheating of the sample could affect the integrity of chromatin and associated proteins.
15. Always avoid drying of beads by closing lid and handling liquids quickly.
16. This input sample will be used for sequencing as it reflects the chromatin state at the point of the ChIP start.
17. Bead blocking can be done for up to eight ChIP reactions at once and stored for one week at 4 °C with 0.01% NaAzide. Adjust volumes accordingly.
18. These options to isolate chromatin associated DNA exist due to the varying yield received from different starting material and/or the abundance of precipitated protein of interest. For first-time ChIP samples always use phenol-chloroform extraction.
19. Be careful not to disrupt the pellet.
20. Index primer sequences are individually added to each sample, allowing their identification from the pooled sequencing reactions.
21. Do not let the gel run for too long to avoid excessive separation of DNA and to concentrate the DNA in a small gel piece as much as possible.
22. The choice between single- or paired-end sequencing depends on the individual research questions to be addressed in the study, on the depth of sequencing coverage, and also the budget. Single-read sequencing involves sequencing from only one end of the DNA fragment and is a simpler way to utilize sequencing whereas paired-end sequencing allows reading both ends of a fragment. Paired end reads can provide more reliable information about the protein locations or mapping along repetitive sequence elements. However, this degree of accuracy may not be required for all experiments.
23. These parameters are just a recommendation and can be changed depending on the biological question in mind.
24. To avoid biases introduced by incorrect genome annotations, different genetic backgrounds, repetitive elements, etc., we recommend removing 1 kb intervals that overlap with satellite repeats, simple repeats or so-called "blacklisted regions" from ENCODE [34]. Furthermore, removal of genomic intervals with insufficient coverage in the input sample helps to reduce false positives.

25. Reads per kilo base pair (RPK) calculation is required for coverage normalization between differently sized genomic intervals and can be achieved in the following way, where RMI = reads mapped to interval and IL = interval length in base pairs.

$$\mathrm{RPK} = \mathrm{RMI}/\mathrm{IL} \times 1000$$

Acknowledgments

We thank Isabel Schwarz and Joël Wirz for carefully reading the manuscript prior to submission. Research in the Baubeclab is supported by an SNSF Professorship (SNF157488) and Systems-X.ch Special Opportunities Grant (2015_322) to T.B., and by the University of Zurich.

References

1. Okano M, Bell DW, Haber DA, Li E (1999) DNA methyltransferases Dnmt3a and Dnmt3b are essential for de novo methylation and mammalian development. Cell 99(3):247–257
2. Smith ZD, Meissner A (2013) DNA methylation: roles in mammalian development. Nat Rev Genet 14(3):204–220. https://doi.org/10.1038/nrg3354
3. Suzuki MM, Bird A (2008) DNA methylation landscapes: provocative insights from epigenomics. Nat Rev Genet 9(6):465–476. https://doi.org/10.1038/nrg2341
4. Landt SG, Marinov GK, Kundaje A, Kheradpour P, Pauli F, Batzoglou S, Bernstein BE, Bickel P, Brown JB, Cayting P, Chen Y, DeSalvo G, Epstein C, Fisher-Aylor KI, Euskirchen G, Gerstein M, Gertz J, Hartemink AJ, Hoffman MM, Iyer VR, Jung YL, Karmakar S, Kellis M, Kharchenko PV, Li Q, Liu T, Liu XS, Ma L, Milosavljevic A, Myers RM, Park PJ, Pazin MJ, Perry MD, Raha D, Reddy TE, Rozowsky J, Shoresh N, Sidow A, Slattery M, Stamatoyannopoulos JA, Tolstorukov MY, White KP, Xi S, Farnham PJ, Lieb JD, Wold BJ, Snyder M (2012) ChIP-seq guidelines and practices of the ENCODE and modENCODE consortia. Genome Res 22(9):1813–1831. https://doi.org/10.1101/gr.136184.111
5. Einhauer A, Jungbauer A (2001) The FLAG peptide, a versatile fusion tag for the purification of recombinant proteins. J Biochem Biophys Methods 49(1–3):455–465
6. Kolodziej KE, Pourfarzad F, de Boer E, Krpic S, Grosveld F, Strouboulis J (2009) Optimal use of tandem biotin and V5 tags in ChIP assays. BMC Mol Biol 10:6. https://doi.org/10.1186/1471-2199-10-6
7. Wilbanks EG, Larsen DJ, Neches RY, Yao AI, Wu CY, Kjolby RA, Facciotti MT (2012) A workflow for genome-wide mapping of archaeal transcription factors with ChIP-seq. Nucleic Acids Res 40(10):e74. https://doi.org/10.1093/nar/gks063
8. Kidder BL, Hu G, Zhao K (2011) ChIP-Seq: technical considerations for obtaining high-quality data. Nat Immunol 12(10):918–922. https://doi.org/10.1038/ni.2117
9. de Boer E, Rodriguez P, Bonte E, Krijgsveld J, Katsantoni E, Heck A, Grosveld F, Strouboulis J (2003) Efficient biotinylation and single-step purification of tagged transcription factors in mammalian cells and transgenic mice. Proc Natl Acad Sci U S A 100(13):7480–7485. https://doi.org/10.1073/pnas.1332608100
10. Green NM (1990) Avidin and streptavidin. Methods Enzymol 184:51–67
11. Baubec T, Colombo DF, Wirbelauer C, Schmidt J, Burger L, Krebs AR, Akalin A, Schubeler D (2015) Genomic profiling of DNA methyltransferases reveals a role for DNMT3B in genic methylation. Nature 520(7546):243–247. https://doi.org/10.1038/nature14176
12. Lindqvist Y, Schneider G (1996) Protein-biotin interactions. Curr Opin Struct Biol 6(6):798–803

13. Schatz PJ (1993) Use of peptide libraries to map the substrate specificity of a peptide-modifying enzyme: a 13 residue consensus peptide specifies biotinylation in Escherichia coli. Biotechnology (N Y) 11(10):1138–1143
14. Kim J, Cantor AB, Orkin SH, Wang J (2009) Use of in vivo biotinylation to study protein--protein and protein–DNA interactions in mouse embryonic stem cells. Nat Protoc 4:506–517
15. Flemr M, Buhler M (2015) Single-step generation of conditional knockout mouse embryonic stem cells. Cell Rep 12(4):709–716. https://doi.org/10.1016/j.celrep.2015.06.051
16. Weber M, Hellmann I, Stadler MB, Ramos L, Paabo S, Rebhan M, Schubeler D (2007) Distribution, silencing potential and evolutionary impact of promoter DNA methylation in the human genome. Nat Genet 39(4):457–466. https://doi.org/10.1038/ng1990
17. Meyer CA, Liu XS (2014) Identifying and mitigating bias in next-generation sequencing methods for chromatin biology. Nat Rev Genet 15(11):709–721. https://doi.org/10.1038/nrg3788
18. FASTX-Toolkit (2010) http://hannonlab.cshl.edu/fastx_toolkit/
19. Langmead B, Trapnell C, Pop M, Salzberg SL (2009) Ultrafast and memory-efficient alignment of short DNA sequences to the human genome. Genome Biol 10(3):R25. https://doi.org/10.1186/gb-2009-10-3-r25
20. Gaidatzis D, Lerch A, Hahne F, Stadler MB (2015) QuasR: quantification and annotation of short reads in R. Bioinformatics 31 (7):1130–1132. https://doi.org/10.1093/bioinformatics/btu781
21. Lawrence M, Huber W, Pages H, Aboyoun P, Carlson M, Gentleman R, Morgan MT, Carey VJ (2013) Software for computing and annotating genomic ranges. PLoS Comput Biol 9 (8):e1003118. https://doi.org/10.1371/journal.pcbi.1003118
22. Robinson MD, McCarthy DJ, Smyth GK (2010) edgeR: a Bioconductor package for differential expression analysis of digital gene expression data. Bioinformatics 26 (1):139–140. https://doi.org/10.1093/bioinformatics/btp616
23. Love MI, Huber W, Anders S (2014) Moderated estimation of fold change and dispersion for RNA-seq data with DESeq2. Genome Biol 15(12):550. https://doi.org/10.1186/s13059-014-0550-8
24. Gentleman R, Carey V, Huber W, Hahne F (2016) genefilter: methods for filtering genes from high-throughput experiments. https://www.bioconductor.org/packages/release/bioc/html/genefilter.html
25. Karolchik D, Hinrichs AS, Furey TS, Roskin KM, Sugnet CW, Haussler D, Kent WJ (2004) The UCSC Table Browser data retrieval tool. Nucleic Acids Res 32(Database issue): D493–D496. https://doi.org/10.1093/nar/gkh103
26. Gardiner-Garden M, Frommer M (1987) CpG islands in vertebrate genomes. J Mol Biol 196 (2):261–282
27. Takai D, Jones PA (2002) Comprehensive analysis of CpG islands in human chromosomes 21 and 22. Proc Natl Acad Sci U S A 99 (6):3740–3745. https://doi.org/10.1073/pnas.052410099
28. Hackenberg M, Carpena P, Bernaola-Galvan P, Barturen G, Alganza AM, Oliver JL (2011) WordCluster: detecting clusters of DNA words and genomic elements. Algorithms Mol Biol 6:2. https://doi.org/10.1186/1748-7188-6-2
29. Stadler MB, Murr R, Burger L, Ivanek R, Lienert F, Scholer A, van Nimwegen E, Wirbelauer C, Oakeley EJ, Gaidatzis D, Tiwari VK, Schubeler D (2011) DNA-binding factors shape the mouse methylome at distal regulatory regions. Nature 480(7378):490–495. https://doi.org/10.1038/nature10716
30. Baubec T, Ivanek R, Lienert F, Schubeler D (2013) Methylation-dependent and -independent genomic targeting principles of the MBD protein family. Cell 153(2):480–492. https://doi.org/10.1016/j.cell.2013.03.011
31. Pagès H, Aboyoun P, Gentleman R, DebRoy S (2016) Biostrings: string objects representing biological sequences, and matching algorithms. https://bioconductor.org/packages/release/bioc/html/Biostrings.html
32. Illingworth RS, Gruenewald-Schneider U, Webb S, Kerr AR, James KD, Turner DJ, Smith C, Harrison DJ, Andrews R, Bird AP (2010) Orphan CpG islands identify numerous conserved promoters in the mammalian genome. PLoS Genet 6(9):e1001134. https://doi.org/10.1371/journal.pgen.1001134
33. Akalin A, Franke V, Vlahovicek K, Mason CE, Schubeler D (2015) Genomation: a toolkit to summarize, annotate and visualize genomic intervals. Bioinformatics 31(7):1127–1129. https://doi.org/10.1093/bioinformatics/btu775
34. mod/mouse/humanENCODE: blacklisted genomic regions for functional genomics analysis (2014) https://sites.google.com/site/anshulkundaje/projects/blacklists
35 Manzo M, et al (2017) Isoform-specific localization of DNMT3A regulates DNA methylation fidelity at bivalent CpG islands. EMBO J 36:3421–3434

Check for updates

Chapter 10

Experimental Design and Bioinformatic Analysis of DNA Methylation Data

Yulia Medvedeva and Alexander Shershebnev

Abstract

DNA methylation is a crucial regulatory mechanism of gene expression, affected in many human pathologies. Therefore, it is not surprising that nowadays, in the era of high-throughput methods, a lot of data sets representing DNA methylation in various conditions are available and the amount of such data keeps growing. In this chapter, we discuss those aspects of experiment planning and data analysis, which we consider the most important for reliability and reproducibility of DNA methylation studies: usage of replicates, data quality control at various stages, selection of a statistical model, and incorporation of DNA methylation into the multi-omics analysis.

Key words DNA methylation, Next generation sequencing, Data analysis, Quality control, Experiment planning

1 Introduction

DNA methylation is an epigenetic mechanism associated with many normal and pathological biological processes: cell differentiation, cell identity and pluripotency maintenance (reviewed in [1–3]), aging [4], memory formation [5], responses to environmental exposures, stress, and diet [6, 7]. Abnormalities of DNA methylation have been reported for various diseases, including metabolic [8], cardiovascular [9], neurodegenerative [10, 11] diseases and cancers [12]. Profiles of DNA methylation can be inherited through cell division and in some cases through generations [13, 14]. On the other hand, DNA methylation can also be removed with the help of TET proteins [15] and chemical hypomethylating agents (Azacitidine [16]; Decitabine [17]). Hypomethylating agents are already used in the clinic for treatment of acute myeloid leukemia (AML) and myelodysplastic syndrome (MDS). Recently, targeted DNA demethylation techniques have been developed [18, 19] which demonstrates an increasing potential of DNA methylation to become a target for noninvasive

Tanya Vavouri and Miguel A. Peinado (eds.), *CpG Islands: Methods and Protocols*, Methods in Molecular Biology, vol. 1766, https://doi.org/10.1007/978-1-4939-7768-0_10, © Springer Science+Business Media, LLC, part of Springer Nature 2018

therapies for multiple diseases. As a result, there are a lot of genome-wide studies of DNA methylation in different conditions and their number is increasing rapidly. These studies usually search for a set of differentially methylated areas (positions, CpG islands or longer regions) between two or more sample sets. In this chapter, we discuss challenges at various steps of DNA methylation analysis. At the stage of experimental design, we discuss pros and cons for the selection of microarray-based vs sequencing-based approaches and how many replicates to choose. Also, we cover various quality control computational techniques. Then we focus on the most common analysis type: comparison of two groups of samples. For this experimental setup, we investigate the role of the selected statistical model and type of data representation. At the last part of this review, we provide our suggestions about which genomic loci are most interesting in terms of DNA methylation and how to incorporate DNA methylation data into the multi-omics analysis.

2 Technologies for DNA Methylation Profiling

Technologies for quantifying DNA methylation have evolved rapidly (reviewed in [20]). Currently, sequencing of bisulfite-converted DNA [21] and microarrays (mostly Illumina Infinium Methylation 27K, 450K, and EPIC) are the methods of choice since all of them provide single CpG resolution and sufficient genomic coverage.

2.1 Illumina Infinium Methylation Arrays

Illumina Infinium Methylation Kits are custom designed microarrays covering a predefined set of mostly CpGs but also some CHHs identified in human stem cells. The most enriched microarray, Infinium MethylationEPIC, also covers a fraction of differentially methylated sites identified in tumor versus normal tissues, FANTOM5 and ENCODE enhancers, ENCODE open chromatin and DNase hypersensitive sites as well as miRNA promoter regions. This array covers about 850 thousand cytosine positions, including more than 90% of the CpGs from the Infinium 450K and additional 400 thousand sites. The additional probes definitely improve the coverage of regulatory elements, including 58% of FANTOM5 enhancers, 7% distal and 27% proximal ENCODE regulatory elements [22]. Unfortunately, a single probe is not always representative of the methylation level of a regulatory element. The latter quite often demonstrates variable methylation across the region [23]. However, overall data from the EPIC array are highly reproducible across technical and biological replicates and are highly correlated with BS-sequencing data [23]. Apart from the reduced cost (as compared to genome-wide bisulfite sequencing (WGBS)) the usage of the arrays can be favorable for a study with multiple samples (dozens to hundreds) since this approach guarantees a minimum of missing values among all CG/CHH sites covered, while a coverage of a particular CpG in bisulfite sequencing (BS) data can be less stable across samples.

2.2 Bisulfite Sequencing

Sodium bisulfite treatment of DNA following next-generation sequencing (BS-seq) is an alternative technology for investigation of DNA methylation at single-nucleotide resolution. One of the advantages of BS-seq over the array-based technique is the use the same sequencing machine as for other experiments (RNA sequencing, for example). Unfortunately, due to the cost of whole-genome bisulfite sequencing, it is still not widely used. The reduced representation bisulfite sequencing (RRBS) method allows cost reduction [24]. This technique includes a restriction step followed by a fragment size selection in order to enrich for GC-rich regions of the genome. The most commonly used restriction enzyme is MspI (with a target site 5′-CCGG-3′). Although initially considered as an advantage, the ability to capture DNA methylation mostly in GC-rich promoters and CpG islands (CGIs), nowadays might not be considered as such. Recently it has been shown that high GC-content prevents CpGs from being methylated in any condition [25, 26]. Also, promoters of lncRNAs—a novel functional class of transcripts that greatly outnumbers protein-coding genes [27, 28]—are mostly depleted of CGIs [29]. In a comprehensive study of 42 WGBS methylomes in 30 diverse human cell and tissue types, dynamic methylation was reported for only about 20% of autosomal CpGs, mostly located in distal regulatory elements, particularly enhancers and transcription factor binding sites [30]. In this regard, the sequencing of GC-rich regions may give little information about the dynamic changes in DNA methylation. A combination of restriction enzymes (AluI, BfaI, HaeIII, HpyCH4V, MluCI, MseI, and MspI) has been demonstrated to improve CpG coverage up to 12% of the genome [31].

2.3 Single-Molecule Sequencing

Single molecule real-time (SMRT) sequencing is a recently developed technique, that allows to sequence single molecules of DNA yielding reads that are significantly longer (>10 kb on average) compared to second generation sequencing methods such as those offered by Illumina. SMRT sequencing employs special nanostructures (zero-mode waveguides), which enables to isolate and monitor a single molecule of DNA polymerase bound to a template DNA. Each nucleotide used in the DNA synthesis contains a distinct fluorophore which produces a pulse during the incorporation. It has been shown that various nucleotide modifications can affect the kinetics of that process changing the width and time between pulses allowing for reliable identification of a modified base [32].

Thus obtained long reads combine the information on both genomic and epigenomic levels and at the same time provide the possibility to detect modifications of DNA with strand-specificity at a single nucleotide resolution on a genome scale. Unfortunately, as of now, SMRT sequencing capabilities are limited in size, so the preferred application area is the sequencing of organisms that have a relatively small genome, such as prokaryotes. Another drawback of SMRT is its high error rate. Also, the cell epigenetic profile varies due to different cell state, phase or for stochastic reasons [33, 34]. There are, however, several papers reporting the usage of SMRT in conjunction with Illumina short reads in order to benefit from long SMRT reads while preserving an accuracy of Illumina reads [35, 36]. For all of these experiments, the main bioinformatics challenge is how to process and analyze extremely sparse data. With such data imputation using contemporary machine learning techniques might become necessary [35] as, for example, the deep learning methods which recently have demonstrated promising results in the analysis of single cell methylomes [36].

3 Experimental Design

Experimental design is—no doubt—the most critical step of every research project. Even complicated data analysis techniques cannot compensate for poor experimental design. It is extremely important to have a clear set of questions the researchers are interested in before the beginning of the study, for example, whether they are searching for biomarkers of a disease/condition or mechanistic insights into the role of DNA methylation on gene regulation. From our perspective, all designs could be used for the search of biomarkers/correlations given that a proper technology, a number of replicates and a coverage depth is chosen. Yet, for the mechanistic insights, our main advice would be to go for a longitudinal (time-course) or gradient study, since it allows not only to capture stable trends but also to infer missing data in an accurate manner.

3.1 Case–Control Design

A case–control study is a widespread type of study in which two existing groups of patients, animals, cells, etc. are identified and compared on the basis of some attribute. Case–control studies are often used to identify factors that may contribute to a response to stimuli or other conditions by comparing subjects in one condition (the "cases") with subjects in a different, usually more typical condition (the "controls"). The advantage of such design is that the size of cases/controls groups can be relatively large increasing the statistical power of an intergroup comparison. On the other hand, different groups will most likely have a complex set of individual phenotypes which cannot be controlled, reducing the potential to discover correlations (and likely also mechanical insights). The study of twins discordant for some trait or disease, or inbred animals from the same litter is probably the best scheme, as it excludes genetic variability, age-related issues, and, to some extent, environmental factors. Family studies benefit from reduced genetic variability of individuals and can uncover de novo (epi)mutations or transgenerational effects, although they cannot avoid age-related bias—a well-known feature of DNA methylation. In this experimental setup, a lot of factors can be controlled, but this design usually suffers from the small data sets available. For general case–-control experiments a set of well-developed statistical instruments reviewed elsewhere (for example, [37]) allows the researchers to estimate the size of case and control groups, which can provide expected levels of Type I and Type II errors based on the type of the tested hypothesis, previously estimated variance of the attribute of interest in the general population and effect size.

For a case–control design in a lab experiment (for example, treated vs nontreated animals or cells) only 1–3 replicates per condition are typically used. In the case of sequencing-based techniques, it is widely believed that the depth of sequencing can compensate for the lack of replicates. For example, the US National Institutes of Health (NIH) Roadmap Epigenomics Project recommends the use of two replicates with a combined total coverage of 30× for genome-wide bisulfite sequencing experiments (http://www.roadmapepigenomics.org/protocol (2011)). Yet Ziller et al. [38] have shown on both real and simulated data that an experimental set of two replicates with 15× coverage each has approximately the same power as a set of three replicates with 5× coverage each (15× coverage in total) in the detection of differentially methylated regions (DMR) with 20% difference in methylation. Based on multiple tests they recommend per-sample coverage in the range of 5–15×, depending on the magnitude of the intergroup methylation difference that needs to be detected and the downstream analysis that is planned to perform. Ziller and colleagues also show that an experiment with more replicates benefits more from the increased depth of coverage than an experiment with one replicate. According to their results, a single replicate experiment cannot get an accuracy much higher than 0.6 even with the highest possible coverage. All in all, Ziller et al. argue that sequencing at levels higher than 5–15× leads mostly to wasted resources that would be much better spent on an increased number of biological replicates. In a more recent saturation analysis, Libertini et al. [39] support the observation that methylome of 30× coverage in a single replicate is not adequate for quantitative identification of differentially methylated positions, suggesting that increasing the number of replicates rather than coverage can increase the accuracy of the downstream analysis.

3.2 Longitudinal (Time Course) or Gradual Design

In this design, DNA methylation should be measured over several time points or with a gradual change of a parameter of interest (concentration of a drug or time passed from the point of induction, for example). This design allows one to observe the trends in the establishment of DNA methylation, filter out the noise and increase the reliability of the results. For instance, a clear increase or decrease in DNA methylation over time or with a gradual change of drug concentration is more reliable as compared to a change when comparing only two conditions.

Unfortunately, there are not many tools developed specifically to process time course DNA methylation data. One of the possible approaches is to model DNA methylation dynamics with Markov chain models [40], assuming that methylation state over tissue differentiation or over a response to a stimulus can be modeled similarly to DNA sequence changes in evolution. Since DNA methylation is relatively slow changing, such approach allows for determination of nonrandom patterns. Moreover, a missing CpG methylation value in a sample can be more reliably imputed from methylation values at the same site in other samples of a longitudinal/gradual study than from neighboring CpG sites in the same sample [40].

3.3 Analysis of Samples that Contain Multiple Cell Types

It is essential to think carefully about the aim of the study—is it to seek mechanistic insights into the phenotype/disease or biomarkers. In the first case, samples should be purified to contain only one cell type (or as close to this aim as possible). In the case of a lab experiment, cell cycle synchronization is desirable. DNA methylation differs from one cell type to another and a mixture of cells can change the overall picture dramatically. In the second case, if the goal is to find biomarkers (of an exposure or of a predictive/prognostic parameter), a surrogate, accessible, and most used in the clinical samples may be of more help. Yet in this case, the researcher should keep in mind that cell composition varies from one sample to another.

Independently of whether purified cells or clinical samples (such as blood) are used, it is extremely important to experimentally or analytically account [41] for any epigenetic variability that is due to cell subtype, transcriptional or sequence variability, as well as any identifiable technical factors that occur during the experiments. These factors should be captured as metadata. We discuss the ways to analyze such data later in the review.

4 Quality Control of DNA Methylation Data

4.1 Quality Control (QC) of BS-seq Reads

As with every other type of sequencing data, the first standard step is assessing the quality of raw reads. There are several tools used for this task (FastQC, Prinseq, etc.) with FastQC (www.bioinformatics.

babraham.ac.uk/projects/fastqc/) being one of the most popular. Read quality sometimes drops in the later cycles and wrong base calls can lead to incorrect methylation calls. Therefore it is important to consider "Per base" quality metric and use the only high-quality portion of reads (quality trimming). Several metrics in FastQC output will most likely report errors due to C to T conversion. "Per base sequence content" metric will show an error as cytosine content will be really low, while thymine content will be elevated. Moreover, if cytosine content increases toward the read end, it is most likely a sign of adapter contamination. For RRBS, the type of a restriction enzyme used will also affect this metric. For instance, when using MspI enzyme, a near 100% peak at nucleotides 2 and 3 for cytosine is expected due to the sequence of a restriction site (CCGG). "GC-content" in BS libraries could also report an error, since the most typical values for human samples are in the range of 20–30%, while values closer to standard genomic GC content (40–50%) might indicate adapter contamination.

For BS data it is critical to perform not only adapter but also quality trimming (using Prinseq, Trim Galore, or other tools). Our recommendation is to avoid removal of low complexity reads since the majority of BS reads have only three nucleotides (A, G, T). For example, Prinseq (http://prinseq.sourceforge.net) complexity filter -lc_method entropy with a standard threshold (-lc_threshold 70) can remove up to 50% of otherwise high quality reads.

4.2 Duplicates

Removal of duplicated reads in BS-seq data is debatable. It is recommended for WGBS data as for any other NGS-seq data to avoid PCR duplicates. On the contrary for the RRBS, the decision is not that simple. Due to the use of restriction enzymes and the fixed read length, it is reasonable to have piles of reads originating from the same starting point. These might not be duplicates. Several lab kits provide a way to filter PCR-duplicates from real but identical reads. One such kit is Nugen Ovation RRBS Methyl-Seq System, where each original DNA fragment is marked using the random N6 sequence attached to each read at the step of adapter ligation. Those with the same starting point and N6 sequence are assumed to be PCR-duplicates. The number of duplicates (if this can be determined unambiguously from the data) can become a useful metric to assess library complexity. According to the Epi-GeneSys BS-seq data protocol (www.epigenesys.eu), 10% duplication is a sign of a diverse library, while 80% duplication is a strong indication of problems at library preparation or sequencing levels. Samples with high duplication levels should be removed from the downstream analysis.

4.3 Read Mapping and Methylation Calling

There are currently several dozen tools available for alignment of BS-reads (extensively reviewed elsewhere [42, 43]). Yet there are only two different categories: wild-card aligners and three-letter aligners [44]. Wild-card aligners (BSMAP, RMAP, MethPipe, etc.) replace cytosines in the sequenced reads with wild-card Y, which corresponds to both C and T. On the other hand, three-letter aligners (Bismark, BS-Seeker, etc.) convert all C to T in both sequenced reads and a reference genome before mapping with conventional fast aligners such as Bowtie. In the case of Bismark, the advantage of fast alignment usage is diminished by the fact that several alignments should be performed for all combinations of converted reads and genome sequence. Without going into details, it should be noted that both approaches reduce the information content of the reads leading to an increased fraction of multi-mapped reads. The standard fraction of uniquely mapped reads in the case of three-letter aligners is about 70–80%.

There are also a few noteworthy ideas specific for paired-end reads. First, read 2 contains information about the methylation from the same strand as read 1 but it will be aligned to the complementary strand. Usually, all aligners can deal with it automatically if appropriate filters for paired-end reads are used. Yet, this should be kept in mind when reads 1 and 2 are aligned separately, for example, to increase mappability. Also, at the stage of extracting methylation information (methyl calling), we would recommend avoiding independent consideration of read 1 and 2 from the same fragment in the case when the fragment length is smaller than the sum of reads lengths (-no-overlap filter in Bismark).

4.4 Post mapping QC

As soon as reads are mapped and methylation calling is performed, several simple tests can assess the quality of data from a biological and data-mining perspective.

4.4.1 BS-Conversion Efficiency

Since at the level of reads it is impossible to tell whether a particular cytosine is unmethylated or just not converted by sodium bisulfite, the estimation of the conversion rate is one of the major QC steps. Indirectly, it can be estimated from non-CpG conversion rate, assuming that non-CpGs are mostly not methylated. The assumption is pretty solid for most cell types, except stem and brain cells, where non-CpG methylation is relatively common. Therefore, cell type should be taken into account when planning the experiment. If there is a concern that the cell types used have a significant amount of non-CpG methylation, a small portion of definitely unmethylated DNA (DNA of phage lambda, for example) should be added to the sequencing library allowing for the direct calculation of conversion rate in the reads aligned to the foreign genome.

4.4.2 Coverage of CpG Positions

BS reads are nonuniformly distributed along the genome, which is particularly true for RRBS. For a specific genomic CpG site, the data from BS-seq can be summarized as counts of methylated and total reads. The methylated read counts are often modeled by a beta-binomial distribution, which accounts for both biological and sampling variations [45, 46]. To make a reliable methylation call one should have at least several reads covering a particular CpG site. In this regard checking the total fraction of the CpGs covered by more than N reads ($N = 10$ seems to be a reasonable threshold) might also be a very informative test. It is difficult to say how many CpGs are expected to be covered with at least N reads, but if this metric varies dramatically between studied samples while the total number of mapped reads is comparable, samples with the lowest CpG coverage may be best excluded.

4.4.3 Coverage of CpG Islands

For RRBS specifically, it is a good idea to assess the number of CpG-islands (CGIs) covered with at least one CpG position for which methylation level is determined with a reasonable level of significance. Coverage of too few CGIs might be interpreted as a sign of an experimental bias. Again if coverage varies between samples, the samples with the lowest coverage might be best excluded.

4.4.4 Methylation of Imprinted Genes

QC of the data might also include checking for the methylation status of known regions in the epigenome, such as promoters of known imprinted genes. For all diploid cell types, one would expect promoters of imprinted genes to be semimethylated, while for haploid cells the values must be around 0 or 1 depending on the parent of origin of the imprinted gene.

4.4.5 SNPs

In BS-seq data it is almost impossible to tell the difference between BS-converted C > T and C > T polymorphism. Heterozygous C > T variation can be detected from the aligned reads if the coverage is deep enough. Homozygous C > T variations are possible to detect only if for a given position enough reads from both strands supporting T-A pairing are detected. Unfortunately, C > T variation is the most common SNP in mammals due to a frequent spontaneous mutation of methylated C to T. Although it has been shown that C > T mutation level is decreased in CpG islands—the regions which are targeted by RRBS—even when being methylated [47] we suggest filtering out all known C > T SNPs prior to downstream analysis unless working with samples with extremely high coverage.

4.4.6 QQ-Plots

The QQ-plot (Fig. 1) is a nice and quick way to assess that there are no hidden associations in the data. It shows the comparison between the observed and expected distributions of p-values of the differentially methylated positions (DMPs) or regions (DMRs). If there is no systematic bias in the data the QQ-plot should be very close to a diagonal line $Y = X$ with a small curve in

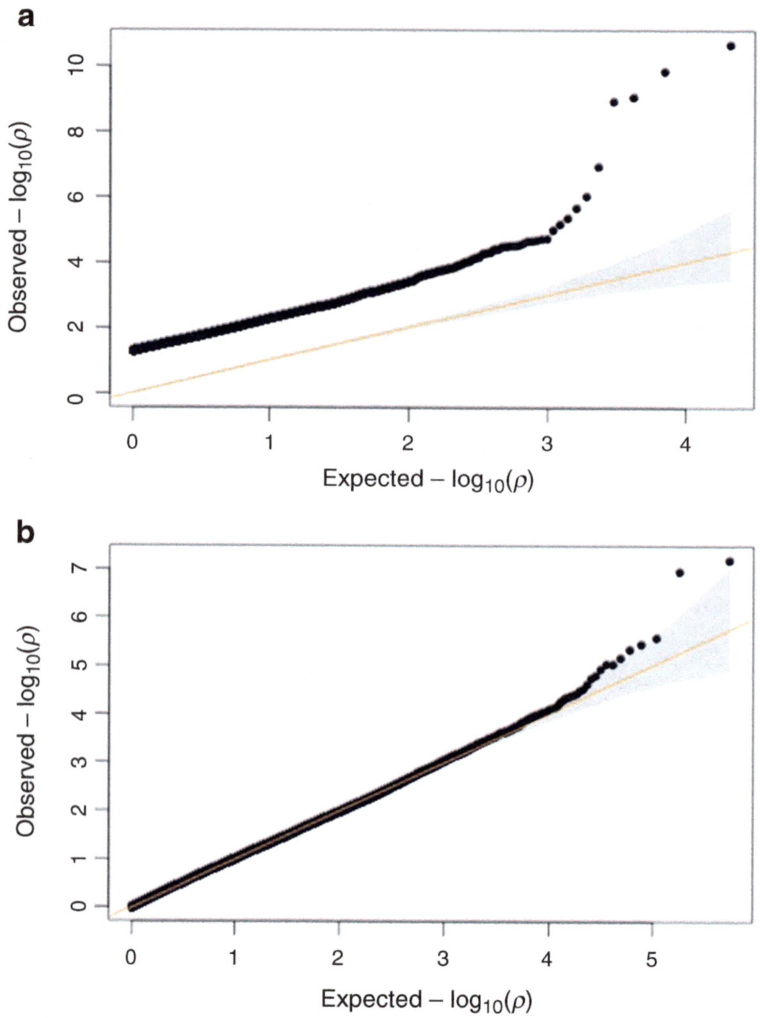

Fig. 1 (**a**) Sample QQ plot suggesting problems with data; (**b**) QQ plot for high-quality data

the upright corner which corresponds to the true associations between the two conditions (for example, DMRs related to a disease or a response to a stimulus). If on the other hand, the QQ-plot shows severe deviation from the diagonal line, this suggests some unaccounted confounder or other problem. Related is a genomic inflation factor (lambda) which is the ratio between the medians of observed and expected *p*-values. Deviations of lambda from 1 again strongly suggest existing biases in the data.

5 Search for Differential Methylation: Go Deeper, Broader, or Aside

There are two basic ways to analyze differential methylation between two samples based on BS-seq data (reviewed in [42, 43, 48]). The most widely used one is to sum reads over predefined regions or sliding windows and then detect differentially methylated regions (DMR). A similar approach can also be used for microarray data. The main flaw of such an approach is the following. Combining many separate CpGs (and reads for BS-seq) into a putative DMR can increase the statistical significance even of small absolute differences between methylation levels. If the aim of the study is to understand the mechanism of epigenetic regulation a small change in DNA methylation is hard to interpret. Since in a single cell at a particular genomic CpG site only three methylation levels are possible (fully unmethylated, fully methylated or semi-methylated, which represents two alleles with different methylation status), a small change in methylation represents a change in a small cell population. If the aim is to find a biomarker a small change is difficult to detect in the clinic. To compensate for this issue, the vast majority of programs use an explicit threshold for the maximum length of DMRs (or maximum distance between putative DMRs before merging). Although this approach sounds reasonable, it cannot find a relatively recently discovered structure—DNA methylation valleys—long (up to several Mb) stretches of low methylated CpGs that can be frequently found in cancer [49, 50] or COMETs—blocks of comethylation, which can be fairly large [51].

An alternative approach is to look for a single significantly differentially methylated CpG and then if needed, merge neighboring CpGs into a DMR. This approach is proven to outperform those based on merging of reads [52]. It also benefits more from the increase of genomic read coverage [38]. In addition, it was shown for several genes that methylation of a single CpG nearby can be a reliable representative marker of its expression [53, 54] and therefore should not be neglected by averaging methylation levels of all neighboring CpGs into DMRs.

Many statistical methods have been applied to the detection of differentially methylated CpGs (DMCs) between two conditions. The ratio-based data, which can be obtained from both microarrays and bisulfite sequencing, offer the flexibility of fitting to many linear models. The raw read count data, which can be obtained only from bisulfite sequencing takes into consideration the information about coverage. The ratio-based tests are usually less sensitive in comparison with the count-based tests. On the other hand, count-based methods might have higher false-positive rates. The beta-binomial model, that takes into account raw reads and replicates, gives a good balance between sensitivity and specificity, and therefore is the preferred method [55]. One of the tools based on the beta-binomial distribution that performs high in several benchmarks is RADmeth, which is now incorporated into the MethPipe software [56].

5.1 CpG Islands: Friends or Foes?

CpG islands (CGI), regions with a high frequency of CpG dinucleotides, are usually considered as a type of promoter. Their regulatory potential for transcription initiation has been widely accepted, although the definition is still quite vague [57]. CGI initiate transcription even being far away from genes [58]. In this regard, it is not surprising that researchers look at the methylation status of CGIs, giving rise to a concept of CpG methylator phenotype in tumors [59]. For a long time, cancer studies assumed that functionally important DNA methylation should occur in CpG islands, specifically the promoter ones. However, more recently, regions with relatively low CpG density have attracted attention. The majority of CpG rich regions (such as CpG islands) are non-dynamic and less variant between tissues [30, 60]. The most remarkable alterations in DNA methylation in cancer occur outside of CGI, in CGI-flanking sequences up to 2 kb distant, which are called "CpG island shores." Methylation levels of CpG island shores are strongly correlated with gene expression, and are discriminative for a cell type [60]. Methylation levels of CpG shelves (<2 kb flanking outward from a CpG shore) are also more dynamic than the ones of CGIs. Unfortunately, the most widely used protocol for the study of DNA methylation (the RRBS protocol that uses MspI restriction enzyme) has been developed with the specific aim to enrich for CGI regions. Also, 27K and 450K Infinium Methylation Arrays explicitly target promoters and therefore CGIs. Yet even with RRBS data, the analysis of the non-CGI methylation can be at least partially performed and we advise to do so.

5.2 Summary of Recommendation for the Search of Differentially Methylated Regions

Having the above in mind, we recommend not just running one of the commonly used software for DMR search, but also the following:

- Carefully choose the statistical model (we recommend methods based on beta-binomial distribution).
- Check the methylation status of important single CpGs.
- Look for DNA methylation valleys, COMETs, CGI shores and shelves, not only DMRs.
- When analyzing DMRs, select them not only based on the level of statistical significance. Small variation within a cell population cannot be reliably distinguished from a small intergroup difference. Therefore, a reasonable threshold of fold change in methylation level should be used.

6 Novel Aspects of DNA Methylation Analysis

6.1 Cell Population Variability

Variation in DNA methylation in different cells in a population briefly mentioned before, can per se be an interesting topic. Tumor heterogeneity has been accepted as one of the major obstacles to effective treatment and it is becoming more and more clear that DNA methylation contributes to cancer heterogeneity. One of the cancer development models suggests that cancer-specific changes in DNA methylation contribute to epigenetic stochasticity of tumor cells, which in turn is subject to natural selection [61, 62]. Also, the increased heterogeneity of methylation in a population of organisms was suggested to compensate for decreased genetic variation in a population shortly after a "bottleneck" [63]. In this regard, looking for regions of high variation in methylation levels within a group might be as informative as looking for high variation between groups.

6.2 Single-Cell Methylation Variability

Assumptions about the role of epigenetic modifications are usually made based on the correlations obtained from a bulk cell population. However, a growing number of articles suggests that epigenetic regulation is more complex than previously believed [34, 64]. Single-cell variability can be of special interest in a highly heterogeneous environment, such as cancer, for detection of cell subtypes in the population. It was also shown, that single-cell epigenomics can be used to assess heterogeneous responses to drugs [33].

6.3 Non-CpG Methylation

BS reads may detect non-CpG methylation as well, yet at the analysis stage quite often such positions are filtered out [65]. Surprisingly, non-CpG methylation has been recently reported to play an important role in the regulation of transcription factors (TF) GR, ER, and BMAL1 binding [66]. At the same time, the majority of TFs capable of binding unmethylated CpG can bind methylated CpG as well [67].

6.4 Hydroxymethylcytosine (5hmC)

If methylation variability in the region of interest is high, it might be indicative of active demethylation. To investigate this process one can measure 5hmC—the first product in the active demethylation of 5mC. Bisulfite sequencing, the gold standard for detecting 5mC, cannot discriminate between 5mC and 5hmC [68]. Experimental methods such as oxidative bisulfite sequencing (oxBS-seq) have been developed to distinguish between the two [69]. In this case 5hmC is oxidized to 5fC, which in turn is converted to uracil by bisulfite. This way, a comparison to standard BS-data of the same sample can discriminate between 5hmC and 5mC shedding light on the dynamics of DNA methylation.

7 Integrating Various OMICS Data

After detection of DMC, DMR, DMV, CGI, or any other kind of differentially methylated loci the standard approach is to associate them to the closest genes. Although several interesting observations have been made using this simplistic approach, the distance to the closest gene can be huge, sometimes up to several hundred megabases. It is hard to believe in this kind of association without additional information. Below are our recommendations on how to computationally infer the function of differentially methylated loci.

7.1 Enhancer–Gene Association

Enhancers are distal regulatory regions that can be found up to dozens megabases away from their target genes. Although they are usually GC-poor, there are multiple known examples showing that methylation of an enhancer can regulate the expression of an associated gene [69]. Recently a powerful technique for enhancer determination has been developed based on CAGE data (Cap Analysis of Gene Expression) [70]. In brief, enhancers can be detected by locating CAGE clusters that are up to two nucleosomes apart, with balanced bidirectional transcription. This approach allows one to make an association between a promoter and an enhancer based on the correlation between their expression profiles in different cell types [71].

7.2 3D Chromatin Structure

Topologically associated domains (TAD) contain coexpressed genes and most likely their regulatory regions as well. At least the association between CG-rich regions and TAD-structure has been shown in several animals [72]. The presence of a gene and DMR in the same TAD supports their functional association. Since TAD structure is similar for different cell types, one can use any available data for an organism of interest.

7.3 Transcriptomics

At the stage of experimental design, it is important to understand that although DNA methylation is correlated with gene expression, this correlation is not always strong. Therefore transcriptomics data for the same cells that are tested for epigenetic changes contribute to interpretability of methylation data. If an observed epigenetic change is severe it is a good idea to look at the expression levels of enzymes and other proteins that establish the particular epigenetic mark [73].

Recently, a novel experimental technique that combines BS-seq and RNA-Seq has been also developed [74]. The main idea behind this is to physically separate DNA and RNA so that bisulfite conversion does not affect the transcriptome. This allows looking simultaneously at both methylome and transcriptome of the same single cell instead of approximating over a large (potentially heterogeneous) cell population.

7.4 Multifactorial Experiments

With rapid advances in sequencing technologies and a continuous reduction in sequencing costs, an increasing number of experiments are now performed under general, multifactor designs. DNA methylation in such experiments is usually measured by

RRBS [75]. Although we recommend using more simple designs, we expect this trend to continue and data with multiple factors/groups and covariates to be widely available in the near future, even from large-scale population level studies. Flexible and efficient methods are in great demand to comprehensively decipher biological processes in current epigenomics research. Novel statistical models have been developed to detect differentially methylated loci using general experimental designs taking covariates into consideration [41].

8 Conclusions

DNA methylation affects and is affected by various processes. Modern technologies allow one to determine DNA methylation at the level of a single nucleotide at a relatively low price, which has already led to enormous amounts of data for different cell types, developmental stages, and diseases. This chapter represents our view on the challenges of DNA methylation analysis. Our aim is to discuss the most critical steps in DNA methylation analysis and provide some ideas on how to deal with such data based on our experience and literature. From the very beginning, we encourage the researchers to understand that sequencing-based approaches generally have more statistical power, yet microarrays-based approaches guarantee measured methylated values in almost all genomic locations present on a microarray, which in turn ensure the possibility of comparison of several dozens or hundreds of patients. We emphasize that using replicates is not only an important contribution to reliability but also in the case of BS-based methods may reduce the overall cost of the experiment while providing the same or even more information. We focus on the techniques for data quality control, in particular, on post mapping QC for BS-data, which in our experience is often and undeservedly omitted. For the most common analysis type—the selection of differentially methylated loci—we discuss the role of the statistical model and data representation. We conclude that methods based on beta-binomial distribution usually outperform others. The best-performing methods first determine differentially methylated cytosines using raw counts and only then merge several significant CpGs into a DMR. We also provide suggestions about which genomic loci are most promising in terms of the expected DNA methylation change and regulatory role. As a final step, we discuss how to incorporate DNA methylation data into multi-omics studies. We believe that a careful, diverse and comprehensive analysis performed with DNA methylation data can contribute to reliable and reproducible DNA methylation results, which in turn can bring new discoveries.

Acknowledgments

Y.A.M.'s work was supported by RSF grant 15-14-30002, and A. S.'s work was supported by RSF grant 14-45-00065. Y.A.M. wrote the manuscript, and A.S. wrote sections about quality control and contributed to others.

References

1. Jaenisch R, Bird A (2003) Epigenetic regulation of gene expression: how the genome integrates intrinsic and environmental signals. Nat Genet 33(Suppl):245–254. https://doi.org/10.1038/ng1089
2. Messerschmidt DM, Knowles BB, Solter D (2014) DNA methylation dynamics during epigenetic reprogramming in the germline and preimplantation embryos. Genes Dev 28 (8):812–828. https://doi.org/10.1101/gad.234294.113
3. Tomazou EM, Meissner A (2010) Epigenetic regulation of pluripotency. Adv Exp Med Biol 695:26–40. https://doi.org/10.1007/978-1-4419-7037-4_3
4. Horvath S (2013) DNA methylation age of human tissues and cell types. Genome Biol 14 (10):R115. https://doi.org/10.1186/gb-2013-14-10-r115
5. Miller CA, Sweatt JD (2007) Covalent modification of DNA regulates memory formation. Neuron 53(6):857–869. https://doi.org/10.1016/j.neuron.2007.02.022
6. Jirtle RL, Skinner MK (2007) Environmental epigenomics and disease susceptibility. Nat Rev Genet 8(4):253–262. https://doi.org/10.1038/nrg2045
7. Ladd-Acosta C, Fallin MD (2016) The role of epigenetics in genetic and environmental epidemiology. Epigenomics 8(2):271–283. https://doi.org/10.2217/epi.15.102
8. Desai M, Jellyman JK, Ross MG (2015) Epigenomics, gestational programming and risk of metabolic syndrome. Int J Obes 39 (4):633–641. https://doi.org/10.1038/ijo.2015.13
9. Zhong J, Agha G, Baccarelli AA (2016) The role of DNA methylation in cardiovascular risk and disease: methodological aspects, study design, and data analysis for epidemiological studies. Circ Res 118(1):119–131. https://doi.org/10.1161/CIRCRESAHA.115.305206
10. Wüllner U, Kaut O, deBoni L, Piston D, Schmitt I (2016) DNA methylation in Parkinson's disease. J Neurochem 139(Suppl 1):108–120. https://doi.org/10.1111/jnc.13646
11. Sanchez-Mut JV, Gräff J (2015) Epigenetic alterations in Alzheimer's disease. Front Behav Neurosci 9:347. https://doi.org/10.3389/fnbeh.2015.00347
12. Baylin SB, Jones PA (2016) Epigenetic determinants of cancer. Cold Spring Harb Perspect Biol 8(9). https://doi.org/10.1101/cshperspect.a019505
13. Klosin A, Lehner B (2016) Mechanisms, timescales and principles of trans-generational epigenetic inheritance in animals. Curr Opin Genet Dev 36:41–49. https://doi.org/10.1016/j.gde.2016.04.001
14. Klosin A, Casas E, Hidalgo-Carcedo C, Vavouri T, Lehner B (2017) Transgenerational transmission of environmental information in C. elegans. Science 356(6335):320–323. https://doi.org/10.1126/science.aah6412
15. Ito S, Shen L, Dai Q, Wu SC, Collins LB, Swenberg JA, He C, Zhang Y (2011) Tet proteins can convert 5-methylcytosine to 5-formylcytosine and 5-carboxylcytosine. Science 333(6047):1300–1303. https://doi.org/10.1126/science.1210597
16. Garcia-Manero G, Stoltz ML, Ward MR, Kantarjian H, Sharma S (2008) A pilot pharmacokinetic study of oral azacitidine. Leukemia 22(9):1680–1684. https://doi.org/10.1038/leu.2008.145
17. Aribi A, Borthakur G, Ravandi F, Shan J, Davisson J, Cortes J, Kantarjian H (2007) Activity of decitabine, a hypomethylating agent, in chronic myelomonocytic leukemia. Cancer 109(4):713–717. https://doi.org/10.1002/cncr.22457
18. Morita S, Noguchi H, Horii T, Nakabayashi K, Kimura M, Okamura K, Sakai A, Nakashima H, Hata K, Nakashima K, Hatada I (2016) Targeted DNA demethylation in vivo using dCas9–peptide repeat and scFv–TET1 catalytic domain fusions. Nat Biotechnol 34 (10):1060–1065. https://doi.org/10.1038/nbt.3658
19. Xu X, Tao Y, Gao X, Zhang L, Li X, Zou W, Ruan K, Wang F, Xu G-L, Hu R (2016) A

CRISPR-based approach for targeted DNA demethylation. Cell Discov 2:16009. https://doi.org/10.1038/celldisc.2016.9

20. McGregor K, Bernatsky S, Colmegna I, Hudson M, Pastinen T, Labbe A, Greenwood CMT (2016) An evaluation of methods correcting for cell-type heterogeneity in DNA methylation studies. Genome Biol 17:84. https://doi.org/10.1186/s13059-016-0935-y
21. Davis BM, Chao MC, Waldor MK (2013) Entering the era of bacterial epigenomics with single molecule real time DNA sequencing. Curr Opin Microbiol 16(2):192–198. https://doi.org/10.1016/j.mib.2013.01.011
22. Pidsley R, Zotenko E, Peters TJ, Lawrence MG, Risbridger GP, Molloy P, Van Djik S, Muhlhausler B, Stirzaker C, Clark SJ (2016) Critical evaluation of the Illumina MethylationEPIC BeadChip microarray for whole-genome DNA methylation profiling. Genome Biol 17(1):208. https://doi.org/10.1186/s13059-016-1066-1
23. Bock C, Tomazou EM, Brinkman AB, Müller F, Simmer F, Gu H, Jäger N, Gnirke A, Stunnenberg HG, Meissner A (2010) Quantitative comparison of genome-wide DNA methylation mapping technologies. Nat Biotechnol 28(10):1106–1114. https://doi.org/10.1038/nbt.1681
24. Meissner A, Gnirke A, Bell GW, Ramsahoye B, Lander ES, Jaenisch R (2005) Reduced representation bisulfite sequencing for comparative high-resolution DNA methylation analysis. Nucleic Acids Res 33(18):5868–5877. https://doi.org/10.1093/nar/gki901
25. Wachter E, Quante T, Merusi C, Arczewska A, Stewart F, Webb S, Bird A (2014) Synthetic CpG islands reveal DNA sequence determinants of chromatin structure. elife 3:e03397. https://doi.org/10.7554/eLife.03397
26. Krebs AR, Dessus-Babus S, Burger L, Schübeler D (2014) High-throughput engineering of a mammalian genome reveals building principles of methylation states at CG rich regions. elife 3:e04094. https://doi.org/10.7554/eLife.04094
27. Kapranov P, Cheng J, Dike S, Nix DA, Duttagupta R, Willingham AT, Stadler PF, Hertel J, Hackermüller J, Hofacker IL, Bell I, Cheung E, Drenkow J, Dumais E, Patel S, Helt G, Ganesh M, Ghosh S, Piccolboni A, Sementchenko V, Tammana H, Gingeras TR (2007) RNA maps reveal new RNA classes and a possible function for pervasive transcription. Science 316(5830):1484–1488. https://doi.org/10.1126/science.1138341
28. Hon CC, Ramilowski JA, Harshbarger J, Bertin N, Rackham OJ, Gough J, Denisenko E, Schmeier S, Poulsen TM, Severin J, Lizio M, Kawaji H, Kasukawa T, Itoh M, Burroughs AM, Noma S, Djebali S, Alam T, Medvedeva YA, Testa AC, Lipovich L, Yip CW, Abugessaisa I, Mendez M, Hasegawa A, Tang D, Lassmann T, Heutink P, Babina M, Wells CA, Kojima S, Nakamura Y, Suzuki H, Daub CO, de Hoon MJ, Arner E, Hayashizaki Y, Carninci P, Forrest AR (2017) An atlas of human long non-coding RNAs with accurate 5′ ends. Nature 543(7644):199–204. https://doi.org/10.1038/nature21374
29. Alam T, Medvedeva YA, Jia H, Brown JB, Lipovich L, Bajic VB (2014) Promoter analysis reveals globally differential regulation of human long non-coding RNA and protein-coding genes. PLoS One 9(10):e109443. https://doi.org/10.1371/journal.pone.0109443
30. Ziller MJ, Gu H, Müller F, Donaghey J, Tsai LTY, Kohlbacher O, De Jager PL, Rosen ED, Bennett DA, Bernstein BE, Gnirke A, Meissner A (2013) Charting a dynamic DNA methylation landscape of the human genome. Nature 500(7463):477–481. https://doi.org/10.1038/nature12433
31. Martinez-Arguelles DB, Lee S, Papadopoulos V (2014) In silico analysis identifies novel restriction enzyme combinations that expand reduced representation bisulfite sequencing CpG coverage. BMC Res Notes 7:534. https://doi.org/10.1186/1756-0500-7-534
32. Guo H, Zhu P, Wu X, Li X, Wen L, Tang F (2013) Single-cell methylome landscapes of mouse embryonic stem cells and early embryos analyzed using reduced representation bisulfite sequencing. Genome Res 23(12):2126–2135. https://doi.org/10.1101/gr.161679.113
33. Farlik M, Sheffield NC, Nuzzo A, Datlinger P, Schönegger A, Klughammer J, Bock C (2015) Single-cell DNA methylome sequencing and bioinformatic inference of epigenomic cell-state dynamics. Cell Rep 10(8):1386–1397. https://doi.org/10.1016/j.celrep.2015.02.001
34. Smallwood SA, Lee HJ, Angermueller C, Krueger F, Saadeh H, Peat J, Andrews SR, Stegle O, Reik W, Kelsey G (2014) Single-cell genome-wide bisulfite sequencing for assessing epigenetic heterogeneity. Nat Methods 11(8):817–820. https://doi.org/10.1038/nmeth.3035
35. Ernst J, Kellis M (2015) Large-scale imputation of epigenomic datasets for systematic annotation of diverse human tissues. Nat Biotechnol 33(4):364–376. https://doi.org/10.1038/nbt.3157

36. Angermueller C, Lee HJ, Reik W, Stegle O (2017) DeepCpG: accurate prediction of single-cell DNA methylation states using deep learning. Genome Biol 18(1):67. https://doi.org/10.1186/s13059-017-1189-z
37. Noordzij M, Tripepi G, Dekker FW, Zoccali C, Tanck MW, Jager KJ (2010) Sample size calculations: basic principles and common pitfalls. Nephrol Dial Transplant 25(5):1388–1393. https://doi.org/10.1093/ndt/gfp732
38. Ziller MJ, Hansen KD, Meissner A, Aryee MJ (2015) Coverage recommendations for methylation analysis by whole-genome bisulfite sequencing. Nat Methods 12(3):230–232., 231 p. following 232. https://doi.org/10.1038/nmeth.3152
39. Libertini E, Heath SC, Hamoudi RA, Gut M, Ziller MJ, Herrero J, Czyz A, Ruotti V, Stunnenberg HG, Frontini M, Ouwehand WH, Meissner A, Gut IG, Beck S (2016) Saturation analysis for whole-genome bisulfite sequencing data. Nat Biotechnol. https://doi.org/10.1038/nbt.3524
40. Capra JA, Kostka D (2014) Modeling DNA methylation dynamics with approaches from phylogenetics. Bioinformatics 30(17): i408–i414. https://doi.org/10.1093/bioinformatics/btu445
41. Park Y, Wu H (2016) Differential methylation analysis for BS-seq data under general experimental design. Bioinformatics 32(10):1446–1453. https://doi.org/10.1093/bioinformatics/btw026
42. Wright ML, Dozmorov MG, Wolen AR, Jackson-Cook C, Starkweather AR, Lyon DE, York TP (2016) Establishing an analytic pipeline for genome-wide DNA methylation. Clin Epigenetics 8:45. https://doi.org/10.1186/s13148-016-0212-7
43. Stockwell PA, Chatterjee A, Rodger EJ, Morison IM (2014) DMAP: differential methylation analysis package for RRBS and WGBS data. Bioinformatics 30(13):1814–1822. https://doi.org/10.1093/bioinformatics/btu126
44. Bock C (2012) Analysing and interpreting DNA methylation data. Nat Rev Genet 13(10):705–719. https://doi.org/10.1038/nrg3273
45. Park Y, Figueroa ME, Rozek LS, Sartor MA (2014) MethylSig: a whole genome DNA methylation analysis pipeline. Bioinformatics 30(17):2414–2422. https://doi.org/10.1093/bioinformatics/btu339
46. Dolzhenko E, Smith AD (2014) Using beta-binomial regression for high-precision differential methylation analysis in multifactor whole-genome bisulfite sequencing experiments. BMC Bioinformatics 15:215. https://doi.org/10.1186/1471-2105-15-215
47. Panchin AY, Makeev VJ, Medvedeva YA (2016) Preservation of methylated CpG dinucleotides in human CpG islands. Biol Direct 11(1):11. https://doi.org/10.1186/s13062-016-0113-x
48. Robinson MD, Kahraman A, Law CW, Lindsay H, Nowicka M, Weber LM, Zhou X (2014) Statistical methods for detecting differentially methylated loci and regions. Front Genet 5:324. https://doi.org/10.3389/fgene.2014.00324
49. Xie W, Schultz MD, Lister R, Hou Z, Rajagopal N, Ray P, Whitaker JW, Tian S, Hawkins RD, Leung D, Yang H, Wang T, Lee AY, Swanson SA, Zhang J, Zhu Y, Kim A, Nery JR, Urich MA, Kuan S, Yen C-A, Klugman S, Yu P, Suknuntha K, Propson NE, Chen H, Edsall LE, Wagner U, Li Y, Ye Z, Kulkarni A, Xuan Z, Chung W-Y, Chi NC, Antosiewicz-Bourget JE, Slukvin I, Stewart R, Zhang MQ, Wang W, Thomson JA, Ecker JR, Ren B (2013) Epigenomic analysis of multilineage differentiation of human embryonic stem cells. Cell 153(5):1134–1148. https://doi.org/10.1016/j.cell.2013.04.022
50. Jeong M, Sun D, Luo M, Huang Y, Challen GA, Rodriguez B, Zhang X, Chavez L, Wang H, Hannah R, Kim S-B, Yang L, Ko M, Chen R, Göttgens B, Lee J-S, Gunaratne P, Godley LA, Darlington GJ, Rao A, Li W, Goodell MA (2014) Large conserved domains of low DNA methylation maintained by Dnmt3a. Nat Genet 46(1):17–23. https://doi.org/10.1038/ng.2836
51. Libertini E, Heath SC, Hamoudi RA, Gut M, Ziller MJ, Czyz A, Ruotti V, Stunnenberg HG, Frontini M, Ouwehand WH, Meissner A, Gut IG, Beck S (2016) Information recovery from low coverage whole-genome bisulfite sequencing. Nat Commun 7:11306. https://doi.org/10.1038/ncomms11306
52. Klein HU, Hebestreit K (2016) An evaluation of methods to test predefined genomic regions for differential methylation in bisulfite sequencing data. Brief Bioinform 17(5):796–807. https://doi.org/10.1093/bib/bbv095
53. Medvedeva YA, Khamis AM, Kulakovskiy IV, Ba-Alawi W, MSI B, Kawaji H, Lassmann T, Harbers M, ARR F, Bajic VB, Consortium F (2014) Effects of cytosine methylation on transcription factor binding sites. BMC Genomics 15:119. https://doi.org/10.1186/1471-2164-15-119

54. Pardo LM, Rizzu P, Francescatto M, Vitezic M, Leday GGR, Sanchez JS, Khamis A, Takahashi H, van de Berg WDJ, Medvedeva YA, van de Wiel MA, Daub CO, Carninci P, Heutink P (2013) Regional differences in gene expression and promoter usage in aged human brains. Neurobiol Aging 34 (7):1825–1836. https://doi.org/10.1016/j.neurobiolaging.2013.01.005
55. Zhang Y, Baheti S, Sun Z (2016) Statistical method evaluation for differentially methylated CpGs in base resolution next-generation DNA sequencing data. Brief Bioinform. https://doi.org/10.1093/bib/bbw133
56. Song Q, Decato B, Hong EE, Zhou M, Fang F, Qu J, Garvin T, Kessler M, Zhou J, Smith AD (2013) A reference methylome database and analysis pipeline to facilitate integrative and comparative epigenomics. PLoS One 8 (12):e81148. https://doi.org/10.1371/journal.pone.0081148
57. Medvedeva YA (2011) Algorithms for CpG islands search: new advantages and old problems. In: Mahdavi MM (ed) Bioinformatics – trends and methodologies. InTech, Rijeka. https://doi.org/10.5772/22883
58. Medvedeva YA, Fridman MV, Oparina NJ, Malko DB, Ermakova EO, Kulakovskiy IV, Heinzel A, Makeev VJ (2010) Intergenic, gene terminal, and intragenic CpG islands in the human genome. BMC Genomics 11:48. https://doi.org/10.1186/1471-2164-11-48
59. Issa J-P (2004) CpG island methylator phenotype in cancer. Nat Rev Cancer 4 (12):988–993. https://doi.org/10.1038/nrc1507
60. Irizarry RA, Ladd-Acosta C, Wen B, Wu Z, Montano C, Onyango P, Cui H, Gabo K, Rongione M, Webster M, Ji H, Potash JB, Sabunciyan S, Feinberg AP (2009) The human colon cancer methylome shows similar hypo- and hypermethylation at conserved tissue-specific CpG island shores. Nat Genet 41(2):178–186. https://doi.org/10.1038/ng.298
61. Feinberg AP (2014) Epigenetic stochasticity, nuclear structure and cancer: the implications for medicine. J Intern Med 276(1):5–11. https://doi.org/10.1111/joim.12224
62. Hansen KD, Timp W, Bravo HC, Sabunciyan S, Langmead B, McDonald OG, Wen B, Wu H, Liu Y, Diep D, Briem E, Zhang K, Irizarry RA, Feinberg AP (2011) Increased methylation variation in epigenetic domains across cancer types. Nat Genet 43 (8):768–775. https://doi.org/10.1038/ng.865
63. Artemov AV, Mugue NS, Rastorguev SM, Zhenilo S, Mazur AM, Tsygankova SV, Boulygina ES, Kaplun D, Nedoluzhko AV, Medvedeva YA, Prokhortchouk EB (2017) Genome-wide DNA methylation profiling reveals epigenetic adaptation of stickleback to marine and freshwater conditions. Mol Biol Evol 5:msx156
64. Gravina S, Dong X, Yu B, Vijg J (2016) Single-cell genome-wide bisulfite sequencing uncovers extensive heterogeneity in the mouse liver methylome. Genome Biol 17 (1):150. https://doi.org/10.1186/s13059-016-1011-3
65. Krueger F, Andrews SR (2011) Bismark: a flexible aligner and methylation caller for bisulfite-Seq applications. Bioinformatics 27 (11):1571–1572. https://doi.org/10.1093/bioinformatics/btr167
66. Jin J, Lian T, Gu C, Yu K, Gao YQ, Su X-D (2016) The effects of cytosine methylation on general transcription factors. Sci Rep 6:29119. https://doi.org/10.1038/srep29119
67. Yin Y, Morgunova E, Jolma A, Kaasinen E, Sahu B, Khund-Sayeed S, Das PK, Kivioja T, Dave K, Zhong F, Nitta KR, Taipale M, Popov A, Ginno PA, Domcke S, Yan J, Schubeler D, Vinson C, Taipale J (2017) Impact of cytosine methylation on DNA binding specificities of human transcription factors. Science 356:6337. https://doi.org/10.1126/science.aaj2239
68. Jin S-G, Kadam S, Pfeifer GP (2010) Examination of the specificity of DNA methylation profiling techniques towards 5-methylcytosine and 5-hydroxymethylcytosine. Nucleic Acids Res 38(11):e125. https://doi.org/10.1093/nar/gkq223
69. Aran D, Hellman A (2013) DNA methylation of transcriptional enhancers and cancer predisposition. Cell 154(1):11–13. https://doi.org/10.1016/j.cell.2013.06.018
70. Booth MJ, Ost TWB, Beraldi D, Bell NM, Branco MR, Reik W, Balasubramanian S (2013) Oxidative bisulfite sequencing of 5-methylcytosine and 5-hydroxymethylcytosine. Nat Protoc 8 (10):1841–1851. https://doi.org/10.1038/nprot.2013.115
71. Andersson R, Gebhard C, Miguel-Escalada I, Hoof I, Bornholdt J, Boyd M, Chen Y, Zhao X, Schmidl C, Suzuki T, Ntini E, Arner E, Valen E, Li K, Schwarzfischer L, Glatz D, Raithel J, Lilje B, Rapin N, Bagger FO, Jørgensen M, Andersen PR, Bertin N, Rackham O, Burroughs AM, Baillie JK, Ishizu Y, Shimizu Y, Furuhata E, Maeda S, Negishi Y, Mungall CJ, Meehan TF,

Lassmann T, Itoh M, Kawaji H, Kondo N, Kawai J, Lennartsson A, Daub CO, Heutink P, Hume DA, Jensen TH, Suzuki H, Hayashizaki Y, Müller F, Consortium F, Forrest ARR, Carninci P, Rehli M, Sandelin A (2014) An atlas of active enhancers across human cell types and tissues. Nature 507 (7493):455–461. https://doi.org/10.1038/nature12787

72. Babenko VN, Chadaeva IV, Orlov YL (2017) Genomic landscape of CpG rich elements in human. BMC Evol Biol 17(Suppl 1):19. https://doi.org/10.1186/s12862-016-0864-0

73. Medvedeva YA, Lennartsson A, Ehsani R, Kulakovskiy IV, Vorontsov IE, Panahandeh P, Khimulya G, Kasukawa T, Consortium F, Drabløs F (2015) EpiFactors: a comprehensive database of human epigenetic factors and complexes. Database 2015:bav067. https://doi.org/10.1093/database/bav067

74. Angermueller C, Clark SJ, Lee HJ, Macaulay IC, Teng MJ, Hu TX, Krueger F, Smallwood SA, Ponting CP, Voet T, Kelsey G, Stegle O, Reik W (2016) Parallel single-cell sequencing links transcriptional and epigenetic heterogeneity. Nat Methods 13(3):229–232. https://doi.org/10.1038/nmeth.3728

75. Jeddeloh JA, Greally JM, Rando OJ (2008) Reduced-representation methylation mapping. Genome Biol 9(8):231. https://doi.org/10.1186/gb-2008-9-8-231

Part III

Chromatin and Related Genomics Methods Used for Understanding the Functional Role of CpG Islands and Other Regulatory Regions

Chapter 11

Assay for Transposase Accessible Chromatin (ATAC-Seq) to Chart the Open Chromatin Landscape of Human Pancreatic Islets

Helena Raurell-Vila, Mireia Ramos-Rodríguez, and Lorenzo Pasquali

Abstract

The regulatory mechanisms that ensure an accurate control of gene transcription are central to cellular function, development and disease. Such mechanisms rely largely on noncoding regulatory sequences that allow the establishment and maintenance of cell identity and tissue-specific cellular functions.

The study of chromatin structure and nucleosome positioning allowed revealing transcription factor accessible genomic sites with regulatory potential, facilitating the comprehension of tissue-specific cis-regulatory networks. Recently a new technique coupled with high-throughput sequencing named Assay for Transposase Accessible Chromatin (ATAC-seq) emerged as an efficient method to chart open chromatin genome wide. The application of such technique to different cell types allowed unmasking tissue-specific regulatory elements and characterizing cis-regulatory networks. Herein we describe the implementation of the ATAC-seq method to human pancreatic islets, a tissue playing a central role in the control of glucose metabolism.

Key words Open chromatin, Pancreatic islets, Gene transcription, Epigenetics

1 Introduction

Transcription regulation is central to cellular function, development, and disease. Thus, the regulatory landscape controlling gene expression is highly dynamic, cell type-specific, and varies by individual genomes. Much of the transcriptional regulation of a cell is orchestrated by transcription factors that, by binding the DNA at proximal and distal regulatory sites, fine tune the amount of RNA produced.

In eukaryotic cells, DNA is packed within a nucleus through a hierarchical folding of 147 bp of DNA around a histone octamer to form nucleosomes, and their further compaction into chromatin.

Helena Raurell-Vila and Mireia Ramos-Rodríguez contributed equally to this work.

Tanya Vavouri and Miguel A. Peinado (eds.), *CpG Islands: Methods and Protocols*, Methods in Molecular Biology, vol. 1766, https://doi.org/10.1007/978-1-4939-7768-0_11,

Major insights into the epigenetic information encoded within the nucleoprotein structure of chromatin have come from high-throughput, genome-wide methods for assaying the accessibility of DNA to the machinery of gene expression also referred as chromatin "openness" [1, 2]. In fact, regulatory DNA often coincides with regions of remodeled chromatin, resulting in genomic sites open or accessible to transcription factor binding.

The use of techniques such as FAIRE and DNase I hypersensitive coupled with high-throughput sequencing, enabled the identification genome-wide of active transcription start sites, enhancers, and insulators in a wide variety of cell lines and tissue samples including the pancreatic islets [2–5]. Despite these successes, such methods employ multiple biochemical steps and substantial amounts of starting sample material. Recently, Buenrostro et al. reported on a robust and sensitive technology for profiling open chromatin by direct transposition of sequencing adapters into native chromatin [6] termed Assay of transposase Accessible Chromatin or "ATAC-seq". This novel assay takes advantage of a hyperactive Tn5 transposase that can simultaneously fragment and tag with sequencing adaptors genomic regions of accessible chromatin allowing an open chromatin library construction in a single enzymatic step.

Importantly ATAC-seq assay and library construction is achieved with considerably less starting material than conventional methods, thus being especially well-suited for the study of rare or difficult to obtain cell types, such as human pancreatic islet.

In this chapter we describe the implementation of the ATAC-seq protocol to chart open chromatin in isolated human pancreatic islets. We also provide a short technical description of the computational procedure to analyze ATAC-seq data. The experiments and analysis here described may be generally applicable to cell lines and primary tissue cellular aggregates.

2 Materials

2.1 Tissue Culture

1. Dithizone (diphenylthiocarbazone) stock solution: add 10 mg dithizone (Sigma) to 2 ml dimethylsulfoxide (DMSO) (Sigma). Store at −20 °C.
2. Islet medium: Ham's F10 medium supplemented with 10% FBS, 2 mM GlutaMAX, 50 U/ml penicillin and 50 μg/ml streptomycin (GIBCO), 6.1 mM glucose, 50 μM 3-isobutyl-1-methylxanthine, 1% BSA (Sigma).
3. Hanks' balanced salt solution (Sigma).
4. Phosphate buffered saline (PBS).

2.2 ATAC-Seq

1. Lysis buffer (10 mM Tris–HCl pH 7.4, 10 mM NaCl, 3 mM $MgCl_2$, 0.1% Igepal CA-630). Store up to 1 week at +4 °C.
2. Nuclear staining: Methyl Green-Pyronin Staining (#HT70116, Sigma).
3. Insulin needle (29G×1/2″).
4. Tn5 transposase and TD buffer (Nextera DNA Library Prep Kit, #15028212, Illumina).
5. Clean up buffer (900 mM NaCl, 300 mM EDTA).
6. Sodium Dodecyl Sulfate (SDS) 20% Solution (Omnipur).
7. Proteinase K (Thermo Scientific).
8. SPRI beads cleanup (Agencourt AMPure XP—5 ml, #A63880, Beckman Coulter).
9. 25 μM PCR Primer 1 (sequences provided in Buenrostro et al. [6]).
10. 25 μM Barcoded PCR Primer 2 (sequences provided in Buenrostro et al. [6]).
11. NEBNext High-Fidelity 2× PCR Master Mix (#M0541, New England Biolabs).
12. SYBR Green (Roche).
13. MinElute PCR Purification Kit (Qiagen).
14. Qubit® dsDNA HS Assay (#Q32851, Invitrogen).

3 Methods

The ATAC-seq method here described is based on that developed by Buenrostro et al. [6] and further modified by other authors [7, 8]. These modifications are aimed to improve the technique efficiency in purified primary human islets and may be used to enhance this assay in other cell lines and primary tissue cellular aggregates.

A successful ATAC-seq experiment relies on the preservation of the native chromatin architecture and the original nucleosome distribution patterns [7]. Thus, ATAC-seq is most efficient when applied to freshly isolated nuclei, in this case freshly human pancreatic endocrine islets, while fixed or frozen cells may reduce the sensitivity of the methodology [8]. Human pancreatic islets are obtained from organ donors after a laborious isolation procedure involving collagenase digestion and gradient purification to separate endocrine pancreatic islets from the exocrine tissue [9, 10]. Human pancreatic islets are then cultivated for 48 h to recover from the stress and minimize the environment induced variation. Freshly isolated pancreatic islets are tissue fragments of typically ~500–2000 cells that can be cultured in suspension.

This protocol is optimized for 50,000 cells corresponding to approximately 50 human pancreatic islets. The number of starting cells to be processed is crucial, as the transposase-to-cell ratio determines the distribution of DNA fragments generated [6]. A correct proportion between the quantity of Tn5 transposase and the number of cells is clue to a successful ATAC-seq experiment.

3.1 Tissue Culture

Prior to culture, the ratio of islets to exocrine tissue is ascertained by dithizone staining. Dithizone binds to zinc ions present in the secretory granules of the β-cells allowing to differentiate them from the acinar contaminants, thus to estimate islet size and purity.

Collect a representative volume of the sample, containing a minimum of 50 islets, to a 1.5 ml eppendorf tube. Let the islet sediment, remove the supernatant and resuspend with 100 μl of culture medium. Add 900 μl of dithizone solution and incubate for 15 min at 37 °C. Dithizone solution is obtained by mixing 200 μl dithizone stock solution (*see* Subheading 2) to 800 μl Hanks' balanced salt solution and centrifuging at maximum speed at room temperature for 5 min. After rinsing the islets with phosphate buffered saline twice, stained islets can be counted under a microscope. The sample purity is extrapolated by calculating the ratio of stained islets to the total tissue aggregates.

In a primary tissue culture hood, transfer the islets to 50 ml falcon tubes and to spin the sample for 1 min at 100 × *g* at room temperature. Discard the supernatant and resuspend, in 75 ml flasks, ~5 ml islet medium (*see* Subheading 2) every 1000 islets and, place in an incubator at 37 °C for 48 h.

3.2 Islet Preparation

Preheat a thermo-block to +37 °C, prepare 5 ml of cold PBS and a table-top centrifuge at +4 °C.

Carefully hand pick 50 healthy and acinar-free islets corresponding approximately to 50,000 cells and transfer them to a 1.5 ml eppendorf tube and keep on ice (*see* Subheading 5, **Note 1**).

All steps are to be done swiftly, to prevent cell stress and disturbance of chromatin.

Centrifuge the islets aliquot for 1 min at 100 × *g*, +4 °C, remove the medium and rinse with 500 μl of ice-cold PBS (1 min, 100 × *g*, +4 °C), remove all supernatant.

3.3 Nuclei Preparation

Add to the islets pellet 300 μl cold lysis buffer (*see* Subheading 2), resuspend, by gentle pipetting, and incubate 25 min on ice. In order to optimize the nuclei isolation, while incubating on ice, resuspend after 5 and 15 min the lysed islets using a syringe with a 29G needle.

During the lysis reaction, check the number and integrity of the nuclei obtained by instant nucleus staining (*see* Subheading 2) 5% of the sample aliquot.

Spin down the nuclei for 15 min at 500 × *g*, +4 °C using a swing rotor with low acceleration and brake settings. Carefully discard the supernatant (the pellet should be translucent).

Gently wash the pellet in 100 μl of lysis buffer and centrifuge 15 min at 500 × *g*, +4 °C using a swing rotor with low acceleration and brake settings and carefully to remove the supernatant.

While washing, prepare the transposase reaction mix described in the next step and keep at room temperature.

3.4 Transposition DNA Purification

Resuspend nuclei pellet in a 25 μl transposase reaction mix containing 2 μl of Tn5 transposase, 12.5 μl of TD buffer (*see* Subheading 2) and 10.5 μl DEPC treated water, per reaction and incubate at 37 °C for 1 h. Gentle mixing may increase fragment yield.

Add 5 μl of clean up buffer (*see* Subheading 2), 2 μl of 5% SDS and 2 μl of Proteinase K and incubate for 30 min at 40 °C.

3.5 DNA Purification

Right after the transposition reaction, place the sample on ice and proceed immediately with the isolation of the tagmented DNA using 2× SPRI beads cleanup (*see* Subheading 2), following kit instructions. Elute in 20 μl DEPC treated water. The isolated DNA samples can be stored at −20 °C before library amplification or used immediately for the next steps.

3.6 Library Amplification and Purification

Two sequential 9-cycle PCR are performed in order to enrich for small tagmented DNA fragments.

Prepare the PCR mix consisting of 2 μl of PCR Primer 1 (25 μM working stock), 2 μl of Barcoded PCR Primer 2 (25 μM working stock), 25 μl of NeBNext High-Fidelity 2× PCR Master Mix, 1μl of DEPC treated water and 20 μl of the eluted sample (add DEPC water to compensate in case the volume of eluted DNA is less than 20 μl).

Amplify the library in a thermocycler using the following program (leave preheated to 72 °C): 72 °C for 5 min; 98 °C for 30 s; 9 cycles of 98 °C for 10 s, 63 °C for 30 s; and 72 °C for 1 min; and at 4 °C hold. After the first PCR round, select for fragments smaller than 600 bp using SPRI cleanup beads (*see* Subheading 2).

Perform a second PCR applying the same conditions in order to obtain the final library.

Finally purify the DNA library using the MinElute PCR Purification Kit, following kit instructions and eluting 2 × 10μl with the elution buffer. The purified libraries can then be stored at −20 °C.

3.7 Library Quality Control and Quantification

Accurate assessment of library quality and concentration is critical to successful sequencing and data analysis. A sample aliquot of 1.2 μl is used to check, by TapeStation or Bioanalyzer, the library quality and the distribution of the fragment size. The library obtained from a successfully tagmented sample will display a nucleosomal pattern (Fig. 1) (*see* Subheading 5, **Note 1**).

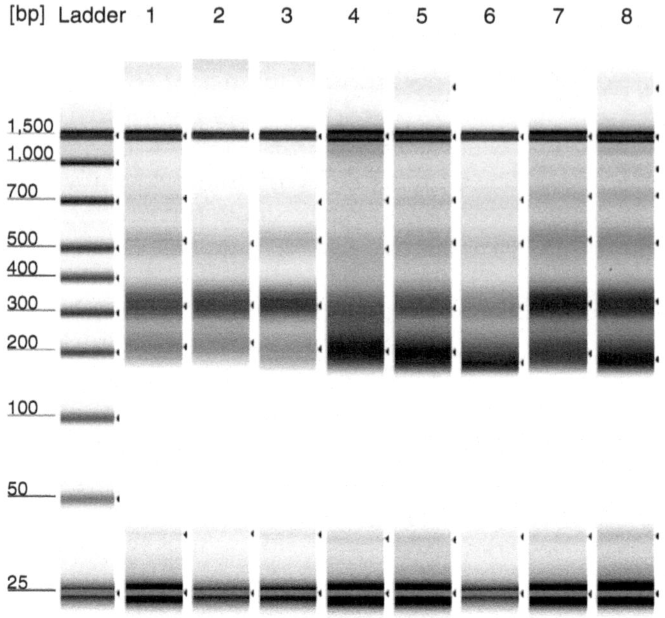

Fig. 1 Agilent TapeStation profiles showing the laddering pattern of ATAC-seq libraries. The band sizes correspond to the expected nucleosomal pattern obtained by chromatin tagmentation of human islets samples (lanes 1–8)

Fragments lower than 100 bp may represent mitochondrial fragments not properly tagged for sequencing, thus should not be included in the calculation of the molar concentration of the library. The library concentration can be measured by QuBit dsDNA HS Assay Kit following the instructions provided with the kit. As in chromatin immunoprecipitation experiments, semiquantitative PCR assays that target specific genomic sites may be carried on to estimate the efficiency of the ATAC-seq experiment in enriching for open chromatin regions. To this end we designed oligonucleotides targeting presumably open chromatin sites as well as negative control sites not expected to harbor open chromatin.

3.8 Sequencing

A minimum DNA concentration of 3.25 ng/μl in a minimum volume of 15 μl is preferred for next generation sequencing. Typically 100 million 50 bp long single or paired-end reads per library are sufficient to chart open chromatin regions genome-wide with a good enrichment over background resolution. Data yield is impacted by a fraction of mitochondrial reads that we report, for this protocol in human pancreatic islets, varying between 20% and 40%.

4 ATAC-Seq Data Analysis

The sequencing data generated is subsequently analyzed using a variety of analytical tools that allow resolving the ATAC-seq genome wide open chromatin profile. Data analysis may be carried out by applying computational procedures analogous to those applied for chromatin immunoprecipation sequencing experiments (ChIP-seq). While data analysis may require computational expertise and progressive increase of computing power and storage capacity, several ATAC-seq and ChIP-seq pipeline have been developed for researchers with limited computational experience (https://usegalaxy.org or http://cistrome.org/ap/root).

We here briefly describe the basic steps of a computational analysis addressed to resolve the ATAC-seq chromatin accessibility genome wide.

4.1 Data Quality Control and Alignment

Quality examination of the raw sequencing data is essential prior to data analysis in order to rule out sources of error and sequencing biases.

FASTQC [11] is a quality control tool which uses the raw FASTQ files to perform several simple quality control analyses. This program returns an HTML file comprehensive of a detailed analysis of the raw sequenced reads including the Phread score (per base sequencing error probability) GC content and overrepresented sequences or kmer content. Notice that for ATAC-seq experiments a "per base sequence content" warning may be due to a preference for the transposase binding [12, 13] that does not affect the quality of the experiment.

After data quality assessment sequenced reads are aligned to a reference genome. There are several mapping programs available including Maq [14], RMAP [15], Cloudburst [16], GEM [17], SHRiMP [18], BWA [19], and Bowtie [20].

Bowtie2 [21] is optimized to work with large reference genomes, such as mammalian, and reads from 50 bp up to 1 kb. This aligner software characterizes the degree of confidence of each alignment as nonnegative integer $Q = -10\ log10\ p$, where p is an estimate of the probability that the alignment does not correspond to the read's true point of origin. The "*Q*" value can be used, postalignment, to filter out reads with a poor mapping quality. While some reads might be mapping to several genomic locations at the same time, with default settings, bowtie2 searches for distinct valid alignments for each read and will report only the best alignment.

In order to align ATAC-seq single end raw reads bowtie2 may be run via command line. Default options can be used and the following useful flags can be specified: "*-t*" for printing to the "stderr", the time taken to load the indexes and align the reads

("*2>*" can be used to keep the statistics of the alignment for posterior usage); "*-S*" to return a SAM format file output; "*-p*" to use a specified number of cores.

4.2 Postalignment Processing

Prior to downstream data analysis several steps are needed to convert file formats and reduce the impact of potential artifact reads.

The typical output of an alignment software is a SAM or a BAM format file, the latter being a binary compressed format for storing sequence data. Conversion between these two file formats and sorting by genomic coordinates, necessary for downstream analyses, can be easily performed by using the SAMtools [22] programs suite.

Reads aligning to mitochondrial genome are discarded as unrelated to the scope of the experiment. Reads must also be filtered to remove overrepresented areas of the genome due to technical bias. A collection of signal artifact blacklist regions in the human genome is provided by the ENCODE project (ENCODE blacklist: https://personal.broadinstitute.org/anshul/projects/encode/rawdata/blacklists/) and can be used for this purpose. Reads filtering can be performed with SAMtools or Picard tools (http://broadinstitute.github.io/picard). Additionally, tools such as Picard allow easily tagging and eventually removing duplicates reads that may represent PCR amplification artifacts (*see* Subheading 5, **Note 2** and **3**).

A number of publicly available genome browser tools, can be used to visualize ATAC-seq profiles relative to available annotation tracks. Such tools include Artemis [23], EagleView [24], MapView [25], Apollo [26], and the Islet Regulome Browser [27]. The University of California Santa Cruz (UCSC) [28] and the Integrative Genomics Viewer (IGV) [29] are the most widely used genome browser tools. Density profile tracks can be loaded to the UCSC browser (Fig. 2) in BigWig format, indexed binary files which associate a genomic location with number of aligning reads. Bedtools utilities [30] and UCSC provide a series of tools to convert different file formats to bedgraph and BigWig.

Finally, identification of those regions that are significantly enriched of mapped reads compared to the background allows charting genome wide the predicted open chromatin sites. Several peak calling algorithms are available such as ZINBA [31], F-seq [32], HOMER [33], or MACS [34]. Differently from ChIP-seq experiments, ATAC-seq cannot rely on input DNA or mock-IP as control background to identify the reads enriched regions. The peak caller MACS2 [34] allows to fine tune the algorithm to the ATAC-seq experiment, typical parameter added to the default arguments are (*—no model --shift – 100 --exsize-200)* that allow to center the peaks to the Tn5 cutting sites.

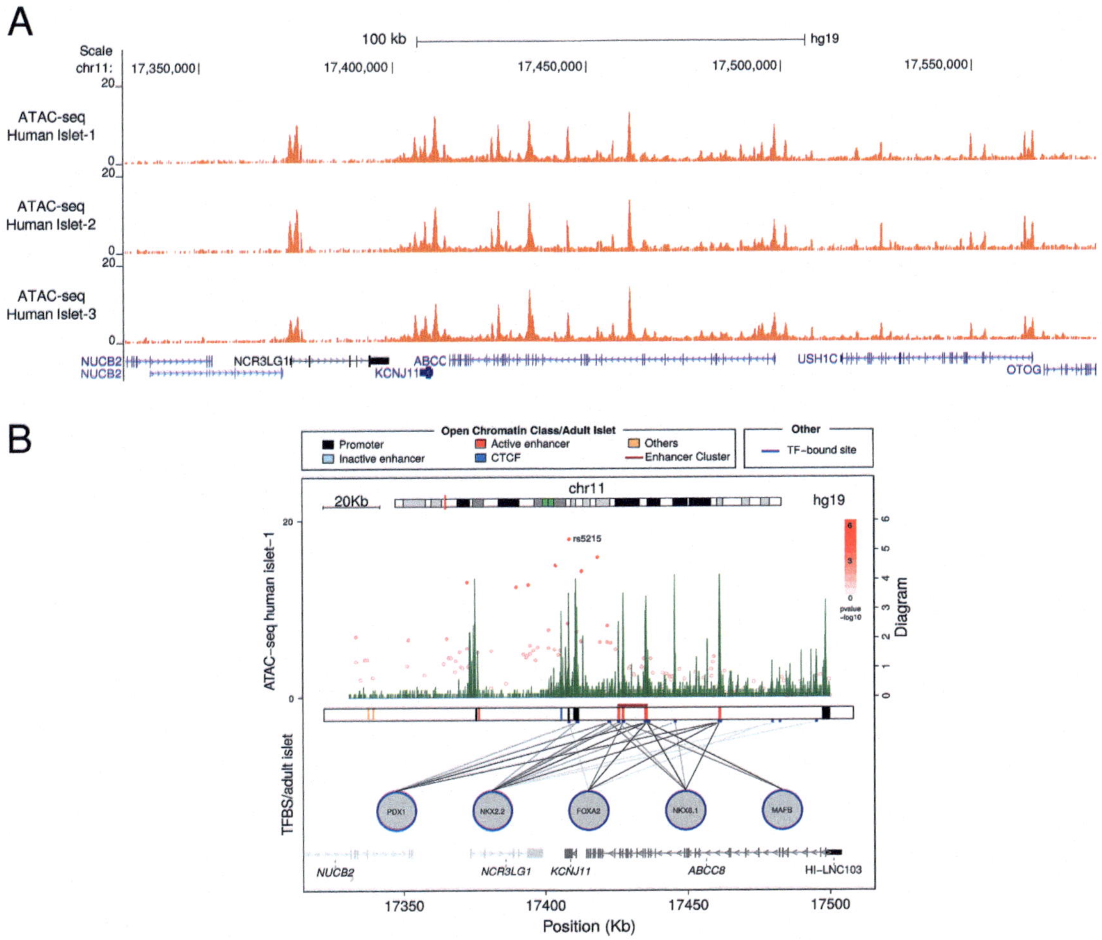

Fig. 2 Genome browser views of ATAC-seq signal in the proximity of the ATP-sensitive potassium ion channels genes, *ABCC8* and *KCNJ11*. (**a**) UCSC genome browser view of 3 human islets ATAC-seq libraries. ATAC-seq tracks display high concordance and a strong resolution over background of the signal. Notice that each ATAC-seq library was prepared by processing only ~50 human pancreatic islets. (**b**) Islet Regulome Browser [27] integrative view (www.isletregulome.com) showing that ATAC-seq open chromatin enrichment are often found to coincide with regulatory elements, promoters and transcription factor binding sites previously identified in human pancreatic islets. The plot also depicts the signal of Type 2 Diabetes Genome Wide Association Studies (GWAS), each red dot represents a genomic variant, being the color intensity of the dot proportional to -Log *p*-value of association, as indicated on the side of the plot

5 Notes

1. Variations in the number of cells processed may result in "under or over transposition" of the DNA. Such variations will be reflected in the library fragment size distribution measured by TapeStation or Bioalnalyzer. An overrepresentation of large fragments may mirror an "under tagmentation" of the DNA due to an excess of cells relative to the transposase. For this reason we recommend hand picking approximately

50 fragments of the islet prep, taking in account the purity of the sample being processed.

2. Identical reads can arise from independent transposition events or PCR amplification of a single fragment. In order to reduce experimental noise we thus recommend removing identical reads. Alternatively a solution proposed for ChIP-seq experiments is that of retaining a fixed number of tags per genomic location according to the sequencing depth [35].
3. In case of using ATAC-seq data to infer the transcription factor binding by footprint analysis [36], we recommend to adjust the read start sites to the center of the transposon's binding event. This can be done by offsetting the reads 4 bp on the + strand and 5 bp on the – strand since Tn5 transposase was shown to bind as a dimer inserting two adaptors separated by 9 bp [12].

Acknowledgment

This work was supported by a grant from the Spanish Ministry of Economy and Competiveness (BFU2014-58150-R), the Spanish Diabetes Society and Fundació La Marató de TV3. LP is a recipient of a Ramon y Cajal contract from the Spanish Ministry of Economy and Competitiveness (RYC 2013-12864). Helena Raurell-Vila and Mireia Ramos-Rodríguez contributed equally to this work.

References

1. Thurman RE, Rynes E, Humbert R, Vierstra J, Maurano MT, Haugen E, Sheffield NC, Stergachis AB, Wang H, Vernot B, Garg K, John S, Sandstrom R, Bates D, Boatman L, Canfield TK, Diegel M, Dunn D, Ebersol AK, Frum T, Giste E, Johnson AK, Johnson EM, Kutyavin T, Lajoie B, Lee BK, Lee K, London D, Lotakis D, Neph S, Neri F, Nguyen ED, Qu H, Reynolds AP, Roach V, Safi A, Sanchez ME, Sanyal A, Shafer A, Simon JM, Song L, Vong S, Weaver M, Yan Y, Zhang Z, Lenhard B, Tewari M, Dorschner MO, Hansen RS, Navas PA, Stamatoyannopoulos G, Iyer VR, Lieb JD, Sunyaev SR, Akey JM, Sabo PJ, Kaul R, Furey TS, Dekker J, Crawford GE, Stamatoyannopoulos JA (2012) The accessible chromatin landscape of the human genome. Nature 489(7414):75–82. https://doi.org/10.1038/nature11232
2. Gaulton KJ, Nammo T, Pasquali L, Simon JM, Giresi PG, Fogarty MP, Panhuis TM, Mieczkowski P, Secchi A, Bosco D, Berney T, Montanya E, Mohlke KL, Lieb JD, Ferrer J (2010) A map of open chromatin in human pancreatic islets. Nat Genet 42(3):255–259. https://doi.org/10.1038/ng.530
3. Pasquali L, Gaulton KJ, Rodriguez-Segui S, Mularoni L, Miguel-Escalada I, Akerman I, Tena JJ, Moran I, Gómez-Marín C, van de Bunt M, Ponsa-Cobas J, Castro N, Nammo T, Cebola I, Garcia-Hurtado J, Maestro MA, Pattou F, Piemonti L, Berney T, Gloyn AL, Ravassard P, Muller F, McCarthy MI, Ferrer J (2014) Pancreatic islet enhancer clusters enriched in type 2 diabetes risk–associated variants. Nat Genet 46(2):136–143. https://doi.org/10.1038/ng.2870
4. Stitzel ML, Sethupathy P, Pearson DS, Chines PS, Song L, Erdos MR, Welch R, Parker SC, Boyle AP, Scott LJ, Margulies EH, Boehnke M, Furey TS, Crawford GE, Collins FS (2010) Global epigenomic analysis of primary human pancreatic islets provides insights into type 2 diabetes susceptibility loci. Cell Metab 12(5):443–455. https://doi.org/10.1016/j.cmet.2010.09.012
5. Nammo T, Rodriguez-Segui SA, Ferrer J (2011) Mapping open chromatin with

formaldehyde-assisted isolation of regulatory elements. Methods Mol Biol 791:287–296. https://doi.org/10.1007/978-1-61779-316-5_21
6. Buenrostro JD, Giresi PG, Zaba LC, Chang HY, Greenleaf WJ (2013) Transposition of native chromatin for fast and sensitive epigenomic profiling of open chromatin, DNA-binding proteins and nucleosome position. Nat Methods 10(12):1213–1218. https://doi.org/10.1038/nmeth.2688
7. Lavin Y, Winter D, Blecher-Gonen R, David E, Keren-Shaul H, Merad M, Jung S, Amit I (2014) Tissue-resident macrophage enhancer landscapes are shaped by the local microenvironment. Cell 159(6):1312–1326. https://doi.org/10.1016/j.cell.2014.11.018
8. Milani P, Escalante-Chong R, Shelley BC, Patel-Murray NL, Xin X, Adam M, Mandefro B, Sareen D, Svendsen CN, Fraenkel E (2016) Cell freezing protocol suitable for ATAC-Seq on motor neurons derived from human induced pluripotent stem cells. Sci Rep 6:25474. https://doi.org/10.1038/srep25474
9. Shapiro AM, Pokrywczynska M, Ricordi C (2016) Clinical pancreatic islet transplantation. Nat Rev Endocrinol. https://doi.org/10.1038/nrendo.2016.178
10. Piemonti L, Pileggi A (2013) A 25 years of the Ricordi automated method for islet isolation. CellR 4(1):e128
11. Andrews S (2010) FastQC: a quality control tool for high throughput sequence data. http://www.bioinformatics.babraham.ac.uk/projects/fastqc
12. Adey A, Morrison HG, Asan XX, Kitzman JO, Turner EH, Stackhouse B, MacKenzie AP, Caruccio NC, Zhang X, Shendure J (2010) Rapid, low-input, low-bias construction of shotgun fragment libraries by high-density in vitro transposition. Genome Biol 11(12):R119. https://doi.org/10.1186/gb-2010-11-12-r119
13. Goryshin IY, Miller JA, Kil YV, Lanzov VA, Reznikoff WS (1998) Tn5/IS50 target recognition. Proc Natl Acad Sci U S A 95 (18):10716–10721
14. Li H, Ruan J, Durbin R (2008) Mapping short DNA sequencing reads and calling variants using mapping quality scores. Genome Res 18 (11):1851–1858. https://doi.org/10.1101/gr.078212.108
15. Smith AD, Chung WY, Hodges E, Kendall J, Hannon G, Hicks J, Xuan Z, Zhang MQ (2009) Updates to the RMAP short-read mapping software. Bioinformatics 25 (21):2841–2842. https://doi.org/10.1093/bioinformatics/btp533
16. Schatz MC (2009) CloudBurst: highly sensitive read mapping with MapReduce. Bioinformatics 25(11):1363–1369. https://doi.org/10.1093/bioinformatics/btp236
17. Marco-Sola S, Sammeth M, Guigo R, Ribeca P (2012) The GEM mapper: fast, accurate and versatile alignment by filtration. Nat Methods 9(12):1185–1188. https://doi.org/10.1038/nmeth.2221
18. Rumble SM, Lacroute P, Dalca AV, Fiume M, Sidow A, Brudno M (2009) SHRiMP: accurate mapping of short color-space reads. PLoS Comput Biol 5(5):e1000386. https://doi.org/10.1371/journal.pcbi.1000386
19. Li H, Durbin R (2009) Fast and accurate short read alignment with burrows-wheeler transform. Bioinformatics 25(14):1754–1760. https://doi.org/10.1093/bioinformatics/btp324
20. Langmead B, Trapnell C, Pop M, Salzberg SL (2009) Ultrafast and memory-efficient alignment of short DNA sequences to the human genome. Genome Biol 10(3):R25. https://doi.org/10.1186/gb-2009-10-3-r25
21. Langmead B, Salzberg SL (2012) Fast gapped-read alignment with bowtie 2. Nat Methods 9 (4):357–359. https://doi.org/10.1038/nmeth.1923
22. Li H, Handsaker B, Wysoker A, Fennell T, Ruan J, Homer N, Marth G, Abecasis G, Durbin R, Genome Project Data Processing S (2009) The sequence alignment/map format and SAMtools. Bioinformatics 25 (16):2078–2079. https://doi.org/10.1093/bioinformatics/btp352
23. Carver T, Harris SR, Berriman M, Parkhill J, McQuillan JA (2012) Artemis: an integrated platform for visualization and analysis of high-throughput sequence-based experimental data. Bioinformatics 28(4):464–469. https://doi.org/10.1093/bioinformatics/btr703
24. Huang W, Marth G (2008) EagleView: a genome assembly viewer for next-generation sequencing technologies. Genome Res 18 (9):1538–1543. https://doi.org/10.1101/gr.076067.108
25. Wolfsberg TG (2011) Using the NCBI Map Viewer to browse genomic sequence data. Curr Protoc Hum Genet Chapter 18:Unit18 15. doi:https://doi.org/10.1002/0471142905.hg1805s69
26. Lee E, Helt GA, Reese JT, Munoz-Torres MC, Childers CP, Buels RM, Stein L, Holmes IH, Elsik CG, Lewis SE (2013) Web Apollo: a web-based genomic annotation editing

platform. Genome Biol 14(8):R93. https://doi.org/10.1186/gb-2013-14-8-r93

27. Mularoni L, Ramos-Rodríguez M, Pasquali L (2017) The pancreatic islet regulome browser. Front Genet 8(13). https://doi.org/10.3389/fgene.2017.00013
28. Speir ML, Zweig AS, Rosenbloom KR, Raney BJ, Paten B, Nejad P, Lee BT, Learned K, Karolchik D, Hinrichs AS, Heitner S, Harte RA, Haeussler M, Guruvadoo L, Fujita PA, Eisenhart C, Diekhans M, Clawson H, Casper J, Barber GP, Haussler D, Kuhn RM, Kent WJ (2016) The UCSC genome browser database: 2016 update. Nucleic Acids Res 44 (D1):D717–D725. https://doi.org/10.1093/nar/gkv1275
29. Thorvaldsdottir H, Robinson JT, Mesirov JP (2013) Integrative genomics viewer (IGV): high-performance genomics data visualization and exploration. Brief Bioinform 14 (2):178–192. https://doi.org/10.1093/bib/bbs017
30. Quinlan AR, Hall IM (2010) BEDTools: a flexible suite of utilities for comparing genomic features. Bioinformatics 26(6):841–842. https://doi.org/10.1093/bioinformatics/btq033
31. Rashid NU, Giresi PG, Ibrahim JG, Sun W, Lieb JD (2011) ZINBA integrates local covariates with DNA-seq data to identify broad and narrow regions of enrichment, even within amplified genomic regions. Genome Biol 12 (7):R67. https://doi.org/10.1186/gb-2011-12-7-r67
32. Boyle AP, Guinney J, Crawford GE, Furey TS (2008) F-Seq: a feature density estimator for high-throughput sequence tags. Bioinformatics 24(21):2537–2538. https://doi.org/10.1093/bioinformatics/btn480
33. Heinz S, Benner C, Spann N, Bertolino E, Lin YC, Laslo P, Cheng JX, Murre C, Singh H, Glass CK (2010) Simple combinations of lineage-determining transcription factors prime cis-regulatory elements required for macrophage and B cell identities. Mol Cell 38 (4):576–589. https://doi.org/10.1016/j.molcel.2010.05.004
34. Zhang Y, Liu T, Meyer CA, Eeckhoute J, Johnson DS, Bernstein BE, Nusbaum C, Myers RM, Brown M, Li W, Liu XS (2008) Model-based analysis of ChIP-Seq (MACS). Genome Biol 9(9):R137. https://doi.org/10.1186/gb-2008-9-9-r137
35. Chen Y, Negre N, Li Q, Mieczkowska JO, Slattery M, Liu T, Zhang Y, Kim TK, He HH, Zieba J, Ruan Y, Bickel PJ, Myers RM, Wold BJ, White KP, Lieb JD, Liu XS (2012) Systematic evaluation of factors influencing ChIP-seq fidelity. Nat Methods 9 (6):609–614. https://doi.org/10.1038/nmeth.1985
36. Neph S, Vierstra J, Stergachis AB, Reynolds AP, Haugen E, Vernot B, Thurman RE, John S, Sandstrom R, Johnson AK, Maurano MT, Humbert R, Rynes E, Wang H, Vong S, Lee K, Bates D, Diegel M, Roach V, Dunn D, Neri J, Schafer A, Hansen RS, Kutyavin T, Giste E, Weaver M, Canfield T, Sabo P, Zhang M, Balasundaram G, Byron R, MacCoss MJ, Akey JM, Bender MA, Groudine M, Kaul R, Stamatoyannopoulos JA (2012) An expansive human regulatory lexicon encoded in transcription factor footprints. Nature 489 (7414):83–90. https://doi.org/10.1038/nature11212

Check for updates

Chapter 12

Defining Regulatory Elements in the Human Genome Using Nucleosome Occupancy and Methylome Sequencing (NOMe-Seq)

Suhn Kyong Rhie, Shannon Schreiner, and Peggy J. Farnham

Abstract

NOMe-seq (nucleosome occupancy and methylome sequencing) identifies nucleosome-depleted regions that correspond to promoters, enhancers, and insulators. The NOMe-seq method is based on the treatment of chromatin with the M.CviPI methyltransferase, which methylates GpC dinucleotides that are not protected by nucleosomes or other proteins that are tightly bound to the chromatin (GpC^m does not occur in the human genome and therefore there is no endogenous background of GpC^m). Following bisulfite treatment of the M.CviPI-methylated chromatin (which converts unmethylated Cs to Ts and thus allows the distinction of GpC from GpC^m) and subsequent genomic sequencing, nucleosome-depleted regions can be ascertained on a genome-wide scale. The bisulfite treatment also allows the distinction of CpG from C^mpG (most endogenous methylation occurs at CpG dinucleotides) and thus the endogenous methylation status of the genome can also be obtained in the same sequencing reaction. Importantly, open chromatin is expected to have high levels of GpC^m but low levels of C^mpG; thus, each of the two separate methylation analyses serve as independent (but opposite) measures which provide matching chromatin designations for each regulatory element.

NOMe-seq has advantages over ChIP-seq for identification of regulatory elements because it is not reliant upon knowing the exact modifications on the surrounding nucleosomes. Also, NOMe-seq has advantages over DHS (DNase hypersensitive site)-seq, FAIRE (Formaldehyde-Assisted Isolation of Regulatory Elements)-seq, and ATAC (Assay for Transposase-Accessible Chromatin)-seq because it also gives positioning information for several nucleosomes on either side of each open regulatory element. Here, we provide a detailed protocol for NOMe-seq that begins with the isolation of chromatin, followed by methylation of GpCs with M.CviPI and treatment with bisulfite, and ending with the creation of next generation sequencing libraries. We also include sequencing QC analysis metrics and bioinformatics steps that can be used to identify nucleosome-depleted regions throughout the genome.

Key words NOMe-seq, Nucleosome-depleted regions, Enhancers, Promoters, Insulators, Open chromatin, DNA methylation

---

Electronic supplementary material: The online version of this chapter (https://doi.org/10.1007/978-1-4939-7768-0_12) contains supplementary material, which is available to authorized users.

Tanya Vavouri and Miguel A. Peinado (eds.), *CpG Islands: Methods and Protocols*, Methods in Molecular Biology, vol. 1766, https://doi.org/10.1007/978-1-4939-7768-0_12, © Springer Science+Business Media, LLC, part of Springer Nature 2018

1 Introduction

Regulatory elements such as promoters, enhancers, and insulators are regions of open chromatin that are created and maintained by the binding of site-specific transcription factors (TFs) and their associated protein complexes. These genomic landing platforms are delineated by nucleosome-depleted regions (NDRs), flanked on either side by a series of phased nucleosomes. At promoters and enhancers, the flanking nucleosomes can harbor one or more modifications, such as acetylation of lysine 27 on histone H3 (H3K27ac) at enhancers or methylation of lysine 4 on histone H3 (H3K4me3) at promoters [1–5], that provide additional information about the specific functional state of a particular NDR. These histone modifications are created by the recruitment of histone-modifying enzymes (e.g., acetylases and methylases) to the NDR via interaction with site-specific transcription factors bound to the DNA [6, 7]. Insulators, on the other hand, are characterized by the presence of site-specific DNA binding components of the cohesin complex, such as CTCF and RAD21, often in the absence of marks associated with active enhancers or promoters [8].

NOMe-seq (nucleosome occupancy and methylome sequencing) identifies NDRs that correspond to promoters, enhancers, and insulators (Fig. 1) [9]. The NOMe-seq method is based on the treatment of chromatin with the M.CviPI methyltransferase. This enzyme, which is isolated from Chlorella virus, methylates Cs in the context of GpC dinucleotides. GpC^{m} does not occur in the human genome (the vast majority of DNA methylation in the human genome is at CpG dinucleotides, not GpC dinucleotides) and therefore there is no endogenous background of GpC^{m}. The enzyme can only methylate GpC dinucleotides that are accessible in the context of chromatin, i.e., not protected by nucleosomes or other proteins that are tightly bound to the chromatin. Following bisulfite treatment of the M.CviPI-methylated chromatin (which converts unmethylated Cs to Ts and thus allows the distinction of GpC from GpC^{m}) and subsequent genomic sequencing, the status of GpC-containing regions can be ascertained on a genome-wide scale. Using this method, NDRs are defined as regions having increased GpC^{m} methylation over background (i.e., they were in open regions and thus were methylated by the M.CviPI enzyme) that are at least 140 bp in length. The bisulfite treatment also allows the distinction of CpG from $C^{m}pG$ and thus the endogenous methylation status of the genome can also be obtained in the same sequencing reaction. It is important to note that in contrast to the induced GpC^{m} which represents nucleosome-free, open chromatin that is available for TF binding, the endogenous $C^{m}pG$ represents nucleosome-bound chromatin that is not available for TF binding. We note that GCG trinucleotides cannot be used to

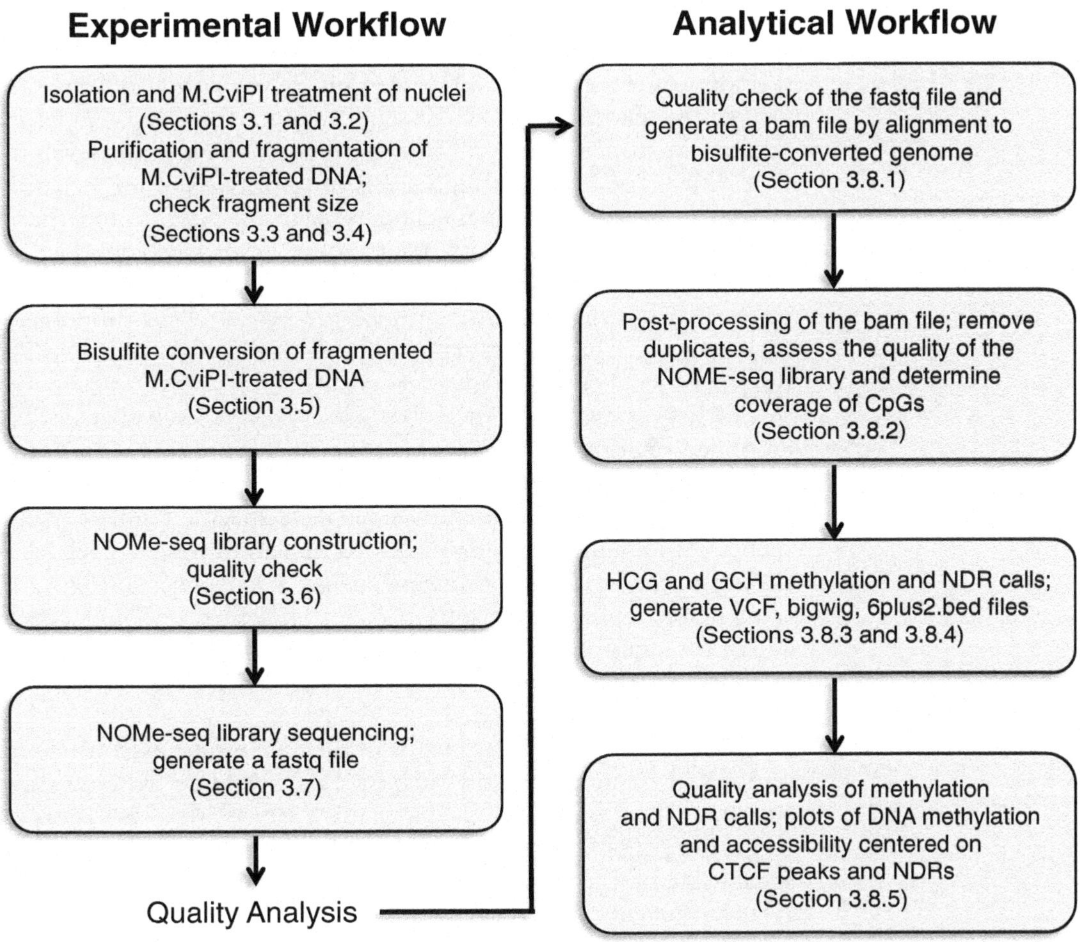

Fig. 1 Schematic overview of NOMe-seq. Left: Experimental workflow, representing Subheadings 3.1–3.7 in the Subheading 3. Right: Analytical workflow, representing Subheadings 3.8.1–3.8.5 in the Subheading 3

distinguish between enforced GpC methylation and endogenous CpG methylation (on the other strand); therefore, in the analysis of NOMe-seq datasets GCH (H = A, C, or T) trinucleotides are selected and analyzed for nucleosome positioning whereas HCG trinucleotides are selected and analyzed for endogenous DNA methylation. As reported earlier, GCG trinucleotides are not frequent in the genome and are almost always within 20 bp of a GCH [9], thus allowing an NDR containing a GCG to be identified by nearby GCH sequences. Importantly, open chromatin is expected to have high levels of GpC^{m} but low levels of $C^{m}pG$; thus, each of the two separate methylation analyses serve as independent (but opposite) measures which should provide matching chromatin designations (open vs. closed) [10].

Although ChIP-seq performed using antibodies to specifically modified histones can also be used to identify regulatory elements [11], NOMe-seq has advantages over ChIP-seq because it is not reliant upon knowing the exact modifications on the surrounding nucleosomes. NOMe-seq may also provide information not easily gained from ChIP-seq. As noted above, regulatory regions identified using NOMe-seq should have high levels of GpC^m or C^mpG, but not high levels of both types of methylation. However, previous analyses using NOMe-seq have found that a small number of regions of the genome have been identified as having both types of methylation in the same cell population [9]. It has been suggested that these regions represent allelic differences, with one allele having an active regulatory element (high GpC^m) but the other allele being in a closed state (high C^mpG). In support of this hypothesis, Kelly et al. [9] previously showed that doubly identified regions (i.e., NDRs identified as having high GpC^m and high C^mpG) are enriched for known imprinted promoters. Thus, NOMe-seq can help to identify new allele-specific regulatory elements without the need for a SNP to be within the element (as is the case for analysis of allele-specific ChIP-seq). Of course, sequencing depth is important in such analyses because high coverage of the examples of "nucleosome-depleted" and "DNA methylated" reads in the same region is needed to be certain that the regions are doubly marked.

NOMe-seq has similarities to other techniques used to detect regions of open chromatin such as DHS (DNase hypersensitive site)-seq [12] and FAIRE (Formaldehyde-Assisted Isolation of Regulatory Elements)-seq [13], both of which rely on the physical separation of nucleosome-free vs. nucleosome-bound DNA, or ATAC (Assay for Transposase-Accessible Chromatin)-seq [14, 15] which identifies regions of open chromatin using transposon integration. However, because treatment with M.CViPI is performed prior to DNA fragmentation, there is less bias toward open chromatin in NOMe-seq and there may be fewer false positive identified regions. Two other advantages of NOMe-seq are that, unlike the other methods, it also gives positioning information of several nucleosomes on either side of each open regulatory element and it provides information concerning the endogenous methylation state of every CpG dinucleotide in the genome.

It is also important to consider the size of the regulatory element identified by the different techniques. For example, the average width of the set of H3K27ac peaks is quite large and it is not reasonable to simply define the center of a H3K27ac-covered area as the functional (i.e., the TF binding platform) region. On the other hand, the NDRs called by NOME-seq are smaller in width, corresponding to inter-nucleosomal regions, and therefore more closely match the region containing TF binding sites (Fig. 2). The ability to refine the functional compartment within open chromatin domains to a small region can have considerable influence on the quality of downstream analyses, such as motif finding and

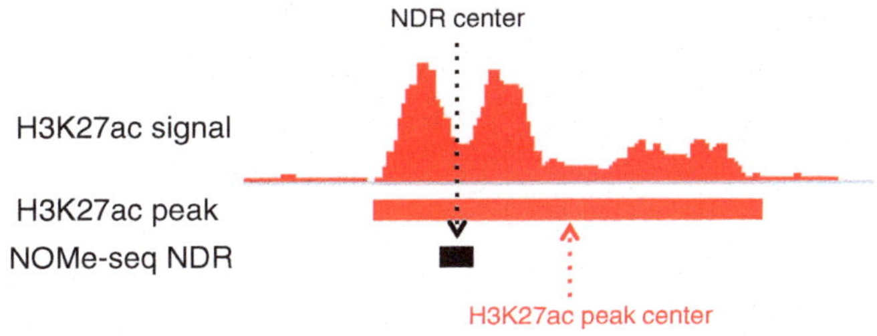

Fig. 2 Refinement of a regulatory element using NOMe-seq. Shown is a nucleosome-depleted region (NDR) flanked by nucleosomes harboring the histone modification H3K27ac; the centers of the NDR and the region covered by H3K27ac are also indicated

interpretation of noncoding variants identified by GWAS. It is also important to precisely delineate the functional compartment of an open regulatory region when using DNA methylation status to link activity of an element to gene expression. For example, DNA methylation levels may be high throughout a large H3K27ac peak, only showing a small hypomethylated region that corresponds to the NDR; averaging methylation levels over a large region may obscure the presence of a differentially active enhancer when comparing different tissue types or disease states.

To date, NOMe-seq has been performed in IMR90 lung cells and glioblastoma cells [9], normal (PREC) and cancer (PC3) prostate cells, normal (HMEC) and cancer (MCF7) breast cells [16, 17], and HCT116 and DKO colon cancer cells [18]. However, due to technology improvements, our current protocol has changed as compared to that used in those initial studies. Here, we provide a detailed protocol for NOMe-seq which differs from that used in previous studies in several important steps, such as the order in which the DNA is treated with bisulfite in the library protocol, which can have a considerable influence in the yield of DNA in the resultant library.

2 Materials

2.1 Isolation of Nuclei

1. 1× Dulbecco's phosphate-buffered saline (DPBS): sterile, no calcium, no magnesium.
2. Trypsin or dispase (if needed for your cell type).
3. Trypan Blue and hemocytometer.
4. Lysis Buffer: 10 mM Tris pH 7.4, 10 mM NaCl, 3 mM $MgCl_2$, 0.1 mM EDTA, 0.5% NP-40.
5. Wash Buffer: 10 mM Tris pH 7.4, 10 mM NaCl, 3 mM $MgCl_2$, 0.1 mM EDTA.

2.2 Treatment of Nuclei with M.CviPI

1. 10× GpC Buffer (New England Biolabs).
2. 32 mM S-adenosylhomocysteine (SAM) (New England Biolabs).
3. 50 U/μL M.CviPI (New England Biolabs).
4. 1 M sucrose.
5. Nuclease-free water.
6. Stop Buffer: 20 mM Tris pH 7.4, 600 mM NaCl, 1% SDS, 10 mM EDTA.

2.3 Isolation of M.CviPI-Treated DNA

1. 5 M NaCl.
2. Proteinase K (Promega).
3. 1:1 phenol–chloroform.
4. 100% ethanol.
5. TE buffer: 10 mM Tris pH 8, 10 mM EDTA pH 8.
6. NanoDrop spectrophotometer.

2.4 Fragmentation of M.CviPI-Treated DNA

1. Covaris sonicator (S220, formerly S2).
2. Covaris MicroTUBE AFA Pre-slit Snap-Cap 6 × 16 mm.
3. NanoDrop spectrophotometer.
4. DNA High Sensitivity Kit (Agilent) for use with Agilent 2100 Bioanalyzer.

2.5 Bisulfite Conversion of M.CviPI-Treated DNA

1. EZ DNA Methylation Kit #D5001 (Zymo Research).

2.6 NOMe-Seq Library Construction

1. Accel-NGS Methyl-Seq DNA Library Kit for Illumina Platforms (Swift Biosciences #30024).
2. Methyl-Seq Set A Indexing Kit (Swift Biosciences).
3. SPRIselect Magnetic Beads (Beckman Coulter).
4. Qubit dsDNA HS (High Sensitivity) Assay Kit (Thermo Fisher Scientific).
5. DNA High Sensitivity Kit (Agilent) for use with Agilent 2100 Bioanalyzer.

3 Methods

3.1 Isolation of Nuclei

(Note: stopping points throughout the experimental protocol are indicated by [Stopping Point]).

1. Treat adherent cells with trypsin or dispase or collect suspension cells and place into a prechilled 15 mL tube (*see* **Note 1**). Centrifuge at 250 × *g* at 4 °C for 5 min.
2. Place cells on ice or at 4 °C for the remaining steps in Subheading 3.1.
3. Remove the media and wash cells with 10 mL ice-cold sterile PBS.
4. Remove 10 μL of the cell suspension and combine with 10 μL of trypan blue in a 1.5 mL tube; mix well.
5. Pipette 10 μL of the cell–trypan blue mixture onto the hemocytometer. Count the number of intact cells (i.e., cells that are not blue) in each of the four quadrants. Take the average of these four counts, multiply by a dilution factor of 2 and multiply by 10,000 to get the number of cells per milliliter.
6. Transfer a volume equivalent to 1 million cells into a new 15 mL conical vial. Centrifuge at 250 × *g* at 4 °C for 5 min, remove PBS wash, and save the pellet.
7. Resuspend the pelleted cells in 1 mL ice-cold Lysis Buffer and let sit undisturbed on ice for 5–10 min to lyse the cells.
8. Check a small aliquot of cells under the microscope using trypan blue and a hemocytometer in the same way as used for counting the cells. The majority of the cells should have blue nuclei, indicating that the cell membrane has been ruptured but the nuclei are intact (*see* **Note 2**).
9. After confirming that most cells (but not most nuclei) are lysed, centrifuge the cells for 5 min at 750 × *g* in 4 °C and discard the supernatant, taking care not to disturb the nuclear pellet.
10. Using a P1000 pipetman, gently resuspend the nuclei in 1 mL ice-cold wash buffer. Centrifuge for 5 min at 750 × *g* in 4 °C, discard supernatant, and immediately proceed to M.CviPI treatment of the pelleted nuclei.

3.2 Treatment of Nuclei with M.CviPI to Methylate Accessible GpCs

1. Prepare at least 378 μL of 1× GpC Buffer (it is recommended that you start with 4 tubes of 250,000 cells and 94.5 μL is needed per 250,000 cells) by diluting the stock 10× GpC buffer in nuclease-free water.
2. Using a P1000 pipetman, resuspend the nuclei obtained from 1 million cells in 378 μL of 1× GpC buffer to obtain a final concentration of 250,000 nuclei per 94.5 μL; keep nuclei on ice.

3. In four prechilled 1.7 mL microcentrifuge tubes, prepare four reaction mixtures containing the following components in the order listed (*see* **Note 3**):

1 M sucrose	45.0 μL
10× GpC buffer	5.0 μL
Nuclei (250,000)	94.5 μL
32 mM SAM	1.5 μL
50 U/μL M.CviPI	4.0 μL
Total	150.0 μL/tube

4. Incubate for 7.5 min at 37 °C, then boost the reaction by adding the following:

32 mM SAM	1.5 μL
50 U/μL M.CviPI	2.0 μL
Total	3.5 μL/tube

5. Incubate for an additional 7.5 min at 37 °C, then stop the reaction by adding 153.5 μL of the Stop Buffer.

3.3 Purification of M. CviPI-Treated DNA

1. Add 200 μg/mL of Proteinase K (3 μL of 20 mg/mL Proteinase K) to each of the four reaction mixtures and incubate for 16 h at 55 °C to inactivate the M.CviPI enzyme and digest proteins present in the treated nuclei preparations.
2. Purify the DNA in the four reaction mixtures using a standard phenol–chloroform extraction method, removing the aqueous layer to a new 1.7 mL tube; note that phase-lock gel can be used to assist the separation of the aqueous and organic phases. Add 2.5 volumes (775 μL) of 100% ethanol to each tube containing the aqueous layer and incubate at −20 °C for overnight or at −80 °C for 1–2 h (*see* **Note 4**).
3. Pellet the DNA by centrifuging at a maximum speed in a microcentrifuge for 15 min. Carefully remove the ethanol and add 300 μL of ice-cold 70% ethanol to the pellet.
4. Pellet the DNA again by centrifuging at a maximum speed in a microcentrifuge for 15 min. Remove the ethanol and allow the pellet to air-dry (~ 20 min).
5. Resuspend the DNA pellet in 20 μL of nuclease-free water or TE buffer.
6. Quantify the DNA and combine the treated DNA from the 4 tubes into a single 1.7 mL tube. In general, a quantity of

100 ng/μL from the starting 1 million cells (~8 μg total) is expected. The DNA can be stored up to 6 months at −20 °C. [Stopping Point]

3.4 Fragmentation of M.CviPI-Treated DNA

1. Dilute M.CviPI-treated DNA to a total volume of 130 μL and transfer into one 6 × 16 mm microTUBE, taking care to avoid air bubbles (*see* **Note 5**).
2. Perform sonication using the Covaris system, producing 150 bp fragments (*see* **Note 6**).
3. Ethanol precipitate the sonicated DNA by adding 2.5 volumes of ice-cold 100% ethanol; incubate at −20 °C for overnight or at −80 °C for 1–2 h.
4. Pellet the DNA by centrifuging at a maximum speed in a microcentrifuge for 15 min. Remove the ethanol and allow the pellet to air-dry (~20 min).
5. Resuspend the DNA pellet in 15 μL of nuclease-free water. Quantify DNA using a NanoDrop spectrophotometer. Check the fragment size using an Agilent Bioanalyzer with a DNA High Sensitivity chip (Fig. 3). The DNA can be stored up to 6 months at −20 °C (*see* **Note 6**). [Stopping Point]

3.5 Bisulfite Treatment of M.CviPI-Methylated DNA to Convert All Unmethylated Cs to Ts

1. Use the EZ DNA Methylation kit from Zymo Research to convert unmethylated Cs in up to 1 μg of M.CviPI-treated and fragmented DNA.
2. Add 5 μL of M-Dilution Buffer to the DNA and adjust total volume to 50 μL with water. Mix the sample by flicking or pipetting up and down.
3. Incubate the sample at 37 °C for 15 min.
4. After the above incubation, add 100 μL of the prepared CT Conversion Reagent to the sample and mix.
5. Incubate the sample in a thermocycler at (95 °C for 30 s, 50 °C for 60 min) for 16 cycles, then hold at 4 °C.
6. Add 400 μL of M-Binding Buffer to a Zymo-Spin IC Column and place the column in the provided collection tube.
7. Transfer the sample being held at 4 °C to the column containing the M-Binding Buffer. Mix by inverting the column in the collection tube several times.
8. Centrifuge at full speed for 30 s. Discard flow through.
9. Add 100 μL of M-Wash Buffer to the column. Centrifuge at full speed for 30 s.
10. Add 200 μL of M-Desulphonation Buffer to the column and let stand at room temperature for 15–20 min. After incubation, centrifuge at full speed for 30 s.

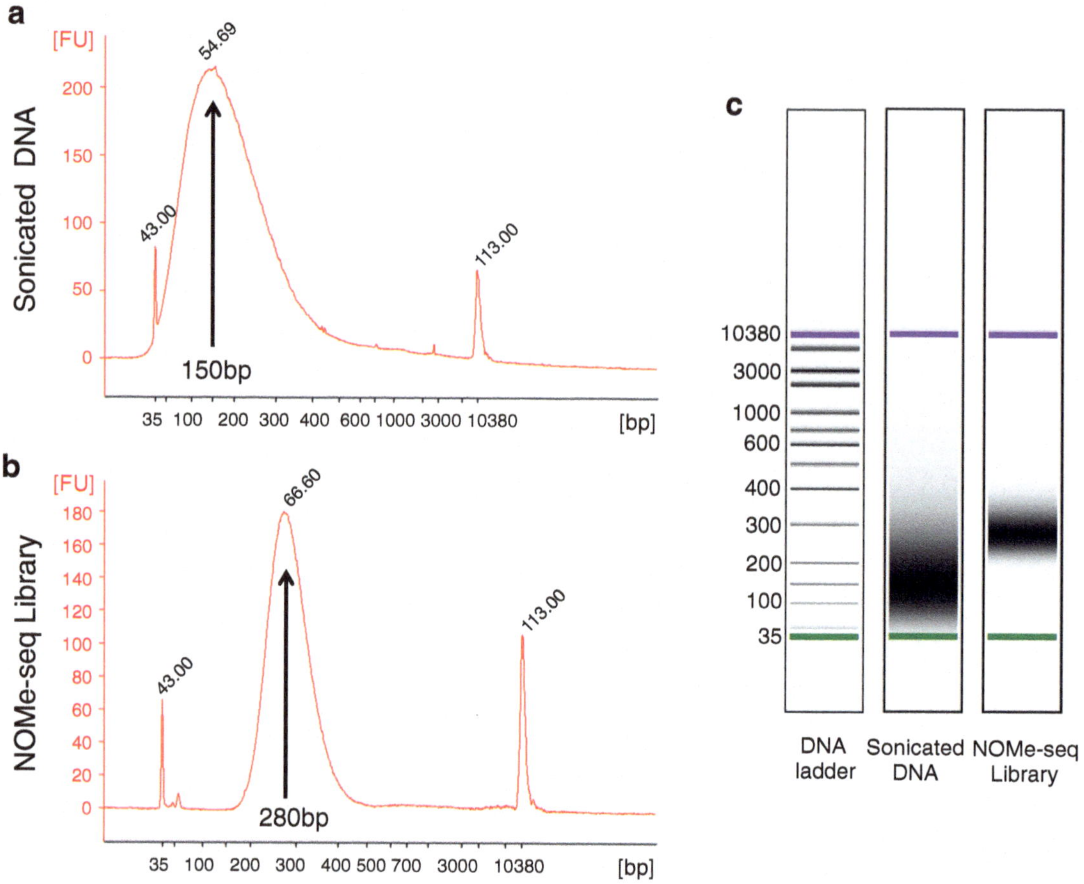

Fig. 3 Size distribution analysis of a NOMe-seq sample and library. Shown is a Bioanalyzer trace, obtained using an Agilent 2100 Bioanalyzer instrument and an Agilent High Sensitivity DNA chip, of the DNA after M. CviPI treatment and fragmentation using a Covaris S220 sonicator **(a)** and of the resultant NOMe-seq library **(b)**. The leftmost and rightmost peaks (labeled 43 and 113) are size markers of 35 bp and 10,380 bp, respectively. The average length of the fragmented DNA is calculated to be 150 bp, whereas the average length of the library fragments is calculated to be 280 bp. **(c)** For comparison to the Bioanalyzer traces, the gel images of the fragmented DNA and the NOMe-seq library are also shown

11. Add 200 μL of M-Wash Buffer to the column. Centrifuge at full speed for 30 s. Add an additional 200 μL of M-Wash Buffer and centrifuge for an additional 30 s.
12. Place the column into a 1.7 mL microcentrifuge tube. Add 20 μL of nuclease-free water to the column matrix. Centrifuge for at full speed for 30 s to collect the DNA solution. Bisulfite-converted DNA can be stored at −20 °C for up to a year. [Stopping Point]

3.6 NOMe-Seq Library Construction

1. Use the bisulfite-converted DNA isolated in the previous step to generate a library using the Accel-NGS Methyl-Seq DNA Library Kit for Illumina Platforms (*see* **Note 7**). The basic steps

in the library preparation include an adaptase step (end repair, tailing of 3′ ends, and ligation of the first truncated sequencing adapter in a single step), extension, ligation of the second truncated adapter, and indexing PCR. A detailed protocol is provided with the kit; however, we note that we generally use the entire amount of converted DNA with 7–10 PCR cycles and that for all steps involving SPRI (Solid Phase Reversible Immobilization) select beads, the volumes indicated for a 165 bp insert size should be used. Importantly, this kit should be purchased along with indexing reagents to barcode your library allowing for the pooling of multiple libraries (*see* **Note 8**). The DNA library can be stored at −20 °C for up to a year. [Stopping Point]

2. Measure the concentration of the library using the Qubit DNA HS assay kit.
3. Check the library size using an Agilent Bioanalyzer with a DNA High Sensitivity chip (Fig. 3) (*see* **Note 6**). The DNA library can be stored at −20 °C for up to a year. [Stopping Point]

3.7 Sequencing a NOMe-Seq Library

NOMe-seq libraries can be sequenced either using single-end or paired-end methods at standard read lengths using Illumina sequencers (e.g., Hi-Seq or NextSeq machines) (see http://www.illumina.com/systems/sequencing.html for details) (*see* **Note 8**). To check the quality of a NOME-seq library, a low pass run should be performed (*see* **Note 9**). After determining that the library is of high quality (*see* Subheading 3.8), a minimum of 200 million reads should be obtained, which corresponds to ~5× coverage of all methylated loci in the human genome.

3.8 Quality Analysis of a NOMe-Seq Library

3.8.1 Genome Alignment

After obtaining fastq files from the Illumina sequencer, the quality of the fastq files is examined using software tools such as FastQC (http://www.bioinformatics.babraham.ac.uk/projects/fastqc/) (*see* **Note 10**). The fastq files must be aligned to a bisulfite-converted genome, which can be done using bisulfite sequencing mapping programs such as BSMAP [19], BWA-METH (https://github.com/brentp/bwa-meth), Bismark [20], or BS-SEEKER [21, 22]. The aligned file produced from the fastq file using the bisulfite sequencing mapping programs should be saved as a bam file format (*see* **Note 8**).

3.8.2 Postprocessing of Bam Files

The bam file generated from Subheading 3.8.1 must be postprocessed for further analyses, such as DNA methylation and NDR calling. To remove duplicate reads caused by PCR from a bam file, the MarkDuplicates function of Picard (http://picard.sourceforge.net) should be used. If multiple sequencing lanes for a given sample were obtained, the bam files from each lane can be combined. Multiple bam files can be merged by using the MergeSamFiles function of Picard or the merge function of SAMTOOLS

[23]. When there are multiple bam files, it is important that each read from multiple sequencing lanes has the proper read group in order to remove duplicates and keep track of the data. By using the AddOrReplaceReadGroups function of Picard, read groups of each read can be added or replaced. The final bam file should be sorted using the sort function of SAMTOOLS and indexed, producing a bai file which contains indices of the bam file required for access to arbitrary genomic coordinates.

To assess the quality of the bam file, the flagstat function of SAMTOOLS or the CollectAlignmentSummaryMetrics function of Picard can be used (*see* **Note 11**). The output file of the flagstat function will list the total number of mapped reads (which includes QC-passed reads and QC-failed reads), the number of duplicates reads, the number of mapped reads, and the number of correctly paired reads if the library was sequenced using a paired-end method. Similarly, using the Picard CollectAlignmentSummary function, statistics such as total reads, aligned reads and percent of aligned pairs can be measured. The coverage of CpGs vs coverage of random regions of the genome can be calculated using the BamToElementEnrichment script from ECWorkflows (https://github.com/uec). This value is critical to assess the quality of the NOMe-seq library. It has been observed that CpG islands can often be poorly represented in bisulfite-converted libraries. Because CpG islands are enriched in promoter regions, it is critical that these regions of the genome be adequately represented in the libraries (*see* **Note 12**).

3.8.3 Methylation Calling

To identify the methylation status of CpG sites (in all HCG trinucleotides) and GpC sites (in all GCH trinucleotides) from the bam file, the Bis-SNP [24] program can be used. The BisulfiteGenotyper function of the Bis-SNP pipeline takes a bam file and generates a VCF file, which contains detailed information about the SNPs in the analyzed genome and provides DNA methylation information. The Vcf2bed6plus2 script in the Bis-SNP pipeline converts vcf files to a 6plus2.bed format file which contains information about each CpG or GpC site, including the chromosome start and end position, status indicating if a SNP or a reference CG is present, a score showing the methylation level (0–1000), the strand orientation, the methylation level (0–100%), and the number of CT reads covered at each locus. The Vcf2wig script in the Bis-SNP pipeline converts vcf files to wiggle files such as bedGraph and bigwig files, which can be used to visualize the DNA methylation levels across the genome by using browsers such as the UCSC genome browser [23], IGV [25], or IGB [26] or to make plots and heatmaps showing the DNA methylation density at regions of interest using the Bis-tools (https://github.com/dnaase/Bis-tools). The MethylSummarizeList.txt file generated from the Bis-SNP pipeline contains statistics of the methylation calling, such as visited bases, callable bases, confidently called bases, and average good reads coverage in all visited and callable loci.

3.8.4 Calling NDRs

For identification of NDRs, the findNDRs function in the aaRon R package can be used (see https://github.com/astatham/aaRon for details). To use the aaRon R package, a GCH.6plus2.bed file, which contains methylation calls from the Bis-SNP program (*see* Subheading 3.8.3), should be transformed to a tsv file, which contains the number of CT reads as methylation levels, multiplying total number of CT reads by the methylation level at each GpC site. For the findNDRs function, different p-value cutoffs and window sizes can be used (*see* **Note 13**). Although the number of NDRs will differ for each NOMe-seq library and for each p-value cutoff, a standard number of NDRs for further analyses of human genomes is 70,000–100,000.

3.8.5 Quality Analysis Methylation and NDR Calls

To determine the quality of the DNA methylation data, the HCG and GCH methylation levels can be visualized at the center of conserved motif-containing CTCF peaks (*see* **Note 14**). A high-quality NOMe-seq library with proper DNA methylation calls will show phasing of HCG and GCH signals; *see* Fig. 4. To determine the quality of the NDR calls, the HCG and GCH methylation levels can be visualized at the center of the called NDRs. As discussed above, the NDRs should have high GCH signals but low HCG signals at their centers; *see* Fig. 5. By generating a heatmap that can visualize methylation signals at each NDR locus, one can remove false positive NDRs and decide *p*-value cutoffs for NDR calls; *see* Fig. 6.

3.8.6 Example Analyses of a NOMe-Seq Library

We generated a NOME-seq library using CNON (Cultured Neuronal cells derived from Olfactory Neuroepithelium) cells [27] from patient sample 45 as part of the PsychENCODE project (https://www.synapse.org/#!Synapse:syn4921369/wiki/235539) [28] using 100 bp paired-end sequencing with the Illumina Hi-Seq 2500. To compare NDRs to regulatory elements defined by the H3K4me3 promoter mark, the H3K27ac enhancer mark, and CTCF binding sites, we also generated ChIP-seq libraries with proper antibodies using our previously published protocols for histones and site-specific DNA binding factors [11, 29]. *See* Supplementary Information for specific cell culture and ChIP-seq protocols for CNON cells; we recommend using MACS2 [30] to call the peaks (*see* also https://github.com/taoliu/MACS/). When we overlapped NDRs with H3K4me3 peaks, H3K27ac peaks and CTCF peaks from CNON cells we found that about 80% of the NDRs were in these regulatory elements. However, we also identified NDRs which are distal from transcription start sites and do not have a significant H3K27ac or CTCF signal (Fig. 7).

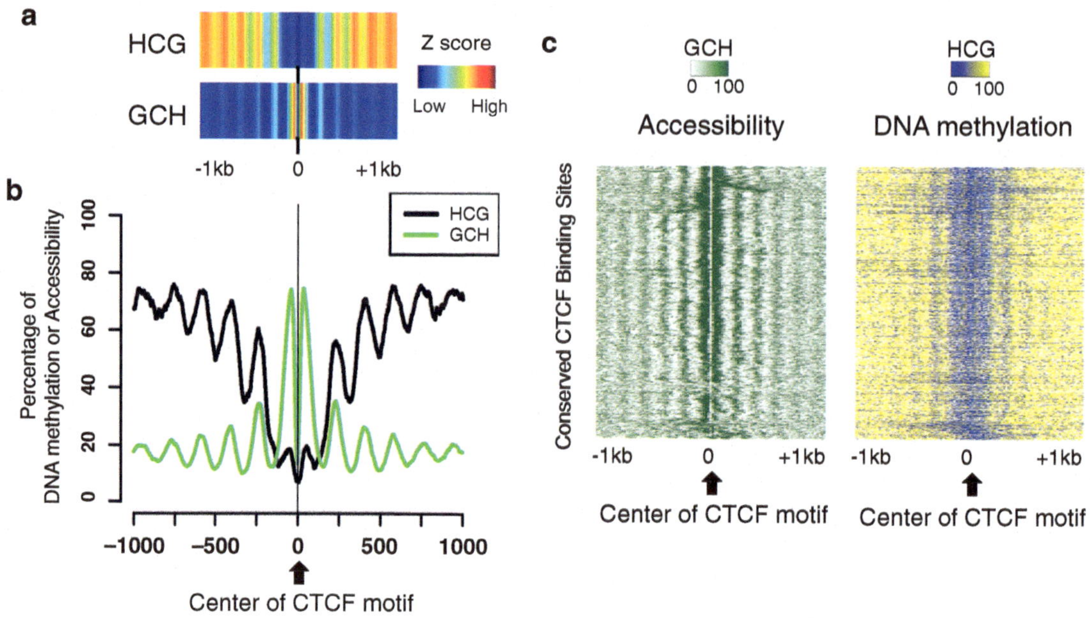

Fig. 4 Quality analysis of NOMe-seq data using CTCF peaks. A set of 3216 CTCF peaks that were commonly identified in 58 different human cell types and that have a CTCF motif were used to compare endogenous DNA methylation (HCG) and accessibility (GCH). In each panel, data is shown for a 2 kb region, centered on the CTCF motifs within the CTCF peaks; the genomic locations of the set of 3216 CTCF sites are provided in Supplementary Table S1. **(a)** The density (Z scores) of HCG methylation and GCH methylation, centered on the CTCF motif, is shown for all common CTCF peaks. **(b)** The average methylation levels of HCG (endogenous DNA methylation) and the average methylation of GCH (accessibility) are shown for all common CTCF peaks. **(c)** A heatmap representing the percentage of GCH methylation (left) and endogenous HCG methylation (right) is shown for all common CTCF peaks. The heatmap was made by first clustering the GCH values at the CTCF peaks, then plotting both the GCH and HCG values in the same order

4 Notes

1. A protocol for growing and processing CNON cells (used to obtain the example NOMe-seq dataset), including the dispase treatment used to collect the cells needed for nuclei isolation, is provided as Supplementary Information. It is recommended that exponentially growing cells be used for these experiments. However, if it is necessary to use tissue samples, then methods such as those used to perform native ChIP (no crosslinking) from tissues should be employed [31, 32]. In addition, other protocols suggest that crosslinked cells can be used as the starting material for NOMe-seq (http://www.activemotif.com/documents/1847.pdf).
2. The time needed to lyse the cells varies based on cell type, so it is recommended to optimize the cell lysis condition for each cell line prior to beginning the NOMe-seq protocol. If intact cells (i.e., cells that do not have blue nuclei) remain after the

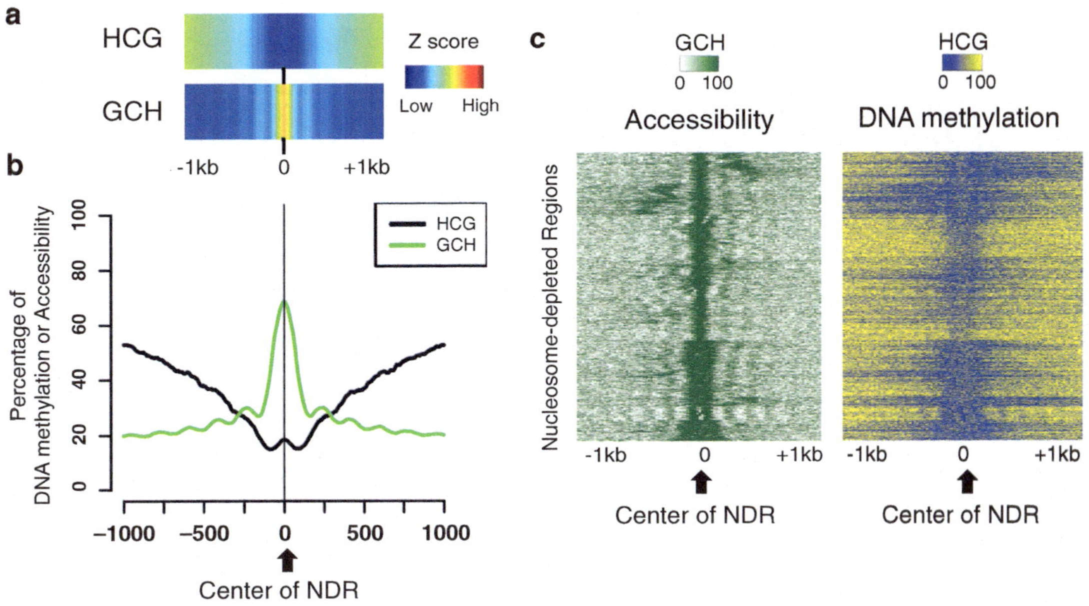

Fig. 5 Quality analysis of called NDRs. Data is shown for a 2 kb region centered on 92,482 called NDRs (P-value cutoff $= 10^{-12}$) for a NOMe-seq dataset. **(a)** The density (Z scores) of HCG methylation (endogenous DNA methylation) and GCH methylation (accessibility), centered on the NDRs, is shown. **(b)** The average methylation levels of HCG (endogenous DNA methylation) and the average methylation of GCH (accessibility) are shown for all NDRs. **(c)** A heatmap representing the percentage of GCH methylation (left) and endogenous HCG methylation (right) is shown for all NDRs. The heatmap was made by first clustering the GCH values at the NDRs, then plotting both the GCH and HCG values in the same order

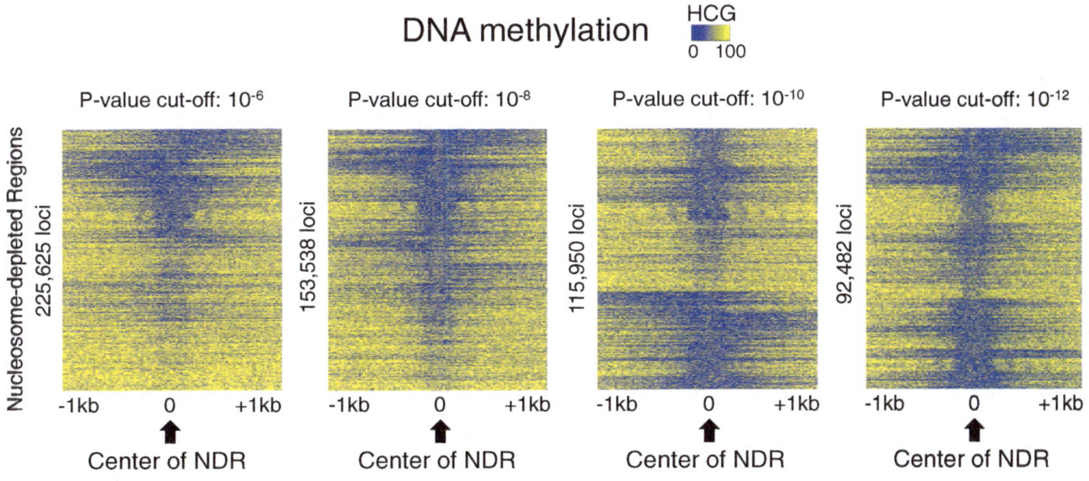

Fig. 6 Comparison of DNA methylation levels at NDRs identified using different *p*-value cutoffs. Shown are heatmaps indicating the percentage of endogenous methylation at HCG sites for a 2 kb region centered on NDRs selected using different *p*-value cutoffs. The heatmaps were made by first clustering the GCH values at each NDR, then plotting the HCG values in the same order

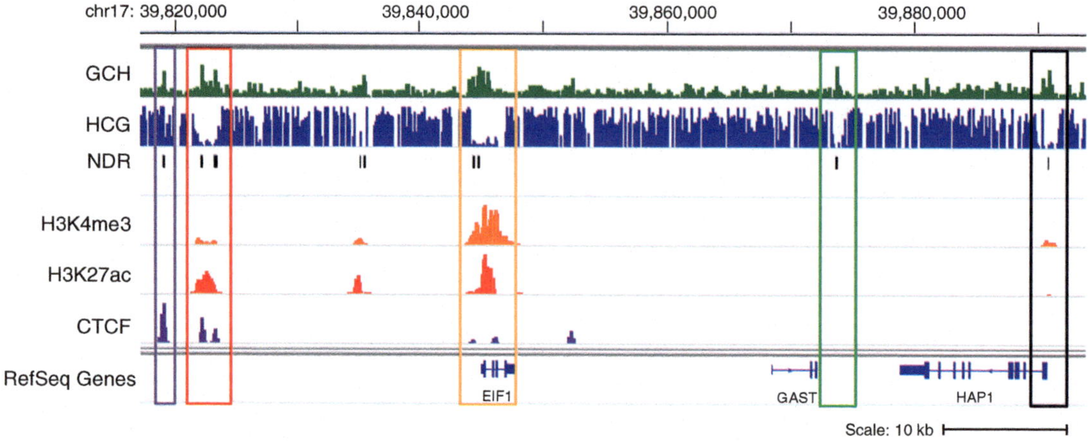

Fig. 7 Examples of NDRs at regulatory elements. Shown is a genome browser screen shot (hg19) of a region from chr17q21.2 with tracks representing accessibility (GCH), endogenous DNA methylation (HCG), called NDRs, and H3K4me3, H3K27ac and CTCF ChIP-seq data; all data is from CNON cells. The purple box highlights an NDR classified as an insulator (a genomic region bound by CTCF that does not have the histone modifications found at promoters or enhancers), the red box highlights an NDR representing an enhancer (a distal genomic region marked by H3K27ac), the orange box highlights an NDR in the promoter of the *EIF1* gene, the green box highlights an NDR that lacks promoter, enhancer, and insulator marks, and the black box highlights an NDR at the promoter of the *HAP1* gene, which is not actively transcribed in these cells (as shown by the small H3K4me3 signal)

initial cell lysis treatment, you will need to extend the time in Lysis Buffer, monitoring progress using trypan blue staining. It is critical that the nuclei remain intact throughout the subsequent wash and reaction steps. Therefore, the shortest amount of time needed to lyse the majority of the cells should be used.

3. The standard concentration of M.CviPI from NewEngland BioLabs is 4 U/μL. However, we recommend that a special, high concentration order of 50 U/μL be purchased from the company; otherwise, the reaction volumes must be adjusted accordingly if the low concentration enzyme is used.

   (a) If using 50 U/μL M.CviPI, resuspend 1 million nuclei in 378 μL 1× GpC Buffer, then:

   For each of 4 tubes:

1 M sucrose	45.0 μL
10× GpC buffer	5.0 μL
Nuclei (250,000)	94.5 μL
32 mM SAM	1.5 μL
50 U/μL M.CviPI	4.0 μL (200 units)
Total	150.0 μL/tube

Incubate for 7.5 minutes at 37 °C, then boost the reaction by adding the following:

32 mM SAM	1.5 μL
50 U/μL M.CviPI	2.0 μL (100 units)
Total	3.5 μL/tube

Incubate for 7.5 min at 37 °C, then stop by adding 153.5 μL of Stop Buffer.

(b) If using 4 U/μl M.CviPI, resuspend 1 million nuclei in 1128 μL 1× GpC Buffer, then:

For each of 4 tubes:

1 M sucrose	150.0 μL
10× GpC buffer	17.0 μL
Nuclei (250,000)	282.0 μL
32 mM SAM	1.5 μL
4 U/μL M.CviPI	50.0 μL (200 units)
Total	500.0 μL/tube

Incubate for 7.5 min at 37 °C, then boost the reaction by adding the following:

32 mM SAM	1.5 μL
4 U/μL M.CviPI	25.0 μL (100 units)
Total	26.5 μL/tube

Incubate for 7.5 min at 37 °C, then stop by adding 526.5 μL of Stop Buffer.

4. In addition to the phenol chloroform extraction method, other methods of isolating human genomic DNA may be used, such as the column-based genomic DNA isolation kit from Zymo (Genomic DNA Clean & Concentrator -25).
5. If using the Covaris S220 sonicator, no more than 10 μg of DNA should be fragmented at a time; if you obtained more than 10 μg of DNA from the treated cells, you should dilute to 100 ng/μL and only use 8–10 μg per sonication tube.
6. Sonication must be optimized for each cell type to produce 100–200 bp fragments. If using a Covaris S220 sonicator, it is recommended that you start by using a 10% duty cycle, an intensity setting of 5, and 200 cycles per burst for 6 min. If the fragments in the resultant sonicated fragments or library are far from the appropriate size when examined on the Bioanalyzer

(Subheading 3.4, **step 5** or Subheading 3.6, **step 3**) it is recommended that the sonication step be optimized and the protocol repeated prior to proceeding with library preparation or sequencing.

7. Although previous NOMe-seq studies [9, 17, 18] have used methods in which adaptors are ligated to the fragmented DNA prior to bisulfite treatment, these methods result in considerable losses of DNA. We recommend using the Accel-NGS Methyl-Seq DNA Library Kit because it enables the preparation of high complexity next-generation sequencing libraries after bisulfite conversion of the DNA. Importantly, this kit is compatible with single-stranded DNA, making it a good choice for use with DNA fragments damaged and denatured by bisulfite conversion. This single-strand compatibility also overcomes the library loss associated with methylated adapter ligation prior to bisulfite conversion.
8. A bisulfite-converted genome is low in complexity, due to the conversion of the vast majority of Cs to Ts, which results in a lower percentage of alignment of sequenced fragments to the genome than obtained from standard genomic libraries. However, the use of paired-end sequencing methods can improve the alignment and therefore this method is preferable rather than single-end sequencing for NOMe-seq libraries; if single-end sequencing is used, longer reads (at least 100 bp) can help increase the alignment percentage. Increasing the number of allowed mismatch parameters in the bisulfite sequencing mapping programs may also improve alignment of reads to the bisulfite genome.
9. The FastQC program can be used to analyze the quality of sequencing libraries. The program outputs QC statistics, focusing on duplicate level, kmer profile, base quality, base GC content, base N content, base sequence content, sequence GC content, sequence quality, and sequence length distribution. It is important to check that your sequenced library has good quality metrics, without any warning or failure/error messages in the each of the statistics sections. Detailed information on FastQC output files, including what is considered appropriate metrics, can be found at http://www.bioinformatics.babraham.ac.uk/projects/fastqc/Help/.
10. It is highly recommended that NOMe-Seq libraries be barcoded so that they can be pooled with other libraries prior to sequencing. It is recommended that a low-pass sequencing run be performed (~10 million reads) for each NOMe-Seq library to assess various quality metrics before proceeding to sequence a library at a high depth. It is important to note that, once a library has passed quality assessment, sequencing data from

multiple NOMe-seq libraries from the same cells can be combined to increase the genomic coverage. Before combining datasets, NDRs called by each dataset should be compared to assess data similarity; we recommend that at least 80% of the top ranked NDRs called at a given p-value overlap for two libraries that will be merged.

11. High-quality NOMe-Seq libraries should have less than 5% duplicate reads, with more than 80% of the total reads mapping to the genome with properly paired ends.
12. A bias for or against CpG islands could be due to size selection of the NOMe-Seq library. Smaller library fragments tend to be enriched for CpG islands. Therefore, if the size of library is too small, CpG islands will be represented but perhaps other regulatory elements, such as distal enhancers, may be lost. On the other hand, if the size of library is too big, there will be low coverage of CpG islands, which will negatively affect the ability to identify NDRs in promoter regions. A ratio of CpG vs random coverage close to 1 is desired; if the ratio is lower than 0.5 the coverage may be biased toward non-CpG islands and if the ratio is larger than 1, the coverage may be biased in favor of CpG islands. Therefore, it is important that the library size be optimized (*see* **Note 6**).
13. We restrict window size to 140 bp for NDRs, which provides a more precise region of the inter-nucleosomal region of open chromatin that can be used for motif analyses.
14. The CTCF protein binds with high affinity to a specific DNA motif, which contains a CpG dinucleotide, which helps to visualize DNA methylation calls. Binding of CTCF is not compatible with high levels of endogenous DNA methylation and therefore the $C^{m}pG$ levels should be very low at these sites. Conversely, because CTCF binds in regions of open chromatin, the levels of GpC^{m} should be high at the sites. To assist investigators who do not have CTCF ChIP-seq data for their particular cell type in which NOMe-seq is being performed, we have generated a file of conserved, motif-containing CTCF sites, which are distal from transcription start sites and commonly found across 114 CTCF ChIP-seq samples from 58 different cell types [1]. This file, which includes 3216 genomic coordinates of CTCF sites (hg19), can be used for quality analysis of any human NOMe-seq library (Supplementary Table S1).

References

1. Consortium EP (2012) An integrated encyclopedia of DNA elements in the human genome. Nature 489(7414):57–74. https://doi.org/10.1038/nature11247
2. Rada-Iglesias A, Bajpai R, Swigut T, Brugmann SA, Flynn RA, Wysocka J (2011) A unique chromatin signature uncovers early developmental enhancers in humans. Nature 470 (7333):279–283. https://doi.org/10.1038/nature09692
3. RoadmapEpigenomicsConsortium (2015) Integrative analysis of 111 reference human epigenomes. Nature 19:317–330
4. Heintzman ND, Ren B (2009) Finding distal regulatory elements in the human genome. Curr Opin Genet Dev 19(6):541–549. https://doi.org/10.1016/j.gde.2009.09.006
5. Heintzman ND, Stuart RK, Hon G, Fu Y, Ching CW, Hawkins RD, Barrera LO, Van Calcar S, Qu C, Ching KA, Wang W, Weng Z, Green RD, Crawford GE, Ren B (2007) Distinct and predictive chromatin signatures of transcriptional promoters and enhancers in the human genome. Nat Genet 39 (3):311–318. https://doi.org/10.1038/ng1966
6. DesJarlais R, Tummino PJ (2016) Role of histone-modifying enzymes and their complexes in regulation of chromatin biology. Biochemistry 55(11):1584–1599. https://doi.org/10.1021/acs.biochem.5b01210
7. Allis CD, Jenuwein T (2016) The molecular hallmarks of epigenetic control. Nat Rev Genet 17(8):487–500. https://doi.org/10.1038/nrg.2016.59
8. Merkenschlager M, Nora EP (2016) CTCF and Cohesin in genome folding and transcriptional gene regulation. Annu Rev Genomics Hum Genet 17:17–43. https://doi.org/10.1146/annurev-genom-083115-022339
9. Kelly TK, Liu Y, Lay FD, Liang G, Berman BP, Jones PA (2012) Genome-wide mapping of nucleosome positioning and DNA methylation within individual DNA molecules. Genome Res 22(12):2497–2506. https://doi.org/10.1101/gr.143008.112
10. Xu M, Kladde MP, Van Etten JL, Simpson RT (1998) Cloning, characterization and expression of the gene coding for a cytosine-5-DNA methyltransferase recognizing GpC. Nucleic Acids Res 26(17):3961–3966
11. O'Geen H, Echipare L, Farnham PJ (2011) Using ChIP-Seq technology to generate high-resolution profiles of histone modifications. Methods Mol Biol 791:265–286. https://doi.org/10.1007/978-1-61779-316-5_20
12. Thurman RE, Rynes E, Humbert R, Vierstra J, Maurano MT, Haugen E, Sheffield NC, Stergachis AB, Wang H, Vernot B, Garg K, John S, Sandstrom R, Bates D, Boatman L, Canfield TK, Diegel M, Dunn D, Ebersol AK, Frum T, Giste E, Johnson AK, Johnson EM, Kutyavin T, Lajoie B, Lee BK, Lee K, London D, Lotakis D, Neph S, Neri F, Nguyen ED, Qu H, Reynolds AP, Roach V, Safi A, Sanchez ME, Sanyal A, Shafer A, Simon JM, Song L, Vong S, Weaver M, Yan Y, Zhang Z, Zhang Z, Lenhard B, Tewari M, Dorschner MO, Hansen RS, Navas PA, Stamatoyannopoulos G, Iyer VR, Lieb JD, Sunyaev SR, Akey JM, Sabo PJ, Kaul R, Furey TS, Dekker J, Crawford GE, Stamatoyannopoulos JA (2012) The accessible chromatin landscape of the human genome. Nature 489 (7414):75–82. https://doi.org/10.1038/nature11232
13. Giresi PG, Kim J, McDaniell RM, Iyer VR, Lieb JD (2007) FAIRE (formaldehyde-assisted isolation of regulatory elements) isolates active regulatory elements from human chromatin. Genome Res 17(6):877–885. https://doi.org/10.1101/gr.5533506
14. Buenrostro JD, Giresi PG, Zaba LC, Chang HY, Greenleaf WJ (2013) Transposition of native chromatin for fast and sensitive epigenomic profiling of open chromatin, DNA-binding proteins and nucleosome position. Nat Methods 10(12):1213–1218. https://doi.org/10.1038/nmeth.2688
15. Buenrostro JD, Wu B, Chang HY, Greenleaf WJ (2015) ATAC-seq: a method for assaying chromatin accessibility genome-wide. Curr Protoc Mol Biol 109(21 29):21–29. https://doi.org/10.1002/0471142727.mb2129s109
16. Taberlay PC, Statham AL, Kelly TK, Clark SJ, Jones PA (2014) Reconfiguration of nucleosome-depleted regions at distal regulatory elements accompanies DNA methylation of enhancers and insulators in cancer. Genome Res 24(9):1421–1432. https://doi.org/10.1101/gr.163485.113
17. Statham AL, Taberlay PC, Kelly TK, Jones PA, Clark SJ (2015) Genome-wide nucleosome occupancy and DNA methylation profiling of four human cell lines. Genom Data 3:94–96. https://doi.org/10.1016/j.gdata.2014.11.012
18. Lay FD, Liu Y, Kelly TK, Witt H, Farnham PJ, Jones PA, Berman BP (2015) The role of DNA

methylation in directing the functional organization of the cancer epigenome. Genome Res 25(4):467–477. https://doi.org/10.1101/gr.183368.114

19. Xi Y, Li W (2009) BSMAP: whole genome bisulfite sequence MAPping program. BMC Bioinformatics 10:232. https://doi.org/10.1186/1471-2105-10-232
20. Krueger F, Andrews SR (2011) Bismark: a flexible aligner and methylation caller for bisulfite-Seq applications. Bioinformatics 27 (11):1571–1572. https://doi.org/10.1093/bioinformatics/btr167
21. Guo W, Fiziev P, Yan W, Cokus S, Sun X, Zhang MQ, Chen PY, Pellegrini M (2013) BS-Seeker2: a versatile aligning pipeline for bisulfite sequencing data. BMC Genomics 14:774. https://doi.org/10.1186/1471-2164-14-774
22. Chen PY, Cokus SJ, Pellegrini M (2010) BS Seeker: precise mapping for bisulfite sequencing. BMC Bioinformatics 11:203. https://doi.org/10.1186/1471-2105-11-203
23. Li H, Handsaker B, Wysoker A, Fennell T, Ruan J, Homer N, Marth G, Abecasis G, Durbin R, Genome Project Data Processing S (2009) The sequence alignment/map format and SAMtools. Bioinformatics 25 (16):2078–2079. https://doi.org/10.1093/bioinformatics/btp352
24. Liu Y, Siegmund KD, Laird PW, Berman BP (2012) Bis-SNP: combined DNA methylation and SNP calling for bisulfite-seq data. Genome Biol 13(7):R61. https://doi.org/10.1186/gb-2012-13-7-r61
25. Robinson JT, Thorvaldsdottir H, Winckler W, Guttman M, Lander ES, Getz G, Mesirov JP (2011) Integrative genomics viewer. Nat Biotechnol 29(1):24–26. https://doi.org/10.1038/nbt.1754
26. Nicol JW, Helt GA, Jr. Blanchard SG, Raja A, Loraine AE (2009) The integrated genome browser: free software for distribution and exploration of genome-scale datasets. Bioinformatics 25(20):2730–2731. https://doi.org/10.1093/bioinformatics/btp472
27. Evgrafov OV, Wrobel BB, Kang X, Simpson G, Malaspina D, Knowles JA (2011) Olfactory neuroepithelium-derived neural progenitor cells as a model system for investigating the molecular mechanisms of neuropsychiatric disorders. Psychiatr Genet 21(5):217–228. https://doi.org/10.1097/YPG.0b013e328341a2f0
28. ThePsychEncodeConsortium, Akbarian S, Liu C, Knowles JA, Vaccarino FM, Farnham PJ, Crawford GE, Jaffe AE, Pinto D, Dracheva S, Geschwind DH, Mill J, Nairn AC, Abyzov A, Pochareddy S, Prabhakar S, Weissman S, Sullivan PF, State MW, Weng Z, Peters MA, White KP, Gerstein MB, Amiri A, Armoskus C, Ashley-Koch AE, Bae T, Beckel-Mitchener A, Berman BP, Coetzee GA, Coppola G, Francoeur N, Fromer M, Gao R, Grennan K, Herstein J, Kavanagh DH, Ivanov NA, Jiang Y, Kitchen RR, Kozlenkov A, Kundakovic M, Li M, Li Z, Liu S, Mangravite LM, Mattei E, Markenscoff-Papadimitriou E, Navarro FC, North N, Omberg L, Panchision D, Parikshak N, Poschmann J, Price AJ, Purcaro M, Reddy TE, Roussos P, Schreiner S, Scuderi S, Sebra R, Shibata M, Shieh AW, Skarica M, Sun W, Swarup V, Thomas A, Tsuji J, van Bakel H, Wang D, Wang Y, Wang K, Werling DM, Willsey AJ, Witt H, Won H, Wong CC, Wray GA, Wu EY, Xu X, Yao L, Senthil G, Lehner T, Sklar P, Sestan N (2015) The PsychENCODE project. Nat Neurosci 18(12):1707–1712. https://doi.org/10.1038/nn.4156
29. O'Geen H, Frietze S, Farnham PJ (2010) Using ChIP-seq technology to identify targets of zinc finger transcription factors. Methods Mol Biol 649:437–455. https://doi.org/10.1007/978-1-60761-753-2_27
30. Zhang Y, Liu T, Meyer CA, Eeckhoute J, Johnson DS, Bernstein BE, Nusbaum C, Myers RM, Brown M, Li W, Liu XS (2008) Model-based analysis of ChIP-Seq (MACS). Genome Biol 9(9):R137. https://doi.org/10.1186/gb-2008-9-9-r137
31. O'Neill LP, Turner BM (2003) Immunoprecipitation of native chromatin: NChIP. Methods 31:76–82
32. Brind'Amour J, Liu S, Hudson M, Chen C, Karimi MM, Lorincz MC (2015) An ultra-low-input native ChIP-seq protocol for genome-wide profiling of rare cell populations. Nat Commun 6:6033. https://doi.org/10.1038/ncomms7033

Check for updates

Chapter 13

Genome-Wide Mapping of Protein–DNA Interactions on Nascent Chromatin

Chenhuan Xu and Victor G. Corces

Abstract

Chromatin immunoprecipitation (ChIP) is the most widely used method to analyze protein–DNA interactions in vivo. Coupled with next generation sequencing, ChIP-seq experiments map protein–DNA interactions in a genome-wide fashion. Here we describe a novel method called nasChIP-seq for mapping genome-wide occupancy of posttranslationally modified histones or transcription factors on newly replicated DNA.

Key words Chromatin immunoprecipitation, ChIP-seq, DNA replication, Nascent chromatin, Click chemistry

1 Introduction

ChIP assays utilize the specific interaction between an antibody and a protein of interest to pull down the fraction of genomic DNA occupied by this protein, achieving an enrichment of regions occupied by the protein out of all randomly sheared genomic regions [1]. The isolated DNA can be used to prepare a library and sequenced under high-throughput conditions to reveal the underlying sequence content. These sequences can then be mapped to a reference genome to locate the genomic regions harboring the sequence, which can be visualized in a genome browser to show the locations of the enriched occupied regions [2–4]. Tandem ChIP assays (re-ChIP or sequential ChIP) have been used to map the location of protein co-occupancy sites [5, 6], or the location of protein occupied sites on chromatin under a certain context, such as nascent (newly replicated) chromatin [7]. Petruk et al. used BrdU to label newly replicated DNA in fly embryos, followed by ChIP with antibodies targeting Trithorax group proteins and BrdU sequentially. Quantitative PCR (qPCR) allowed these authors to measure the occupancy of proteins on nascent chromatin [7]. Here we describe a method utilizing EdU to label newly replicated DNA

Tanya Vavouri and Miguel A. Peinado (eds.), *CpG Islands: Methods and Protocols*, Methods in Molecular Biology, vol. 1766, https://doi.org/10.1007/978-1-4939-7768-0_13,

followed by click chemistry to transform EdU into biotin on nascent DNA [8–11]. The replacement of BrdU with EdU enabled us to take advantage of the strong interaction between biotin and streptavidin for further isolation of labeled DNA. Furthermore, we carried out the click chemistry in a facilitative aqueous phase of purified ChIP DNA molecules, instead of the classical way to "click" in a suspension of formaldehyde-crosslinked nuclei [8–11], which severely limits the reaction components to access the EdU-labeled DNA. These key improvements allowed us to achieve very high labeling efficiency and pulldown of the nascent chromatin fraction. Here we describe two examples in which this technique is used for genome-wide mapping of histone modifications or protein occupancy on nascent chromatin. In one example we use 20 min of EdU labeling with 10 million synchronized *Drosophila* Kc167 cells to map the distribution of H3K4me3 on nascent chromatin (Fig. 1a). In the second example, we describe the mapping of CTCF occupancy on nascent chromatin with 20 min labeling of 10 million asynchronous H9 human embryonic stem cells (Fig. 1b).

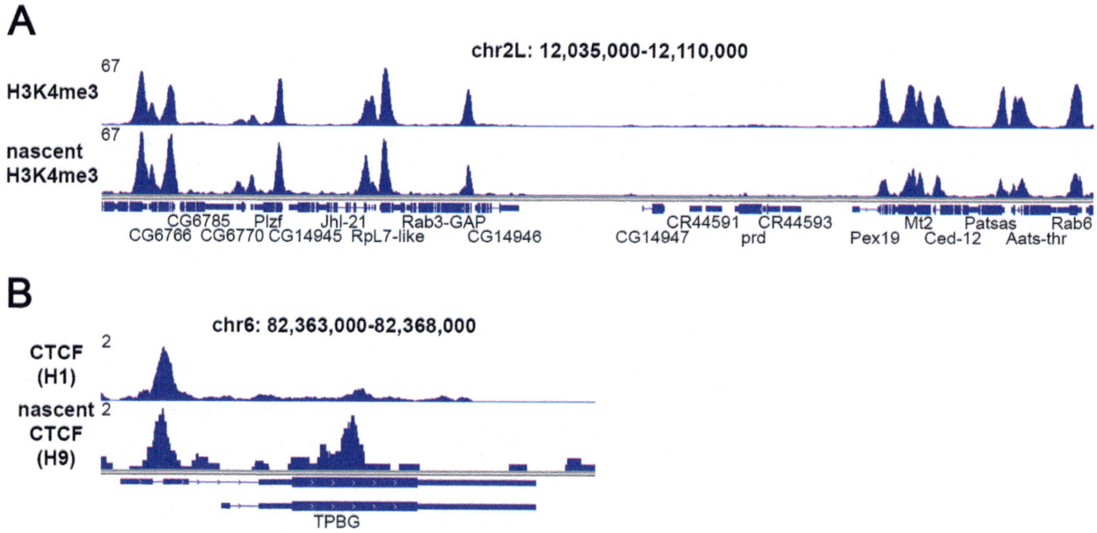

Figure 1, Xu and Corces

Fig. 1 Results from ChIP-seq on nascent chromatin. (**a**) A 75 kb genomic region of the *Drosophila* genome showing a published H3K4me3 ChIP-seq data in asynchronous Kc167 cells (upper track) [12], and our H3K4me3 nasChIP-seq data in synchronized Kc167 cells with 20 min of EdU labeling (lower track). The sequence read density was normalized to *per* one million reads (RPM). The left half of the genomic region shows full restoration of H3K4me3 on nascent chromatin within 20 min, while H3K4me3 is only partially restored in the right half of the region on nascent chromatin. (**b**) A 5 kb genomic region of the human genome showing a published CTCF ChIP-seq data in H1 hES cells (upper track) [13], and our CTCF nasChIP-seq data in H9 hES cells (lower track). The sequence read density was normalized to *per* one million reads (RPM). One conserved CTCF occupied site and one nascent chromatin-specific site are shown around the transcription start site of the TPBG gene

2 Materials

Prepare all solutions every 3 months unless otherwise noted. Molecular biology grade water is the solvent for all solutions unless otherwise noted.

2.1 Cell Synchronization, EdU Labeling, and Chromatin Immunoprecipitation

1. Kc167 cells (ATCC) or H9 cells (WiCell).
2. SFX-Insect cell culture media (HyClone) or StemPro hESC SFM (ThermoFisher).
3. 2 M hydroxyurea solution. Aliquot and store at −20 °C (*see* **Note 1**).
4. 40 mM EdU (5-ethynyl-2′-deoxyuridine) solution. Aliquot and store at −20 °C (*see* **Note 2**).
5. 1× DPBS (Dulbecco's Phosphate Buffered Saline).
6. Formaldehyde solution (37%). Store in the dark.
7. 1.25 M glycine solution. Store at 4 °C.
8. Anti-H3K4me3 antibody. Store at 4 °C.
9. Anti-CTCF antibody. Store at 4 °C.
10. Refer to [14] for other reagents for ChIP.

2.2 Click Chemistry

1. 1 mM Biotin azide in DMSO. Aliquot and store at −20 °C.
2. 100 mM $CuSO_4$ solution. Store at 4 °C.
3. 100 mM sodium ascorbate solution. Aliquot and store at −20 °C.
4. Microcentrifuge tubes.
5. Ethanol. Store at −20 °C.
6. 5 M NaCl solution.
7. 20 mg/ml glycogen solution. Aliquot and store at −20 °C.

2.3 Nascent DNA Pulldown

1. 70% ethanol solution. Store at 4 °C.
2. Dynabeads® MyOne™ Streptavidin C1. Store at 4 °C.
3. 2× binding and washing buffer: 10 mM Tris–HCl (pH 7.5), 1 mM EDTA, 2 M NaCl. Store at 4 °C.
4. 10 mM Tris–HCl (pH 8.0). Store at 4 °C.

2.4 Library Preparation

1. NEB End Repair Enzyme Mix. Store at −20 °C.
2. Klenow Fragment ($3' \rightarrow 5'$ exo$^-$). Store at −20 °C.
3. 1 mM dATP. Store at −20 °C.
4. T4 DNA Ligase. Store at −20 °C.
5. KAPA SYBR FAST qPCR Kit Master Mix (2×). Store at −20 °C.

6. Refer to [15] for sequence information on adaptors and PCR primers.
7. Agencourt AMPure XP beads.

3 Methods

All steps were performed at room temperature (25 °C) unless otherwise noted.

3.1 Cell Synchronization, EdU Labeling, and Chromatin Immunoprecipitation

1. **Kc167:** Cells were grown to 50% confluence in SFX medium at 25 °C in an incubator.

   H9: Cells were grown to 60–70% confluence in StemPro hESC medium at 37 °C in an incubator under 5% CO_2.
2. **Kc167**: Add hydroxyurea to cells at a final concentration of 2 mM, and culture cells in an incubator for 20 h (*see* **Note 3**).
3. **Kc167**: Premix EdU with fresh SFX medium to obtain a final concentration of 40 μM. Discard hydroxyurea-containing medium and wash the cells gently with DPBS once. Add EdU-containing medium to cells (*see* **Note 4**). Leave cells in the incubator for 20 min (*see* **Note 5**).

   H9: Premix EdU with fresh StemPro medium to obtain a final concentration of 50 μM. Prewarm the medium to 37 °C. Add EdU-containing medium to cells. Leave cells in the incubator for 20 min (*see* **Note 6**).
4. Premix formaldehyde with DPBS to obtain a final concentration of 1%. Discard EdU-containing medium, add 1% formaldehyde to cells.
5. After 10 min, add glycine solution to reach 0.125 M final concentration.
6. Proceed to ChIP with H3K4me3 or CTCF antibody respectively, and refer to [14] for details.

3.2 Click Chemistry

1. Centrifuge ethanol-precipitated ChIP DNA at 16,000 × *g* for 30 min at 4 °C.
2. Discard supernatant. Add 70% ice-cold ethanol. Invert the microcentrifuge tube several times to wash the pellet (*see* **Note** 7).
3. Centrifuge at 16,000 × *g* for 5 min.
4. Discard supernatant. Air-dry the pellet (*see* **Note 8**).
5. Add 174 μl 1×DPBS to the pellet. Incubate at 37 °C for 10 min.
6. Add these reagents to 174 μl DNA solution: 2 μl 1 mM biotin-azide (or DMSO, as control, *see* **Note 9**), 4 μl 100 mM $CuSO_4$ solution. Vortex briefly.

7. Add 20 μl 100 mM sodium ascorbate solution. Immediately vortex for 3 s (*see* **Note 10**).
8. Incubate at 37 °C for 1 h (*see* **Note 11**).
9. Add 9 μl 5 M NaCl solution and 2 μl glycogen. Vortex briefly. Add 500 μl ice-cold ethanol. Vortex vigorously. Incubate at −80 °C for 2 h.

3.3 Nascent DNA Pulldown

1. Centrifuge ethanol precipitated DNA at 16,000 × *g* for 30 min at 4 °C.
2. Discard supernatant. Add 70% ice-cold ethanol. Invert the microcentrifuge tube several times to wash the pellet (*see* **Note 12**).
3. Centrifuge at 16,000 × *g* for 5 min.
4. Discard supernatant. Air-dry the pellet (*see* **Note 8**).
5. Add 100 μl 10 mM Tris–HCl (pH 8.0) to the pellet. Incubate at 37 °C for 10 min.
6. Wash 5 μl Streptavidin C1 beads with 1 ml 1× binding and washing buffer. Remove supernatant on a magnetic stand. Resuspend the beads in 100 μl 2× binding and washing buffer. Mix 100 μl DNA solution with 100 μl bead suspension. Rotate on a rotor for 30 min.
7. Remove supernatant on a magnetic stand. Wash the beads with 1 ml 1× binding and washing buffer on a rotor for 10 min. Repeat this step once.
8. Remove supernatant on a magnetic stand. Wash the beads with 1 ml 10 mM Tris–HCl (pH 8.0) on a rotor for 5 min.
9. Gently suspend the beads in 24 μl dH_2O.

3.4 Library Preparation

1. Add 3 μl reaction buffer and 3 μl NEB End Repair Enzyme Mix to the bead suspension. Gently mix. Rotate on a rotor for 30 min (*see* **Note 13**).
2. Remove supernatant on a magnetic stand. Wash the beads with 1 ml 1× binding and washing buffer on a rotor for 10 min.
3. Remove supernatant on a magnetic stand. Wash the beads with 1 ml 10 mM Tris–HCl (pH 8.0) on a rotor for 5 min.
4. Gently resuspend the beads in 19 μl dH_2O. Add 3 μl reaction buffer, 6 μl 1 mM dATP, and 2 μl Klenow Fragment ($3' \rightarrow 5'$ exo^-). Gently mix. Rotate on a rotor at 37 °C for 30 min.
5. Repeat **steps 2** and **3**.
6. Gently resuspend the beads in 19 μl dH_2O. Add 3 μl reaction buffer, 5 μl 10 μM adaptors, and 3 μl T4 DNA ligase. Gently mix. Rotate on a rotor for 4 h.
7. Repeat **steps 2** and **3**.

8. Gently resuspend the beads in 21 μl dH_2O. Add 2 μl each of 10 mM primers, and 25 μl KAPA SYBR FAST qPCR Kit Master Mix (2×). Gently mix. Amplify on a qPCR instrument. Monitor the real-time amplification curve and stop the amplification appropriately (*see* **Note 14**).
9. Size-select and clean the DNA using Agencourt AMPure XP beads (Refer to product manuals.).
10. Send the library to sequencing service providers for quality assessment and high throughput sequencing.

4 Notes

1. Hydroxyurea will decompose in water. Make the stock solution monthly, aliquot into small volumes for single use, and store at −20 °C.
2. Short-term 37 °C incubation and vigorous vortexing are sometimes necessary to fully dissolve the 40 mM EdU stock solution stored at −20 °C.
3. Usually the cells will reach 70%–80% confluence after synchronization with hydroxyurea for 20 h. It should be noted that synchronization is only necessary for DNA replication timing-specific mapping. Our scheme of 20 min labeling after synchronization and release preferentially maps the genomic regions replicated during first 20 min after entry into S phase.
4. Kc167 cells attach loosely after synchronization with hydroxyurea. Be cautious not to perturb the cells when washing and changing the medium. Always pipette the medium through the side surface of the cultureware.
5. The labeling time is highly tunable. For high temporal resolution (the degree of nascentness), choose the shortest labeling time that generates a library with good signal-to-noise ratio after sequencing. For high yield of final DNA product, increase the labeling time appropriately. Also, cell harvesting can be done after a "chase" period following EdU labeling to map dynamic changes in protein occupancy on nascent chromatin.
6. Prewarm the EdU medium to 37 °C to ensure efficient labeling. When labeling cells in multiple pieces of cultureware at the same time, always place individual cultureware back in the incubator immediately after the EdU medium is added.
7. Washing with 70% ethanol helps to remove residual SDS (sodium dodecyl sulfate) carried over from elution buffer in ChIP. SDS is a strong inhibitor of click chemistry. Repeat this step if necessary.

8. 5–10 min are optimal for air-drying the DNA pellet. Over-drying will lead to low solubility of the pellet and overall reduction of final DNA yield.
9. Setting aside a DMSO reaction and making a mock library serves as a negative control for the nascent DNA pulldown step. In our hands, amplification of the DMSO library with the same number of PCR cycles yields very little DNA, ruling out the possibility that the nascent DNA after pulldown contains some carryover ChIP DNA, proving the nascent ChIP DNA faithfully reflects the enrichment of protein occupancy on nascent chromatin.
10. Sodium ascorbate is highly prone to oxidation. Always keep caps on when not using it. Protect from light. Make fresh stock solution monthly. Vortex immediately after adding it to the mixture, ensuring timely reaction with Cu^{2+} before any putative oxidation.
11. Oxygen interferes with click chemistry. Choose microcentrifuge tubes with good airtightness as reaction containers. Place tubes in an airtight container if necessary.
12. Washing with 70% ethanol helps to remove residual biotin azide carried over from click chemistry. Free biotin will reduce the ability of biotin-labeled DNA to bind to the streptavidin beads. Repeat this step if necessary.
13. We determined 30–40 μl as the optimal reaction volume in a 1.5 ml microcentrifuge tube. Volumes lower than this will lead to higher density of beads and higher chance of bead aggregation. Higher volumes will lead to spill of beads to the side surface of the microcentrifuge tube during rotation. Both conditions reduce the overall yield of final DNA.
14. PCR amplification with optimal cycle numbers is critical for the quality of the library. Insufficient amplification will lead to insufficient input material for sequencing and failure of the experiment. Overamplification will lead to higher PCR duplication level (piling up of clonally identical sequence) and reduce the overall cost-efficiency of sequencing.

Acknowledgments

Work in the authors' laboratory was supported by U.S. Public Health Service Award R01 GM035463 from the National Institutes of Health. The content is solely the responsibility of the authors and does not necessarily represent the official views of the National Institutes of Health.

References

1. Struhl K (2007) Interpreting chromatin Immunoprecipitation experiments. In: Zuk D (ed) Evaluating techniques in biochemical research. Cell Press, Cambridge, MA
2. Barski A, Cuddapah S, Cui K, Roh TY, Schones DE, Wang Z et al (2007) High-resolution profiling of histone methylations in the human genome. Cell 129:823–837
3. Johnson DS, Mortazavi A, Myers RM, Wold B (2007) Genome-wide mapping of in vivo protein-DNA interactions. Science 316:1497–1502
4. Robinson JT, Thorvaldsdottir H, Winckler W, Guttman M, Lander ES, Getz G et al (2011) Integrative genomics viewer. Nat Biotechnol 29:24–26
5. Chaya D, Hayamizu T, Bustin M, Zaret KS (2001) Transcription factor FoxA (HNF3) on a nucleosome at an enhancer complex in liver chromatin. J Biol Chem 276:44385–44389
6. Metivier R, Penot G, Hübner MR, Reid G, Brand H, Kos M et al (2003) Estrogen receptor-alpha directs ordered, cyclical, and combinatorial recruitment of cofactors on a natural target promoter. Cell 115:751–763
7. Petruk S, Sedkov Y, Johnston DM, Hodgson JW, Black KL, Kovermann SK et al (2012) TrxG and PcG proteins but not methylated histones remain associated with DNA through replication. Cell 150:922–933
8. Sirbu BM, Couch FB, Feigerle JT, Bhaskara S, Hiebert SW, Cortez D (2011) Analysis of protein dynamics at active, stalled, and collapsed replication forks. Genes Dev 25:1320–1327
9. Sirbu BM, Couch FB, Cortez D (2012) Monitoring the spatiotemporal dynamics of proteins at replication forks and in assembled chromatin using isolation of proteins on nascent DNA. Nat Protoc 7:594–605
10. Lopez-Contreras AJ, Ruppen I, Nieto-Soler M, Murga M, Rodriguez-Acebes S, Remeseiro S et al (2013) A proteomic characterization of factors enriched at nascent DNA molecules. Cell Rep 3:1105–1116
11. Aranda S, Rutishauser D, Ernfors P (2014) Identification of a large protein network involved in epigenetic transmission in replicating DNA of embryonic stem cells. Nucleic Acids Res 42:6972–6986
12. Li L, Lyu X, Hou C, Takenaka N, Nguyen HQ, Ong CT et al (2015) Widespread rearrangement of 3D chromatin organization underlies Polycomb-mediated stress-induced silencing. Mol Cell 58:216–231
13. Gertz J, Savic D, Varley KE, Partridge EC, Safi A, Jain P et al (2013) Distinct properties of cell-type-specific and shared transcription factor binding sites. Mol Cell 52:25–36
14. Lee TI, Johnstone SE, Young RA (2006) Chromatin immunoprecipitation and microarray-based analysis of protein location. Nat Protoc 1:729–748
15. Bowman SK, Simon MD, Deaton AM, Tolstorukov M, Borowsky ML, Kingston RE (2013) Multiplexed Illumina sequencing libraries from picogram quantities of DNA. BMC Genomics 14:466. https://doi.org/10.1186/1471-2164-14-466

Chapter 14

Analysis of Chromatin Interactions Mediated by Specific Architectural Proteins in Drosophila Cells

Masami Ando-Kuri, I. Sarahi M. Rivera, M. Jordan Rowley, and Victor G. Corces

Abstract

Chromosome conformation capture assays have been established, modified, and enhanced for over a decade with the purpose of studying nuclear organization. A recently published method uses in situ Hi-C followed by chromatin immunoprecipitation (HiChIP) to enrich the overall yield of significant genome-wide interactions mediated by a specific protein. Here we applied a modified version of the HiChIP protocol to retrieve the significant contacts mediated by architectural protein CP190 in *D. melanogaster* cells.

Key words Chromatin immunoprecipitation, Chromatin architecture, In situ Hi-C, HiChIP, Epigenetics, Transcription

1 Introduction

Genome-wide architectural landscapes can be generated using Hi-C [1]; this approach shares template creation with its chromosome conformation capture assay predecessors. The major difference is the use of biotinylated nucleotides to fill in the overhangs of restriction fragments, which are then ligated and pulled down to retrieve all the interactions across the genome. The ligation step is performed after nuclear lysis under diluted conditions, and this has been shown to abate the number of meaningful ligation junctions [2]. A variant of the regular Hi-C method is in situ Hi-C, which has the advantage of reducing the number of false contacts arising from ligation proximity [3]. Several variations of the Hi-C technology have been developed to map interacting regions associated with a protein of interest with high resolution. These techniques include Chromatin Interaction Analysis with Paired-End Tag (ChIA-PET) [4] and, more recently, HiChIP [5]. HiChIP combines in situ Hi-C

Masami Ando-Kuri and I. Sarahi M. Rivera contributed equally to this work.

Tanya Vavouri and Miguel A. Peinado (eds.), *CpG Islands: Methods and Protocols*, Methods in Molecular Biology, vol. 1766, https://doi.org/10.1007/978-1-4939-7768-0_14, © Springer Science+Business Media, LLC, part of Springer Nature 2018

(cross-linking, digestion, ligation, and sonication) and chromatin immunoprecipitation with antibodies against a protein of interest to enrich for interactions mediated by this protein. Here we have adapted the HiChIP method to characterize interactions mediated by the architectural protein CP190 in *D. melanogaster* Kc167 cells.

The protocol consists of fixation of DNA–DNA–protein interactions, lysis and digestion of the DNA, biotinylation, and proximity ligation of the restriction fragments, followed by DNA shearing and ChIP. Afterward, the cross-links between DNA and proteins are reversed, DNA is precipitated with ethanol and the ligation junctions containing biotin are pulled down with streptavidin beads. Once the DNA is on the streptavidin beads, the ends of sheared DNA are repaired and biotin is removed from unligated ends. Final steps involve the preparation of an Illumina sequencing library with adaptors for paired-end sequencing and size selection to sequence the DNA of interest. Subsequent computational analysis and visualization of the data was done with Juicer and Juicebox [6, 7]. Significant DNA looping interactions were detected using Capture Hi-C Analysis of Genomic Organization (CHiCAGO) [8]. CHiCAGO is a tool developed to erase bias associated to Capture Hi-C experiments. However, both Capture Hi-C and HiChIP rely on the enrichment of interactions linked to specific regions of the genome. Therefore, the data from these two techniques share similar properties, making CHiCAGO a suitable tool for HiChIP data analysis. CHiCAGO performs larger numbers of tests at regions with smaller number of expected interactions and a background correction with a model that accounts for expected interactions and sequencing artifacts to detect significant interactions [8]. Specific software to account for biases in HiChIP data are not yet available and further development of computational tools remains a challenge in the field [4, 5].

2 Materials

Deionized water is the solvent for all solutions unless otherwise noted. Prepare and store all reagents at room temperature unless indicated otherwise.

2.1 Cross-Linking

1. *D. melanogaster* Kc167 cells.
2. SFX-Insect cell culture media (HyClone).
3. Formaldehyde (37%). Store in dark conditions.
4. 2.5 M glycine solution.

2.2 Lysis and Restriction Digest

1. Hi-C lysis buffer: 10 mM Tris–HCl pH 8.0, 10 mM NaCl, 0.2% Igepal CA-630 (NP-40). Add 500 μL of 1 M Tris–HCl pH 8.0 stock solution, 100 μL of 5 M NaCl stock solution and

500 μL of 20% Igepal CA-630 (NP-40) stock solution in a 50 mL tube. Add water to a volume of 50 mL. Store at 4 °C.

2. 25× Protease Inhibitor stock. Store at 4 °C.
3. 0.5% SDS solution.
4. 10% Triton X-100 Stock solution.
5. 10× DpnII Reaction Buffer.
6. 1× DpnII restriction enzyme (10,000 U/mL).

2.3 End-Repair and Biotinylation

1. Fill-in master mix: Add per tube 22.5 μL of water, 15 μL of 1 mM biotin-16-dCTP, 1.5 μL of 10 mM dTTP, 1.5 μL of 10 mM dATP, 1.5 μL of 10 mM dGTP, and 8 μL of 5 U/μL DNA polymerase I Large (Klenow) fragment.
2. Ligation Master mix: Add per tube 663 μL of water, 120 μL of 10× NEB T4 DNA Ligase buffer, 100 μL of 10% Triton X-100, 12 μL of 10 mg/mL BSA, and 5 μL 400 U/μL T4 DNA Ligase.

2.4 Bead Preparation

1. Magnetic beads with recombinant Protein A and Protein G.
2. Blocking buffer: 0.5% BSA/PBS. Store at 4 °C.
3. Pre-immune rabbit serum or immunoglobulin.
4. Anti-CP190 antibody.
5. IP Dilution Buffer: 0.01% SDS, 1.1% Triton X-100, 1.2 mM EDTA pH 8.0, 16.7 mM Tris–HCl pH 8.0, 16.7 mM NaCl. Add 50 μL of 10% SDS stock solution, 5.5 mL of 10% Triton X-100 stock solution, 120 μL of 0.5 M EDTA pH 8.0 stock solution, 835 μL of 1 M Tris–HCl pH 8.0 stock solution, and 1.67 μL of 1 M NaCl stock solution in a 50 mL tube. Add water to a volume of 50 mL. Store at 4 °C.

2.5 DNA Shearing

1. Nuclei Lysis Buffer: 50 mM Tris–HCl pH 8, 10 mM EDTA pH 8.0, 1% SDS. Add 2.5 mL of 1 M Tris–HCl pH 8.0 stock solution, 1 mL of 0.5 M EDTA pH 8.0 stock solution, and 5 mL of 10% SDS stock solution in a 50 mL tube. Add water to a volume of 50 mL. Store at 4 °C.
2. 25× Protease Inhibitor stock. Store at 4 °C.
3. RNase A 10 mg/mL aliquot. Store at −20 °C.
4. Proteinase K (20 mg/mL). Store at −20 °C.
5. Phenol–chloroform–isoamyl alcohol 25:24:1 Saturated with 10 mM Tris–HCl pH 8.0, 1 mM EDTA pH 8.0. Use only in the laminar flow hood. Store at 4 °C.
6. Glycogen aliquot. Store at −20 °C.
7. 3 M sodium acetate solution.
8. 100% ethanol solution. Store at 4 °C.

9. 70% ethanol solution. Store at 4 °C.
10. 2% agarose gel. Weigh 1 g of agarose for 50 mL of 1× TAE buffer, dissolve until solution is translucent for 3–5 min in microwave.
11. 1× TAE Buffer: 40 mM Tris, 20 mM acetic acid, 1 mM EDTA pH 8.0.
12. 6× gel loading dye, no SDS.
13. 100 bp DNA ladder.
14. Sonicator.

2.6 Chromatin Immunoprecipitation

1. IP dilution buffer (previously described in Subheading 2.4).
2. IP Wash Buffer Low salt: 0.1% SDS, 1% Triton X-100, 2 mM EDTA pH 8.0, 20 mM Tris–HCl pH 8.0, 150 mM NaCl. Add 500 μL of 10% SDS stock solution, 5 mL of 10% Triton X-100 stock solution, 200 μL of 0.5 M EDTA pH 8.0 stock solution, 1 mL of 1 M Tris–HCl pH 8.0 stock solution, and 1.5 mL of 5 M NaCl stock solution in a 50 mL tube. Add water to a volume of 50 mL.
3. IP Wash Buffer High salt: 0.1% SDS, 1% Triton X-100, 2 mM EDTA pH 8, 20 mM, Tris–HCl pH 8.0, 500 mM NaCl. Add 500 μL of 10% SDS stock solution, 5 mL of 10% Triton X-100 stock solution, 200 μL of 0.5 M EDTA pH 8.0 stock solution, 1 mL of 1 M Tris–HCl pH 8.0 stock solution, and 5 mL of 5 M NaCl stock solution in a 50 mL tube. Add water to a volume of 50 mL.
4. LiCl Buffer: 10 mM Tris–HCl pH 8.0, 1 mM EDTA pH 8.0, 0.25 M LiCl, 1% Igepal CA-630 (NP-40), 1% DOC. Add 500 μL of 1 M Tris–HCl pH 8.0 stock solution, 100 μL of 0.5 M EDTA pH 8.0 stock solution, 3.125 mL of 4 M LiCl stock solution, 2.5 mL of 20% Igepal CA-630 (NP-40),, and 5 mL of 10% DOC stock solution in a 50 mL tube. Add water to a volume of 50 mL.
5. IP Elution Buffer: 0.1 M $NaHCO_3$, 1% SDS. Prepare at the time of use. Add 100 μL of 1 M $NaHCO_3$, and 100 μL of 10% SDS. Add water to a total volume of 1.5 mL.
6. 5 M NaCl solution.
7. 0.5 M EDTA pH 8.0 stock solution.
8. Elution Buffer: 10 mM Tris–HCl, pH 8.5.
9. Proteinase K (20 mg/mL). Store at −20 °C.

2.7 Biotin/Streptavidin Pull-Down and Preparation for Sequencing

1. Low-binding 1.5 mL tubes.
2. Magnetic beads with recombinant streptavidin. Store at 4 °C.

3. Tween Wash Buffer (TWB): 5 mM Tris–HCl pH 7.5, 0.5 mM EDTA pH 8.0, 1 M NaCl, 0.05% Tween 20. Add 250 μL of 1 M Tris–HCl pH 7.5 stock solution, 50 μL of 0.5 M EDTA pH 8.0 stock solution, 10 mL of 5 M NaCl stock solution, 250 μL of 10% Tween 20 stock solution in a 50 mL tube. Add water to a volume of 50 mL.
4. Binding buffer (2×): 10 mM Tris–HCl pH 7.5, 1 mM EDTA pH 8.0, 2 M NaCl. Add 150 μL of 1 M Tris–HCl pH 7.5 stock solution, 30 μL of 0.5 M EDTA pH 8.0 stock solution, 6 mL of 5 M NaCl stock solution in a 15 mL tube. Add water to a volume of 15 mL.

2.8 Preparation of Illumina Sequencing Libraries

1. End-repair Master Mix: Add per tube 25 nM dNTPmix, 10 U/μL T4 Polynucleotide Kinase, 3 U/μL T4 DNA polymerase I, 5 U/μL DNA polymerase I large (Klenow) fragment.
2. 1× NEB buffer 2. Store at −20 °C.
3. 10 nM dATP aliquot. Store at −20 °C.
4. Klenow Fragment (3′ → 5′ exo-) DNA Polymerase I (5000 U/mL). Store at −20 °C.
5. 1× Quick ligase reaction buffer: Original stock is 2× concentration, dilute to reach 1× concentration.
6. Quick ligase enzyme.
7. 2× KAPA SYBR FAST qPCR Kit aliquot. Store at −20 °C.
8. Refer to for sequence information on adaptors and PCR primers.
9. Agencourt AMPure XP beads. Store at −20 °C.
10. 80% ethanol solution. Must be made before usage.

3 Methods

3.1 Cross-Linking

1. Kc167 cells were grown to 80% confluence in SFX medium at 25 °C. Use ~100 × 10^6 as total number of cells.
2. In a 15 mL tube, pellet cells at 600 × *g* for 10 min at room temperature (22–25 °C). Discard supernatant.
3. Add 10 mL of SFX medium and pipette to resuspend cells.
4. Divide into two 15 mL tubes, 5 mL in each one.
5. Add formaldehyde to a final concentration of 1% (*see* **Note 1**).
6. Incubate for 7 min at room temperature on a rocker.
7. Add glycine to a final concentration of 0.2 M.
8. Incubate for 5 min at room temperature on a rocker.
9. Pellet cells at 300 × *g* for 5 min at 4 °C.
10. Discard supernatant into an appropriate collection container (*see* **Note 2**).

3.2 Lysis and Restriction Digest

1. Add 500 μL of ice-cold Hi-C lysis buffer. Mix by pipetting and move to a 2 mL tube.
2. Add 20 μL of 25× Protease Inhibitor (PI).
3. Incubate on ice for 1 h.
4. Pellet cells at 2500 × g for 5 min at 4 °C and discard supernatant.
5. Add 100 μL of 0.5% SDS and pipette to resuspend cells.
6. Incubate for 5 min at 65 °C.
7. Add 290 μL of water and 50 μL of 10% Triton X-100.
8. Incubate for 15 min at 37 °C.
9. Add 50 μL of 10× DpnII buffer and 200 U of DpnII.
10. Digest overnight at 37 °C.

3.3 End Repair and Biotinylation

1. Incubate digest reaction at 65 °C for 20 min to inactivate DpnII.
2. Divide each reaction between two tubes of 2 mL.
3. Cool to room temperature.
4. Add 50 μL of the Fill-in Master Mix.
5. Mix by pipetting and incubate at 37 °C for 1.5 h.
6. Add 900 μL of Ligation Master Mix to each pellet.
7. Mix by inverting and incubate for 4 h at room temperature with gentle rocking.

3.4 Bead Preparation

Prepare Pre-clear and Antibody beads during the 1.5 h biotinylation incubation.

1. Pre-Clear beads: Add 10 μL of protein A beads and 10 μL of protein G beads in a 2 mL clean tube.

   Antibody beads: Add 20 μL of protein A beads and 20 μL of protein G beads in a 2 mL clean tube.
2. Wash step:
   (a) Add 1 mL of Blocking Buffer to each tube.
   (b) Rotate for 5 min at room temperature.
   (c) Collect on magnet stand.
   (d) Remove supernatant.
3. Repeat wash step two more times.
4. Add 500 μL of Blocking Buffer to each tube.
5. Pre-Clear beads: Add 10 μL of pre-immune rabbit serum (or IgG).

   Antibody beads: The amount of antibody depends on its efficiency.

6. Incubate the beads at 4 °C on rotator for at least 4 h (*see* **Note 3**).
7. Collect on magnet stand and remove supernatant.
8. Wash with Blocking Buffer:
   (a) Add 1 mL of Blocking Buffer to each tube.
   (b) Rotate for 2 min at room temperature.
   (c) Collect on magnet stand.
   (d) Remove supernatant.
9. Wash with IP Dilution Buffer.
   (a) Add 1 mL of ice-cold IP Dilution Buffer.
   (b) Rotate for 2 min at room temperature.
   (c) Collect on magnet stand.
   (d) Remove supernatant.
10. Repeat wash with IP Dilution Buffer.
11. Add 300 μL of ice-cold IP Dilution Buffer.

    Beads are ready for steps in Subheading 3.6.

3.5 DNA Shearing

1. Pellet nuclei from ligation reaction at 1503 × *g* for 8 min at 4 °C and discard supernatant.
2. Add 200 μL of ice-cold Nuclei Lysis Buffer. Mix by pipetting and move to a 1.5 mL tube.
3. Add 8 μL of 25× PI.
4. Incubate in ice for 20 min.
5. Add 100 μL of ice-cold IP Dilution Buffer and 4 μL of 25× PI.
6. Shear DNA to obtain DNA fragments between 250 and 350 bp (*see* **Notes 4** and **5**).
7. Pellet cell debris at 21,130 × *g* for 10 min at 4 °C.
8. Transfer supernatant (chromatin) of each 1.5 mL tube (4) to a 1.5 mL clean tube.
9. Verify sonication quality (*see* **Note 6**):
   (a) Add 15 μL of postshear supernatant in a 2 mL clean tube.
   (b) Add 185 μL of IP Dilution Buffer.
   (c) Add 0.5 μL of 10 mg/mL RNaseA.
   (d) Reverse cross-link by heating to 95 °C for 5 min.
   (e) Add 4 μL of 10 mg/mL proteinase K.
   (f) Incubate at 50 °C for 2 h.
   (g) Extract DNA with phenol–chloroform–isoamyl alcohol (P:C:I, 25:24:1).
      - Add 1 volume of P:C:I (~205 μL) and vortex well.
      - Spin at 21,130 × *g* for 5 min at 4 °C.
      - Transfer aqueous (top) layer to a 2 mL clean tube.

(h) Add 1 μL of 20 mg/mL glycogen.
(i) Add 1/10 volume of 3 M sodium acetate.
(j) Add 2.5 volume of ice-cold 100% ethanol.
(k) Incubate for 30 min at −80 °C.
(l) Pellet DNA at 21130 × *g* for 10 min at 4 °C and discard supernatant.
(m) Add 1 mL of 70% ethanol.
(n) Pellet DNA at 21130 × *g* for 10 min at 4 °C and discard supernatant.
(o) Let pellet air-dry (5–10 min).
(p) Add 10 μL of water and resuspend the pellet.
(q) Run on a 2% agarose gel. Target is 250–300 bp.

3.6 Chromatin Immunoprecipitation

1. Pool samples back together in a 15 mL tube.
2. Dilute chromatin fivefold with ice-cold IP Dilution Buffer.
3. Add pre-clear beads to sample.
4. Incubate for 1 h at 4 °C with rotation.
5. Take 75 μL out to use as "input" reference. Can store at 4 °C overnight.
6. Place the remaining solution on a magnet stand.
7. Move supernatant to tube with antibody coated beads.
8. Incubate at 4 °C overnight with rotation.
9. Place on magnet stand and remove supernatant.
10. Add 1 mL of Low Salt Wash Buffer. Mix by pipetting and move onto a 1.5 mL tube.
11. Incubate for 5 min at room temperature with rotation.
12. Washes: Place on magnet stand and remove supernatant; add 1 mL of Wash Buffer; incubate for 5 min at room temperature with rotation.
    (a) Wash with Low Salt Wash Buffer two more times.
    (b) Wash with High Salt Wash Buffer two times.
    (c) Wash with LiCl Buffer two times.
    (d) Wash with TE buffer two times.
13. Place on magnet stand and remove supernatant.
14. Elution step:
    (a) Add 150 μL of fresh IP elution buffer.
    (b) Incubate for 10 min at room temperature.
    (c) Incubate for 5 min at 37 °C.
    (d) Place on magnet stand.

15. Move supernatant to a clean tube.
16. Repeat Elution step with the beads one more time.
17. Add supernatant to the same tube for a total of 300 μL of ChIP eluate.
18. Add 20 μL of 5 M NaCl, 8 μL of 0.5 M EDTA pH 8.0 and 16 μL of 1 M Tris–HCl pH 8.0.
19. Incubate for 1.5 h at 68 °C to reverse cross-link.
20. Add 8 μL of proteinase K.
21. Incubate at 50 °C for 2 h.
22. Cool to room temperature.
23. Precipitate the DNA:
    (a) Add 1/10 volume of 3 M sodium acetate.
    (b) Add 2 volumes of ice-cold 100% ethanol.
    (c) Incubate for 30 min at −80 °C. Can go overnight.
24. Pellet DNA at 21,130 × *g* for 10 min at 4 °C and discard supernatant.
25. Add 1 mL of 70% ethanol.
26. Pellet DNA at 21,130 × *g* for 10 min at 4 °C and discard supernatant.
27. Let pellet air-dry (5–10 min).
28. Add 20 μL of Elution Buffer and resuspend the pellet.
29. Incubate at 37 °C for 5 min to make sure pellet is properly dissolved.
30. Quantify the amount of DNA on a spectrophotometer.
31. Bring volume up to 300 μL with Elution Buffer.

3.7 Biotin/Streptavidin Pull-Down and Preparation for Sequencing

Perform remaining steps in 1.5 mL low-binding tubes.

1. Add 400 μL of Tween Wash Buffer (TWB) to a clean tube.
2. Add 20 μL of MyOne Streptavidin T1 beads for every 5 μg of DNA.
3. Mix by pipetting.
4. Separate on magnet stand and discard supernatant.
5. Add 300 μL of 2× Binding Buffer to resuspend the beads and add to the eluate from **step 31** in Subheading 3.6.
6. Incubate at room temperature for 15 min with rotation.
7. Separate on magnet stand and discard supernatant.
8. Wash in TWB:
    (a) Add 600 μL of TWB.
    (b) Rotate for 2 min.
    (c) Separate on magnet stand and discard supernatant.
9. Repeat wash in TWB one more time.

3.8 Preparation of Illumina Sequencing Libraries

1. Add 100 μL of 1× T4 ligase buffer and resuspend beads (*see* **Note 5**).
2. Move to a clean tube. Separate on magnet stand and discard supernatant.
3. Resuspend beads in 100 μL of End-repair Master Mix.
4. Pipette gently to mix.
5. Incubate for 30 min at room temperature.
6. Wash beads two times with TWB as before.
7. Add 100 μL of 1× NEB Buffer 2 and resuspend beads.
8. Move to a clean tube. Separate on magnet stand and discard supernatant.
9. Add 100 μL of dATP Master Mix and resuspend by pipetting.
10. Incubate for 30 min at 37 °C.
11. Wash beads two times with TWB as before.
12. Add 100 μL of 1× Quick ligase buffer and resuspend beads.
13. Move to a clean tube. Separate on magnet stand and discard supernatant.
14. Add 50 μL of 1× Quick Ligase Buffer and resuspend beads.
15. Add 2 μL of Quick Ligase and 2 μL of the Illumina indexed adapter at 10 mM.
16. Incubate at room temperature for 15 min. This incubation can go longer without adverse effects.
17. Wash beads two times with TWB as before.
18. Add 100 μL of Elution Buffer and resuspend beads.
19. Move to a clean tube (*see* **Note 7**). Separate on magnet stand and discard supernatant.
20. Add 23 μL of Elution Buffer and resuspend beads.

    Amplify the library directly on the beads.
21. Transfer the beads to a PCR tube.
22. Add 25 μL of Kapa SYBR FAST qPCR Kit Master Mix (2×), 2 μL of Illumina primers.
23. PCR parameters:
    (a) Initial Denaturation 3 min at 98 °C.
    (b) PCR parameters: Denaturation 15 s at 98 °C; Annealing 30 s at 60 °C; Extension 30 s at 72 °C.
    (c) Final Extension 30 s at 72 °C.
24. Run a total of six PCR cycles on a thermocycler.

25. Separate on a magnet stand and move reaction (supernatant) to a clean real-time PCR (qPCR) tube.
    (a) Add more SYBR Green to a final concentration of 1×.
26. qPCR parameters:
    (a) Initial Denaturation 30 s at 98 °C.
    (b) Cycle parameters: Denaturation 15 s at 98 °C; Annealing 30 s at 60 °C; Extension 30 s at 72 °C.
    (c) Final Extension 30 s at 72 °C.
27. Run 4–6 cycles while monitoring on qPCR (*see* **Note 8**).
28. Move liquid from qPCR tube to a 2 mL clean tube.
29. Add 0.7–0.9× volumes of AMPure XP beads.
30. Incubate for 5 min at room temperature on a rocker.
31. Separate on magnet stand and discard supernatant.
32. Without removing from magnetic stand, quickly add and discard 200 μL of 80% ethanol.
33. Repeat the last step.
34. Let pellet air-dry (5–10 min).
35. Add 15 μL of Elution Buffer and resuspend the DNA.
36. Incubate for 5 min at room temperature.
37. Separate on magnet stand and move the supernatant to a clean tube.
38. Quantify the amount of DNA on a spectrophotometer.
39. Send the library to sequencing by service provider for quality assessment and high throughput sequencing.

3.9 Data Analysis

Most of the pipelines were ran through the terminal in a Linux-based system.

1. Set up
    (a) Install Juicer software from http://aidenlab.org/juicer/.
    (b) Install Juicebox software from http://aidenlab.org/juicebox/.
    (c) Install R, download source file from https://cran.r-project.org/mirrors.html and choose the closest mirror to your location.
    (d) Download CHiCAGO from https://bitbucket.org/chicagoTeam/chicago/downloads.
2. Read mapping and read-pair level filtering: Juicer (*see* **Note 9**).
    (a) Create a directory called "fastq" and move sequence fastq files to this directory.

(b) Change the name of the files for each of the two mates obtained from paired-end sequencing. Each mate should have the name of the sample followed by either "R1" or "R2" to distinguish between the two mates. For example, sample_R1.fastq and sample_R2.fastq) (*see* **Note 10**).

(c) Go to the directory that contains the Juicer program "juicer.sh".

(d) Set the following parameters for Juicer and hit enter (*see* **Note 11**).

-g genome: e.g., dm6.

-d path for "fastq" directory of the sample: e.g., /home/user/exp/.

-s restriction enzyme used: e.g., DpnII.

-p path for chromosome sizes file. This file should have two columns, the first is the name of the chromosome and the second one the length of the chromosome.

-y enter path for restriction site file (locations of restriction sites in genome). This contains each chromosome and the positions of the cutting sites by the previously designated restriction enzyme in a row separated by a space, the last position of each row should be the end of that chromosome.

-z enter path for sequence of the reference genome file in fasta format; the BWA index file must be in the same directory.

We use the following command:

$juicer.sh -g dm6 -d /home/user/exp/ -s DpnII -p /home/user/exp/chromosome-sizes.txt -y /home/user/exp/restriction-sites.txt -z /home/user/exp/reference-sequence.

(e) To verify the success of the alignment, consult that the "aligned" directory from the Juicer output contains the "inter.hic" file (*see* **Note 12**).

3. Normalization to account for experimental bias and extracting significant contacts: Juicebox and CHiCAGO.

(a) Use the "juicebox dump" tool to retrieve a contact map. Set the following parameters and hit enter (*see* **Notes 13** and **14**).

(i) Type of normalization: NONE (*see* **Note 15**).

(ii) Path of the .hic file: e.g., sample/aligned/inter.hic.

(iii) Name of the first chromosome: either 2L, 2R, 3L, 3R, 4, X, or Y.

(iv) Name of the second chromosome: either 2L, 2R, 3L, 3R, 4, X, or Y.

(v) Resolution desired: e.g., BP 5000 (*see* **Note 16**).

(vi) Name of output file: e.g., sample_firstchr_secondchr_obst.txt.

(b) Create design files for CHiCAGO using dumped files from Juicer (*see* **Note 17**).

(i) Create a directory called "designDir".

(ii) Create the .rmap file. This is a BED file with the positions of the restriction fragments with the columns chr, start, end, and fragmentID (*see* **Note 18**).

(iii) Create the .baitmap file. This is also a BED file containing the restriction fragments with a bait, defined as the protein that is immunoprecipitated. It contains five columns: chr, start, end, fragmentID, baitAnnotation (*see* **Note 19**).

(iv) Run Chicago/chicagoTools/makeDesignFiles.py to create .nbpb .poe and .npb files. Set the following parameters and hit enter (*see* **Note 20**).

--designDir = path to the "designDir" directory.

--rmapfile = path to the .rmap file.

--baitmapfile = path to the .baitmap file.

(c) Create .chinput file based on the juicebox dumped files. This 5-column file contains the contact frequency. The format is: baitID, otherendID, N, otherEndLen, and distSign; where otherendID is the ID of the fragment or bin that is in the interacting pair of the bait, N is the number of reads detected for ligation products between the "bait" and "other end", otherEndLen is the length of the "other-end" restriction fragment and distSign is the linear distance between the bait and other-end fragments, respectively [8] (*see* **Note 21**).

(d) Run CHiCAGO following the commands in http://regulatorygenomicsgroup.org/wp-content/uploads/Chicago_vignette.html.

4. Visualization: Juicebox.

(a) Raw data (*see* **Note 22**).

(i) The input file for visualization in Juicebox is inter.hic located inside the "aligned" directory.

(ii) Open Juicebox.

(iii) Click on File > Open > Local and then choose the .hic file.

(b) CHiCAGO's significant interactions: There are two ways to visualize CHiCAGO interactions. To visualize the data as a .hic matrix in juicebox go to **step 4(b)(iii)**, For an annotation track go to **step 4(b)(vii)**.

(i) CHiCAGO's output in the designated directory contains the following directories: "data", "diag_plots", "enrichment_data examples".

(ii) Enter "data" directory and find CP190_CHICAGO_washU_text.txt.

(iii) Create .hic file using "juicebox pre" tool from CP190_CHICAGO_washU_text.txt file. to create a proper "juicebox pre" input file. It should be an 11-column file with the following format: readname, str1, chr1, pos1, frag1, str2, chr2, pos2, frag2, mapq1, mapq2; where str can be set to 0 for forward, anything else for reverse [7] (*see* **Note 23**) (e.g., 1 0 2L 73000 74000 0 2L 84000 85000 50 50).

(iv) Set the parameters of "juicebox pre" tool and hit enter (*see* **Note 24**).

- Pairwise Interaction BED file: CP190_CHICAGO_washU_text_5.txt.
- Output name: cp190_chicago.hic.
- Chromosome size file: path for chrom.sizes file. This file should have two columns, the first one is the name of the chromosome and the second one is the length of the chromosome.
- Restriction map of *D. melanogaster* genome: enter path for restriction site file (locations of restriction sites in genome). This contains each chromosome and its restriction positions by the previously designated enzyme in a row separated by a space.

(v) Open Juicebox.

(vi) Click on File > Open > Local and then choose the CP190_chicago.hic file.

(vii) Modify CP190_CHICAGO_washU_text.txt file to create a proper 2D annotation input file. It should be a 4-column file with the following format: readname, chr1, pos1, pos2, chr2, pos3, pos4, score (e.g., 2L 73000 74000 2L 2569000 2570000 35.6913535863312).

(viii) Open Juicebox.

(ix) Open .hic file of the raw data. Click on File > Open > Local and then choose the CP190_raw.hic file.

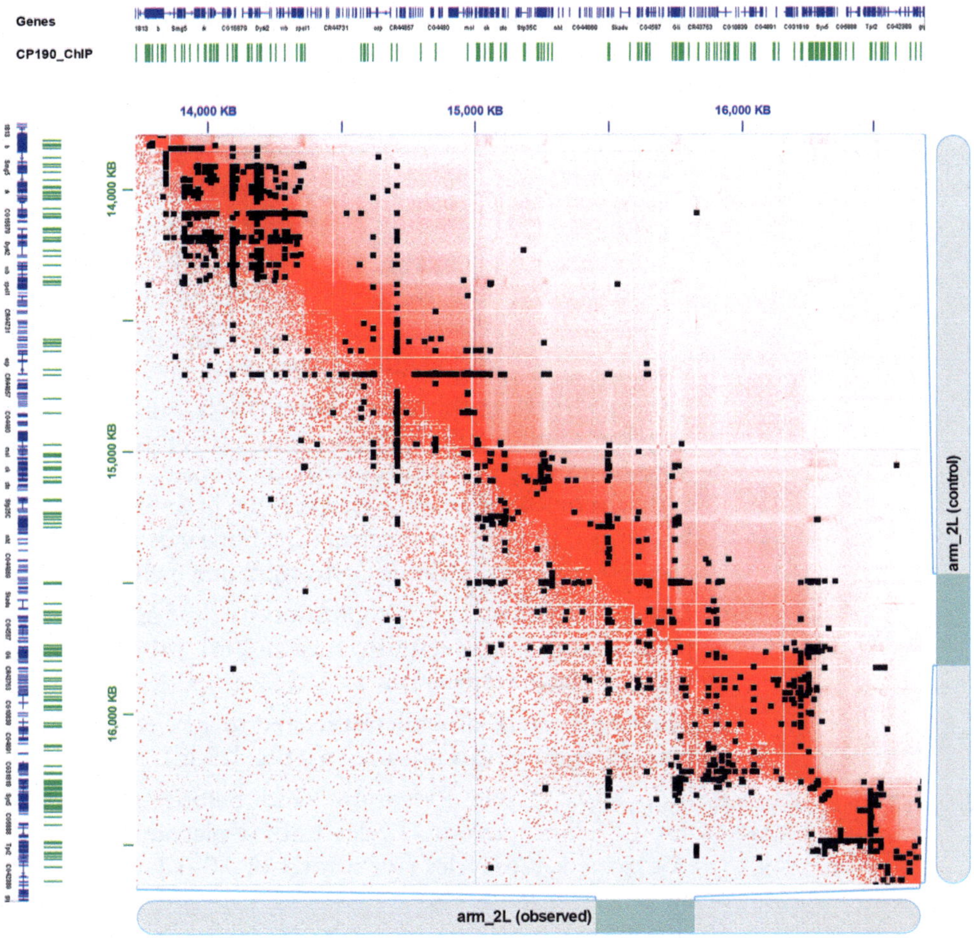

Fig. 1 Significant interactions associated with CP190 retrieved by CHiCAGO. The upper and the lower triangle represent Hi-C and HiChIP raw data respectively with a 5 kb bin resolution. The black dots depict significant interactions obtained from CHiCAGO. The axes show the annotated genes for *D. melanogaster* and CP190 ChIP-seq peaks in this region. Figure was created using Juicebox v1.5

(x) Load the significant interactions called by CHiCAGO using the modified version of the CP190_CHICAGO_washU_text.txt file (Fig. 1). Click on Annotations > Load Basic Annotations > Dataset-specific 2D-features > Add 2D and select local file.

4 Notes

1. Handle in a fume hood. Commercial formaldehyde can be used up to 3 months from the date the bottle was opened.

2. After freezing in liquid nitrogen either proceed to "Lysis and Digest" or store cell pellet at −80 °C and flash freeze in liquid nitrogen.
3. Incubate "Pre-Clear beads" for 15 min before the end of DNA shearing. Incubate "Antibody beads" for 15 min before the end of "Chromatin and Pre-Clear incubation" in Subheading 3.6.
4. We perform DNA shearing in a Diagenode Bioruptor 300 with the following parameters: Intensity: High; ON: 30 s; OFF: 60 s; Cycles: 28–30; Temperature: 4 °C water bath.
5. Tn5 transposase can be used to construct the library [5]. The shearing step must be modified to obtain larger fragments (400–1000 bp), otherwise tagmentation will decrease size of fragments limiting the amplification of the library (described in Subheading 3.8 of this protocol). End-repair, addition of dATP and ligation of the adaptor (**steps 1–16** in Subheading 3.8) will be skipped in order to construct the library using Tn5 transposase. Next steps should be performed normally.
6. Either proceed to verify sonication quality or store 15 μL of postshearing supernatant from each 2 mL tube at −4 °C overnight for later verification.
7. The beads can be directly transferred to a PCR tube. For easier manipulation in this protocol we use 1.5 mL tubes.
8. Stop after "Cycle Extension" and before "Cycle Denaturation." Continue with cycles until a plateau is reached. Do not exceed 4–6 cycles (for a total of 10–12 PCR plus qPCR cycles).
9. Paired-end sequencing is performed on the sample; thus, the sequencing data is obtained in two fastq files.
10. **Steps 1** and **2** can be done in any order. Also, it is possible to compress the fastq files, Juicer can work with both, compressed and uncompressed fastq files.
11. Use the flags described below followed by the option chosen. Additional options are available, use. ./juicer.sh -h to see additional information. The output from Juicer will contain the following directories: "aligned", "errors", "fastq", and "splits".)
12. For deeper sequencing it is advisable to check the duplication rate and determine the number of usable reads. This information can be consulted in the inter.txt file located inside of the directory called "aligned" in the output files. We recommend verifying that the "PCR Duplicates" is below 7% and that the "Hi-C Contacts" is above 40%.
13. Each parameter should be separated by a space; no use of flags is required.

14. The Juicebox dumped file is done per chromosome.
15. As we will normalize the data with CHiCAGO, retrieve raw data. Vanilla Coverage (VC), square root of vanilla coverage (VC_SQRT) and Knight–Ruiz/Balanced normalization (KR) are available normalization options, but are not recommended to be used with this method.
16. To choose the resolution of the contact map you can select the size of the bin determining the base pair resolution with "BP" and designate it from these options: 2,500,000, 1,000,000, 500,000, 250,000, 100,000, 50,000, 25,000, 10,000, 5000; or you can specify the number of fragments by using "FRAG" followed by any of these options: 500, 200, 100, 50, 20, 5, 2, 1.
17. Prior to proceeding, it is highly recommended to read the Vignette explaining the files required to run CHiCAGO: http://regulatorygenomicsgroup.org/wp-content/uploads/Chicago_vignette.html.
18. We created a 1 kb fragment resolution file and assigned sequential numbers as fragmentIDs (e.g., 2R 1 1000 1).
19. To obtain the baits, we overlapped the positions from a CP190 ChIP-seq to the .rmap previously described. The baitAnnotation was assigned as the protein name with sequential numbers (e.g., 2L 65000 66000 66 CP190_1).
20. Use the flags described below followed by the option chosen. Make sure .nbpb .poe and .npb are inside the "designDir" directory. To see additional parameters, use: python makeDesignFiles.py.
21. To create the .chinput file, one must parse dumped files from Juicebox and use a script to generate it. We used a python script for this. The "otherEndLen" column is always 1000 because we set 1 kb resolution for the restriction fragment file. Additionally, CHiCAGO is capable of using several .chinput files as replicates. However, only one replicate for CP190 was processed in this analysis.
22. Juicebox is capable of doing the normalization described in Lieberman-Aiden et al. [1] and KR matrix balancing [9], as well as other statistics. However, the normalization methods implemented do not account for HiCHiP specific biases. For this reason, Juicebox is only used to visualize the data.
23. We did not filter by MAPQ; 50 was used as a default number for mapq1, mapq2. Bash pipelines were used to generate this file.
24. Separate each parameter by one space only; no use of flags required.

Acknowledgments

This work was supported by US Public Health Service Award R01 GM035463 from the National Institutes of Health. The content is solely the responsibility of the authors and does not necessarily represent the official views of the National Institutes of Health. Masami Ando-Kuri and I. Sarahi M. Rivera contributed equally to this work.

References

1. Lieberman-Aiden E, Van Berkum NL, Williams L, Imakaev M, Ragoczy T, Telling A et al (2009) Comprehensive mapping of long-range interactions reveals folding principles of the human genome. Science 326 (5950):289–293
2. Gavrilov AA, Golov AK, Razin SV (2013) Actual ligation frequencies in the chromosome conformation capture procedure. PLoS One 8(3): e60403
3. Rao SS, Huntley MH, Durand NC, Stamenova EK, Bochkov ID, Robinson JT et al (2014) A 3D map of the human genome at kilobase resolution reveals principles of chromatin looping. Cell 159(7):1665–1680
4. Li G, Cai L, Chang H, Hong P, Zhou Q, Kulakova EV et al (2014) Chromatin interaction analysis with paired-end tag (ChIA-PET) sequencing technology and application. BMC Genomics 15(Suppl 12):S11
5. Mumbach MR, Rubin AJ, Flynn RA, Dai C, Khavari PA, Greenleaf WJ, Chang HY (2016) HiChIP: efficient and sensitive analysis of protein-directed genome architecture. Nat Methods 13(11):919–992
6. Durand NC, Shamim MS, Machol I, Rao SS, Huntley MH, Lander ES, Aiden EL (2016) Juicer provides a one-click system for analyzing loop-resolution Hi-C experiments. Cell Syst 3 (1):95–98
7. Durand NC, Robinson JT, Shamim MS, Machol I, Mesirov JP, Lander ES, Aiden EL (2016) Juicebox provides a visualization system for Hi-C contact maps with unlimited zoom. Cell Syst 3(1):99–101
8. Cairns J, Freire-Pritchett P, Wingett SW, Dimond A, Plagnol V, Zerbino D et al (2015) CHiCAGO: robust detection of DNA looping interactions in capture Hi-C data. Genome Biol 17:127

9 Knight PA, Ruiz D (2012) A fast algorithm for matrix balancing. IMA J Numer Anal 33 (3):1029–1047

Check for updates

Chapter 15

High-Throughput Single-Cell RNA Sequencing and Data Analysis

Sagar, Josip Stefan Herman, John Andrew Pospisilik, and Dominic Grün

Abstract

Understanding biological systems at a single cell resolution may reveal several novel insights which remain masked by the conventional population-based techniques providing an average readout of the behavior of cells. Single-cell transcriptome sequencing holds the potential to identify novel cell types and characterize the cellular composition of any organ or tissue in health and disease. Here, we describe a customized high-throughput protocol for single-cell RNA-sequencing (scRNA-seq) combining flow cytometry and a nanoliter-scale robotic system. Since scRNA-seq requires amplification of a low amount of endogenous cellular RNA, leading to substantial technical noise in the dataset, downstream data filtering and analysis require special care. Therefore, we also briefly describe in-house state-of-the-art data analysis algorithms developed to identify cellular subpopulations including rare cell types as well as to derive lineage trees by ordering the identified subpopulations of cells along the inferred differentiation trajectories.

Key words Single cell RNA sequencing, High-throughput, Single cell data analysis, CEL-Seq2, Next-generation sequencing

1 Introduction

Single cell analysis empowers biologists to answer some of the fundamental long-standing questions in the field of developmental and stem cell biology, immunology, and cancer research. In the last few years, there has been a great interest as well as significant progress in the development and improvement of single cell technologies resulting in their widespread use in basic and clinical research. Single cell transcriptome sequencing is one of such techniques which has emerged as an important component of the experimental toolbox of many leading laboratories around the world.

Before the advent of scRNA-seq, protocols for single cell gene expression profiling by single molecule fluorescence in-situ hybridization (smFISH), quantitative PCR (qPCR), and microarrays had been developed [1–3]. However, only a limited number of

Tanya Vavouri and Miguel A. Peinado (eds.), *CpG Islands: Methods and Protocols*, Methods in Molecular Biology, vol. 1766, https://doi.org/10.1007/978-1-4939-7768-0_15,

genes can be analyzed using such techniques. scRNA-seq overcomes this inherent limitation by measuring the expression of thousands of genes simultaneously in single cells, thereby allowing the robust identification of various subpopulations of cells, including rare cell types [4]. Additionally, single cell transcriptomics allows us to understand differentiation dynamics during embryonic development, adult homeostasis, and diseased condition at an unprecedented resolution [5–8].

Several methods have been published for scRNA-seq experiments [9–17]. All these protocols involve amplification of subpicogram amounts of endogenous RNA present in single cells. Here we describe a miniaturized version of the CEL-Seq2 protocol [17] with several modifications (Fig. 1). Using a nanoliter scale pipetting robot, we are able to reduce the reaction volumes for cDNA synthesis by fivefold, thereby significantly reducing the costs. In addition, we describe our bioinformatic pipeline for scRNA-seq data analysis, which includes quality checks and alignment of the reads to the reference transcriptome, quantification of transcript abundance, filtering of low-quality cells, data normalization, identification of subpopulations of cells as well as rare cell types using the RaceID2 algorithm [8] and lineage inference using the StemID algorithm [8].

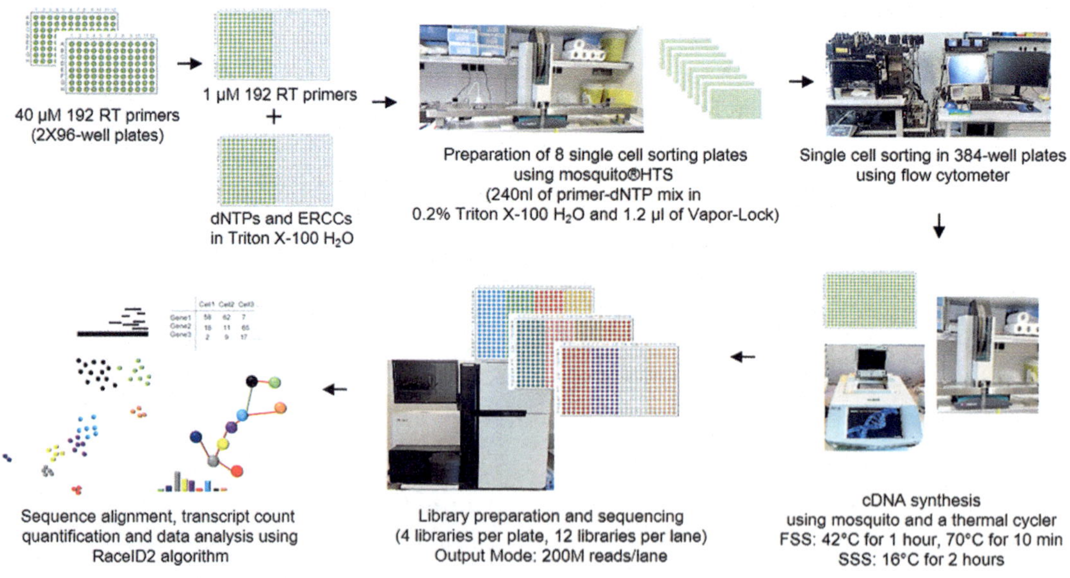
40 µM 192 RT primers
(2X96-well plates)
1 µM 192 RT primers
+
dNTPs and ERCCs
in Triton X-100 H2O
Preparation of 8 single cell sorting plates
using mosquito®HTS
(240nl of primer-dNTP mix in
0.2% Triton X-100 H2O and 1.2 µl of Vapor-Lock)
Single cell sorting in 384-well plates
using flow cytometer
cDNA synthesis
using mosquito and a thermal cycler
FSS: 42°C for 1 hour, 70°C for 10 min
SSS: 16°C for 2 hours
Library preparation and sequencing
(4 libraries per plate, 12 libraries per lane)
Output Mode: 200M reads/lane
Sequence alignment, transcript count
quantification and data analysis using
RaceID2 algorithm

Fig. 1 Schematic showing an overview of the various steps involved in the high-throughput single-cell RNA-sequencing protocol used in our laboratory

2 Materials

Keep the working area clean and RNase-free while making solutions and handling the reagents. Prepare all the solutions in RNase-free water. Use LoBind RNase-free filter tips to perform all the dilutions. Please note that suppliers are mentioned in parentheses after the name of the equipment and reagents. However, readers can use the equipment and reagents from any other suitable or local supplier in accordance with the manufacturer's guidelines.

2.1 Specialized Equipment and General Labware

1. mosquito®HTS (TTP Labtech).
2. Flow cytometer, e.g., BD Influx™, MoFlo XDP Cell Sorter, and BD FACSAria III.
3. Laboratory centrifuge with adaptors for well plates.
4. Minicentrifuge to spin tubes.
5. 384 and 96-well plates, LoBind, DNase/RNase-free.
6. Plate sealers, temperature range: −80–105 °C.
7. 384-well and 96-well thermal cycler or hybridization oven.
8. 96-well plate magnetic stand for bead cleanup.
9. Qubit® Fluorometer (Invitrogen).
10. Agilent 2100 Bioanalyzer.
11. DNA LoBind tubes, PCR Clean—0.5, 1.5, and 2 ml.
12. DNA Low Binding filter tips, DNase- and RNase-free.
13. RNase decontamination solution.
14. RNase-free water.
15. RNase-free ethanol, molecular biology grade.
16. Single-channel and multichannel pipettes.

2.2 Primers

All primers are synthesized by Integrated DNA Technologies (IDT Belgium) as ultramer DNA oligos with standard desalting. We store all the purchased primer plates and tubes at −80 °C. All dilutions are performed in RNase-free water.

1. 192 reverse transcription (RT) primers (sequences provided in Table 1): anchored polyT primers having a 6 bp cell barcode, 6 bp unique molecular identifiers (UMIs), a part of 5′ Illumina adapter and a T7 promoter: 40 μM

   192 RT primers are purchased in 96-well plate format (total 2 plates) at a concentration of 40 μM. We prepare several 384-well primer source plates with 192 RT primers at a final working concentration of 1 μM (total volume: 10 μl, columns: 1–12, 192 wells) from the purchased 40 μM primer plates. 1 μM primer plates are prepared by serially diluting the primers

Table 1
Sequences of 192 RT primers

1	GCCGGTAATACGACTCACTATAGGGAGTTCTACAGTCCGACGATCNNNNNNAGTGTCTTTTTTTTTTTTTTTTTTTTTTTTV
2	GCCGGTAATACGACTCACTATAGGGAGTTCTACAGTCCGACGATCNNNNNNACCATGTTTTTTTTTTTTTTTTTTTTTTTTV
3	GCCGGTAATACGACTCACTATAGGGAGTTCTACAGTCCGACGATCNNNNNNGAGTGATTTTTTTTTTTTTTTTTTTTTTTTV
4	GCCGGTAATACGACTCACTATAGGGAGTTCTACAGTCCGACGATCNNNNNNCACTCATTTTTTTTTTTTTTTTTTTTTTTTV
5	GCCGGTAATACGACTCACTATAGGGAGTTCTACAGTCCGACGATCNNNNNNCATGTCTTTTTTTTTTTTTTTTTTTTTTTTV
6	GCCGGTAATACGACTCACTATAGGGAGTTCTACAGTCCGACGATCNNNNNNACAGGATTTTTTTTTTTTTTTTTTTTTTTTV
7	GCCGGTAATACGACTCACTATAGGGAGTTCTACAGTCCGACGATCNNNNNNGTACCATTTTTTTTTTTTTTTTTTTTTTTTV
8	GCCGGTAATACGACTCACTATAGGGAGTTCTACAGTCCGACGATCNNNNNNACAGACTTTTTTTTTTTTTTTTTTTTTTTTV
9	GCCGGTAATACGACTCACTATAGGGAGTTCTACAGTCCGACGATCNNNNNNACGTTGTTTTTTTTTTTTTTTTTTTTTTTTV
10	GCCGGTAATACGACTCACTATAGGGAGTTCTACAGTCCGACGATCNNNNNNACCAACTTTTTTTTTTTTTTTTTTTTTTTTV
11	GCCGGTAATACGACTCACTATAGGGAGTTCTACAGTCCGACGATCNNNNNNGTGAAGTTTTTTTTTTTTTTTTTTTTTTTTV
12	GCCGGTAATACGACTCACTATAGGGAGTTCTACAGTCCGACGATCNNNNNNCACTTCTTTTTTTTTTTTTTTTTTTTTTTTV
13	GCCGGTAATACGACTCACTATAGGGAGTTCTACAGTCCGACGATCNNNNNNGAGTTGTTTTTTTTTTTTTTTTTTTTTTTTV
14	GCCGGTAATACGACTCACTATAGGGAGTTCTACAGTCCGACGATCNNNNNNGAAGACTTTTTTTTTTTTTTTTTTTTTTTTV
15	GCCGGTAATACGACTCACTATAGGGAGTTCTACAGTCCGACGATCNNNNNNTGCAGATTTTTTTTTTTTTTTTTTTTTTTTV
16	GCCGGTAATACGACTCACTATAGGGAGTTCTACAGTCCGACGATCNNNNNNCTAGGATTTTTTTTTTTTTTTTTTTTTTTTV
17	GCCGGTAATACGACTCACTATAGGGAGTTCTACAGTCCGACGATCNNNNNNACCAGATTTTTTTTTTTTTTTTTTTTTTTTV
18	GCCGGTAATACGACTCACTATAGGGAGTTCTACAGTCCGACGATCNNNNNNGTGACATTTTTTTTTTTTTTTTTTTTTTTTV
19	GCCGGTAATACGACTCACTATAGGGAGTTCTACAGTCCGACGATCNNNNNNCTAGACTTTTTTTTTTTTTTTTTTTTTTTTV
20	GCCGGTAATACGACTCACTATAGGGAGTTCTACAGTCCGACGATCNNNNNNAGCTCATTTTTTTTTTTTTTTTTTTTTTTTV

(continued)

Table 1
(continued)

21	GCCGGTAATACGACTCACTATAGGGAGTTCTACAGTCCGACGATCNNNNNNACTCGA TTTTTTTTTTTTTTTTTTTTTTTTV
22	GCCGGTAATACGACTCACTATAGGGAGTTCTACAGTCCGACGATCNNNNNNCTGTTG TTTTTTTTTTTTTTTTTTTTTTTTV
23	GCCGGTAATACGACTCACTATAGGGAGTTCTACAGTCCGACGATCNNNNNNCATGCA TTTTTTTTTTTTTTTTTTTTTTTTV
24	GCCGGTAATACGACTCACTATAGGGAGTTCTACAGTCCGACGATCNNNNNNCAGAAG TTTTTTTTTTTTTTTTTTTTTTTTV
25	GCCGGTAATACGACTCACTATAGGGAGTTCTACAGTCCGACGATCNNNNNNGTCTCA TTTTTTTTTTTTTTTTTTTTTTTTV
26	GCCGGTAATACGACTCACTATAGGGAGTTCTACAGTCCGACGATCNNNNNNGTGATC TTTTTTTTTTTTTTTTTTTTTTTTV
27	GCCGGTAATACGACTCACTATAGGGAGTTCTACAGTCCGACGATCNNNNNNTGTCTG TTTTTTTTTTTTTTTTTTTTTTTTV
28	GCCGGTAATACGACTCACTATAGGGAGTTCTACAGTCCGACGATCNNNNNNGACAGA TTTTTTTTTTTTTTTTTTTTTTTTV
29	GCCGGTAATACGACTCACTATAGGGAGTTCTACAGTCCGACGATCNNNNNNACTCTG TTTTTTTTTTTTTTTTTTTTTTTTV
30	GCCGGTAATACGACTCACTATAGGGAGTTCTACAGTCCGACGATCNNNNNNTGCAAC TTTTTTTTTTTTTTTTTTTTTTTTV
31	GCCGGTAATACGACTCACTATAGGGAGTTCTACAGTCCGACGATCNNNNNNGAAGGA TTTTTTTTTTTTTTTTTTTTTTTTV
32	GCCGGTAATACGACTCACTATAGGGAGTTCTACAGTCCGACGATCNNNNNNGTTGAG TTTTTTTTTTTTTTTTTTTTTTTTV
33	GCCGGTAATACGACTCACTATAGGGAGTTCTACAGTCCGACGATCNNNNNNAGACCA TTTTTTTTTTTTTTTTTTTTTTTTV
34	GCCGGTAATACGACTCACTATAGGGAGTTCTACAGTCCGACGATCNNNNNNTGGTTG TTTTTTTTTTTTTTTTTTTTTTTTV
35	GCCGGTAATACGACTCACTATAGGGAGTTCTACAGTCCGACGATCNNNNNNGATCTG TTTTTTTTTTTTTTTTTTTTTTTTV
36	GCCGGTAATACGACTCACTATAGGGAGTTCTACAGTCCGACGATCNNNNNNCTAGTG TTTTTTTTTTTTTTTTTTTTTTTTV
37	GCCGGTAATACGACTCACTATAGGGAGTTCTACAGTCCGACGATCNNNNNNCTCAGA TTTTTTTTTTTTTTTTTTTTTTTTV
38	GCCGGTAATACGACTCACTATAGGGAGTTCTACAGTCCGACGATCNNNNNNCTTCGA TTTTTTTTTTTTTTTTTTTTTTTTV
39	GCCGGTAATACGACTCACTATAGGGAGTTCTACAGTCCGACGATCNNNNNNAGCTAG TTTTTTTTTTTTTTTTTTTTTTTTV
40	GCCGGTAATACGACTCACTATAGGGAGTTCTACAGTCCGACGATCNNNNNNGATCGA TTTTTTTTTTTTTTTTTTTTTTTTV

(continued)

Table 1
(continued)

41	GCCGGTAATACGACTCACTATAGGGAGTTCTACAGTCCGACGATCNNNNNNGTACTC TTTTTTTTTTTTTTTTTTTTTTTTV
42	GCCGGTAATACGACTCACTATAGGGAGTTCTACAGTCCGACGATCNNNNNNTGTCGA TTTTTTTTTTTTTTTTTTTTTTTTV
43	GCCGGTAATACGACTCACTATAGGGAGTTCTACAGTCCGACGATCNNNNNNACGTGA TTTTTTTTTTTTTTTTTTTTTTTTV
44	GCCGGTAATACGACTCACTATAGGGAGTTCTACAGTCCGACGATCNNNNNNAGGATC TTTTTTTTTTTTTTTTTTTTTTTTV
45	GCCGGTAATACGACTCACTATAGGGAGTTCTACAGTCCGACGATCNNNNNNCTCATG TTTTTTTTTTTTTTTTTTTTTTTTV
46	GCCGGTAATACGACTCACTATAGGGAGTTCTACAGTCCGACGATCNNNNNNAGACTC TTTTTTTTTTTTTTTTTTTTTTTTV
47	GCCGGTAATACGACTCACTATAGGGAGTTCTACAGTCCGACGATCNNNNNNGACAAC TTTTTTTTTTTTTTTTTTTTTTTTV
48	GCCGGTAATACGACTCACTATAGGGAGTTCTACAGTCCGACGATCNNNNNNAGGACA TTTTTTTTTTTTTTTTTTTTTTTTV
49	GCCGGTAATACGACTCACTATAGGGAGTTCTACAGTCCGACGATCNNNNNNACTCAC TTTTTTTTTTTTTTTTTTTTTTTTV
50	GCCGGTAATACGACTCACTATAGGGAGTTCTACAGTCCGACGATCNNNNNNGTACAG TTTTTTTTTTTTTTTTTTTTTTTTV
51	GCCGGTAATACGACTCACTATAGGGAGTTCTACAGTCCGACGATCNNNNNNAGGAAG TTTTTTTTTTTTTTTTTTTTTTTTV
52	GCCGGTAATACGACTCACTATAGGGAGTTCTACAGTCCGACGATCNNNNNNAGTGCA TTTTTTTTTTTTTTTTTTTTTTTTV
53	GCCGGTAATACGACTCACTATAGGGAGTTCTACAGTCCGACGATCNNNNNNTGGTGA TTTTTTTTTTTTTTTTTTTTTTTTV
54	GCCGGTAATACGACTCACTATAGGGAGTTCTACAGTCCGACGATCNNNNNNAGACAG TTTTTTTTTTTTTTTTTTTTTTTTV
55	GCCGGTAATACGACTCACTATAGGGAGTTCTACAGTCCGACGATCNNNNNNAGCTTC TTTTTTTTTTTTTTTTTTTTTTTTV
56	GCCGGTAATACGACTCACTATAGGGAGTTCTACAGTCCGACGATCNNNNNNTGAGGA TTTTTTTTTTTTTTTTTTTTTTTTV
57	GCCGGTAATACGACTCACTATAGGGAGTTCTACAGTCCGACGATCNNNNNNACGTAC TTTTTTTTTTTTTTTTTTTTTTTTV
58	GCCGGTAATACGACTCACTATAGGGAGTTCTACAGTCCGACGATCNNNNNNTCACAG TTTTTTTTTTTTTTTTTTTTTTTTV
59	GCCGGTAATACGACTCACTATAGGGAGTTCTACAGTCCGACGATCNNNNNNACAGTG TTTTTTTTTTTTTTTTTTTTTTTTV
60	GCCGGTAATACGACTCACTATAGGGAGTTCTACAGTCCGACGATCNNNNNNCGATTG TTTTTTTTTTTTTTTTTTTTTTTTV

(continued)

Table 1
(continued)

61	GCCGGTAATACGACTCACTATAGGGAGTTCTACAGTCCGACGATCNNNNNNTCTTGC TTTTTTTTTTTTTTTTTTTTTTTTV
62	GCCGGTAATACGACTCACTATAGGGAGTTCTACAGTCCGACGATCNNNNNNGGTAAC TTTTTTTTTTTTTTTTTTTTTTTTV
63	GCCGGTAATACGACTCACTATAGGGAGTTCTACAGTCCGACGATCNNNNNNTCATCC TTTTTTTTTTTTTTTTTTTTTTTTV
64	GCCGGTAATACGACTCACTATAGGGAGTTCTACAGTCCGACGATCNNNNNNTAGGAC TTTTTTTTTTTTTTTTTTTTTTTTV
65	GCCGGTAATACGACTCACTATAGGGAGTTCTACAGTCCGACGATCNNNNNNTTCACC TTTTTTTTTTTTTTTTTTTTTTTTV
66	GCCGGTAATACGACTCACTATAGGGAGTTCTACAGTCCGACGATCNNNNNNAACGAG TTTTTTTTTTTTTTTTTTTTTTTTV
67	GCCGGTAATACGACTCACTATAGGGAGTTCTACAGTCCGACGATCNNNNNNGTGGAA TTTTTTTTTTTTTTTTTTTTTTTTV
68	GCCGGTAATACGACTCACTATAGGGAGTTCTACAGTCCGACGATCNNNNNNATGTCG TTTTTTTTTTTTTTTTTTTTTTTTV
69	GCCGGTAATACGACTCACTATAGGGAGTTCTACAGTCCGACGATCNNNNNNATCACG TTTTTTTTTTTTTTTTTTTTTTTTV
70	GCCGGTAATACGACTCACTATAGGGAGTTCTACAGTCCGACGATCNNNNNNGAATCC TTTTTTTTTTTTTTTTTTTTTTTTV
71	GCCGGTAATACGACTCACTATAGGGAGTTCTACAGTCCGACGATCNNNNNNCGATGA TTTTTTTTTTTTTTTTTTTTTTTTV
72	GCCGGTAATACGACTCACTATAGGGAGTTCTACAGTCCGACGATCNNNNNNGAATGG TTTTTTTTTTTTTTTTTTTTTTTTV
73	GCCGGTAATACGACTCACTATAGGGAGTTCTACAGTCCGACGATCNNNNNNGCAACA TTTTTTTTTTTTTTTTTTTTTTTTV
74	GCCGGTAATACGACTCACTATAGGGAGTTCTACAGTCCGACGATCNNNNNNTTCTCG TTTTTTTTTTTTTTTTTTTTTTTTV
75	GCCGGTAATACGACTCACTATAGGGAGTTCTACAGTCCGACGATCNNNNNNATTGCG TTTTTTTTTTTTTTTTTTTTTTTTV
76	GCCGGTAATACGACTCACTATAGGGAGTTCTACAGTCCGACGATCNNNNNNTAGTGG TTTTTTTTTTTTTTTTTTTTTTTTV
77	GCCGGTAATACGACTCACTATAGGGAGTTCTACAGTCCGACGATCNNNNNNAAGCCA TTTTTTTTTTTTTTTTTTTTTTTTV
78	GCCGGTAATACGACTCACTATAGGGAGTTCTACAGTCCGACGATCNNNNNNCTATCC TTTTTTTTTTTTTTTTTTTTTTTTV
79	GCCGGTAATACGACTCACTATAGGGAGTTCTACAGTCCGACGATCNNNNNNTCCGAA TTTTTTTTTTTTTTTTTTTTTTTTV
80	GCCGGTAATACGACTCACTATAGGGAGTTCTACAGTCCGACGATCNNNNNNTGAACC TTTTTTTTTTTTTTTTTTTTTTTTV

(continued)

Table 1
(continued)

81	GCCGGTAATACGACTCACTATAGGGAGTTCTACAGTCCGACGATCNNNNNNTGTACG TTTTTTTTTTTTTTTTTTTTTTTTV
82	GCCGGTAATACGACTCACTATAGGGAGTTCTACAGTCCGACGATCNNNNNNGACGAA TTTTTTTTTTTTTTTTTTTTTTTTV
83	GCCGGTAATACGACTCACTATAGGGAGTTCTACAGTCCGACGATCNNNNNNCCACAA TTTTTTTTTTTTTTTTTTTTTTTTV
84	GCCGGTAATACGACTCACTATAGGGAGTTCTACAGTCCGACGATCNNNNNNCACCAA TTTTTTTTTTTTTTTTTTTTTTTTV
85	GCCGGTAATACGACTCACTATAGGGAGTTCTACAGTCCGACGATCNNNNNNCTAAGC TTTTTTTTTTTTTTTTTTTTTTTTV
86	GCCGGTAATACGACTCACTATAGGGAGTTCTACAGTCCGACGATCNNNNNNGATACG TTTTTTTTTTTTTTTTTTTTTTTTV
87	GCCGGTAATACGACTCACTATAGGGAGTTCTACAGTCCGACGATCNNNNNNACAAGC TTTTTTTTTTTTTTTTTTTTTTTTV
88	GCCGGTAATACGACTCACTATAGGGAGTTCTACAGTCCGACGATCNNNNNNTGAAGG TTTTTTTTTTTTTTTTTTTTTTTTV
89	GCCGGTAATACGACTCACTATAGGGAGTTCTACAGTCCGACGATCNNNNNNTAACGG TTTTTTTTTTTTTTTTTTTTTTTTV
90	GCCGGTAATACGACTCACTATAGGGAGTTCTACAGTCCGACGATCNNNNNNAACCTC TTTTTTTTTTTTTTTTTTTTTTTTV
91	GCCGGTAATACGACTCACTATAGGGAGTTCTACAGTCCGACGATCNNNNNNCGTCTA TTTTTTTTTTTTTTTTTTTTTTTTV
92	GCCGGTAATACGACTCACTATAGGGAGTTCTACAGTCCGACGATCNNNNNNCCATAG TTTTTTTTTTTTTTTTTTTTTTTTV
93	GCCGGTAATACGACTCACTATAGGGAGTTCTACAGTCCGACGATCNNNNNNTTCCAG TTTTTTTTTTTTTTTTTTTTTTTTV
94	GCCGGTAATACGACTCACTATAGGGAGTTCTACAGTCCGACGATCNNNNNNGGACAA TTTTTTTTTTTTTTTTTTTTTTTTV
95	GCCGGTAATACGACTCACTATAGGGAGTTCTACAGTCCGACGATCNNNNNNACTTCG TTTTTTTTTTTTTTTTTTTTTTTTV
96	GCCGGTAATACGACTCACTATAGGGAGTTCTACAGTCCGACGATCNNNNNNTTGTGC TTTTTTTTTTTTTTTTTTTTTTTTV
97	GCCGGTAATACGACTCACTATAGGGAGTTCTACAGTCCGACGATCNNNNNNGGTATG TTTTTTTTTTTTTTTTTTTTTTTTV
98	GCCGGTAATACGACTCACTATAGGGAGTTCTACAGTCCGACGATCNNNNNNCTGCTA TTTTTTTTTTTTTTTTTTTTTTTTV
99	GCCGGTAATACGACTCACTATAGGGAGTTCTACAGTCCGACGATCNNNNNNATGAGG TTTTTTTTTTTTTTTTTTTTTTTTV
100	GCCGGTAATACGACTCACTATAGGGAGTTCTACAGTCCGACGATCNNNNNNGGTAGA TTTTTTTTTTTTTTTTTTTTTTTTV

(continued)

Table 1
(continued)

101	GCCGGTAATACGACTCACTATAGGGAGTTCTACAGTCCGACGATCNNNNNNATCGTGTTTTTTTTTTTTTTTTTTTTTTTTV
102	GCCGGTAATACGACTCACTATAGGGAGTTCTACAGTCCGACGATCNNNNNNATGGACTTTTTTTTTTTTTTTTTTTTTTTTV
103	GCCGGTAATACGACTCACTATAGGGAGTTCTACAGTCCGACGATCNNNNNNAGTAGGTTTTTTTTTTTTTTTTTTTTTTTTV
104	GCCGGTAATACGACTCACTATAGGGAGTTCTACAGTCCGACGATCNNNNNNCCATCATTTTTTTTTTTTTTTTTTTTTTTTV
105	GCCGGTAATACGACTCACTATAGGGAGTTCTACAGTCCGACGATCNNNNNNAGTACCTTTTTTTTTTTTTTTTTTTTTTTTV
106	GCCGGTAATACGACTCACTATAGGGAGTTCTACAGTCCGACGATCNNNNNNCCAGTATTTTTTTTTTTTTTTTTTTTTTTTV
107	GCCGGTAATACGACTCACTATAGGGAGTTCTACAGTCCGACGATCNNNNNNCGTTACTTTTTTTTTTTTTTTTTTTTTTTTV
108	GCCGGTAATACGACTCACTATAGGGAGTTCTACAGTCCGACGATCNNNNNNGAGGTATTTTTTTTTTTTTTTTTTTTTTTTV
109	GCCGGTAATACGACTCACTATAGGGAGTTCTACAGTCCGACGATCNNNNNNTTGGCATTTTTTTTTTTTTTTTTTTTTTTTV
110	GCCGGTAATACGACTCACTATAGGGAGTTCTACAGTCCGACGATCNNNNNNCAATGCTTTTTTTTTTTTTTTTTTTTTTTTV
111	GCCGGTAATACGACTCACTATAGGGAGTTCTACAGTCCGACGATCNNNNNNGCGTTATTTTTTTTTTTTTTTTTTTTTTTTV
112	GCCGGTAATACGACTCACTATAGGGAGTTCTACAGTCCGACGATCNNNNNNTAGCTCTTTTTTTTTTTTTTTTTTTTTTTTV
113	GCCGGTAATACGACTCACTATAGGGAGTTCTACAGTCCGACGATCNNNNNNTTCGACTTTTTTTTTTTTTTTTTTTTTTTTV
114	GCCGGTAATACGACTCACTATAGGGAGTTCTACAGTCCGACGATCNNNNNNGAGCAATTTTTTTTTTTTTTTTTTTTTTTTV
115	GCCGGTAATACGACTCACTATAGGGAGTTCTACAGTCCGACGATCNNNNNNTTGCTGTTTTTTTTTTTTTTTTTTTTTTTTV
116	GCCGGTAATACGACTCACTATAGGGAGTTCTACAGTCCGACGATCNNNNNNTTGCGATTTTTTTTTTTTTTTTTTTTTTTTV
117	GCCGGTAATACGACTCACTATAGGGAGTTCTACAGTCCGACGATCNNNNNNGCAGAATTTTTTTTTTTTTTTTTTTTTTTTV
118	GCCGGTAATACGACTCACTATAGGGAGTTCTACAGTCCGACGATCNNNNNNCCTACATTTTTTTTTTTTTTTTTTTTTTTTV
119	GCCGGTAATACGACTCACTATAGGGAGTTCTACAGTCCGACGATCNNNNNNGCATGATTTTTTTTTTTTTTTTTTTTTTTTV
120	GCCGGTAATACGACTCACTATAGGGAGTTCTACAGTCCGACGATCNNNNNNAACTGGTTTTTTTTTTTTTTTTTTTTTTTTV

(continued)

Table 1
(continued)

121	GCCGGTAATACGACTCACTATAGGGAGTTCTACAGTCCGACGATCNNNNNNCGGTTATTTTTTTTTTTTTTTTTTTTTTTTV
122	GCCGGTAATACGACTCACTATAGGGAGTTCTACAGTCCGACGATCNNNNNNCTAACGTTTTTTTTTTTTTTTTTTTTTTTTV
123	GCCGGTAATACGACTCACTATAGGGAGTTCTACAGTCCGACGATCNNNNNNCACGTATTTTTTTTTTTTTTTTTTTTTTTTV
124	GCCGGTAATACGACTCACTATAGGGAGTTCTACAGTCCGACGATCNNNNNNTTGGAGTTTTTTTTTTTTTTTTTTTTTTTTV
125	GCCGGTAATACGACTCACTATAGGGAGTTCTACAGTCCGACGATCNNNNNNGCAATGTTTTTTTTTTTTTTTTTTTTTTTTV
126	GCCGGTAATACGACTCACTATAGGGAGTTCTACAGTCCGACGATCNNNNNNTATCCGTTTTTTTTTTTTTTTTTTTTTTTTV
127	GCCGGTAATACGACTCACTATAGGGAGTTCTACAGTCCGACGATCNNNNNNATGCAGTTTTTTTTTTTTTTTTTTTTTTTTV
128	GCCGGTAATACGACTCACTATAGGGAGTTCTACAGTCCGACGATCNNNNNNGCTCTATTTTTTTTTTTTTTTTTTTTTTTTV
129	GCCGGTAATACGACTCACTATAGGGAGTTCTACAGTCCGACGATCNNNNNNATTCGCTTTTTTTTTTTTTTTTTTTTTTTTV
130	GCCGGTAATACGACTCACTATAGGGAGTTCTACAGTCCGACGATCNNNNNNTGTTGGTTTTTTTTTTTTTTTTTTTTTTTTV
131	GCCGGTAATACGACTCACTATAGGGAGTTCTACAGTCCGACGATCNNNNNNATGACCTTTTTTTTTTTTTTTTTTTTTTTTV
132	GCCGGTAATACGACTCACTATAGGGAGTTCTACAGTCCGACGATCNNNNNNCCGTAATTTTTTTTTTTTTTTTTTTTTTTTV
133	GCCGGTAATACGACTCACTATAGGGAGTTCTACAGTCCGACGATCNNNNNNTGATCGTTTTTTTTTTTTTTTTTTTTTTTTV
134	GCCGGTAATACGACTCACTATAGGGAGTTCTACAGTCCGACGATCNNNNNNTACAGGTTTTTTTTTTTTTTTTTTTTTTTTV
135	GCCGGTAATACGACTCACTATAGGGAGTTCTACAGTCCGACGATCNNNNNNAGAACGTTTTTTTTTTTTTTTTTTTTTTTTV
136	GCCGGTAATACGACTCACTATAGGGAGTTCTACAGTCCGACGATCNNNNNNGCCATATTTTTTTTTTTTTTTTTTTTTTTTV
137	GCCGGTAATACGACTCACTATAGGGAGTTCTACAGTCCGACGATCNNNNNNACGGTATTTTTTTTTTTTTTTTTTTTTTTTV
138	GCCGGTAATACGACTCACTATAGGGAGTTCTACAGTCCGACGATCNNNNNNAAGCACTTTTTTTTTTTTTTTTTTTTTTTTV
139	GCCGGTAATACGACTCACTATAGGGAGTTCTACAGTCCGACGATCNNNNNNCGAACATTTTTTTTTTTTTTTTTTTTTTTTV
140	GCCGGTAATACGACTCACTATAGGGAGTTCTACAGTCCGACGATCNNNNNNATGCTCTTTTTTTTTTTTTTTTTTTTTTTTV

(continued)

Table 1
(continued)

141	GCCGGTAATACGACTCACTATAGGGAGTTCTACAGTCCGACGATCNNNNNNGGCTTATTTTTTTTTTTTTTTTTTTTTTTTV
142	GCCGGTAATACGACTCACTATAGGGAGTTCTACAGTCCGACGATCNNNNNNATCGCATTTTTTTTTTTTTTTTTTTTTTTTV
143	GCCGGTAATACGACTCACTATAGGGAGTTCTACAGTCCGACGATCNNNNNNGGATCATTTTTTTTTTTTTTTTTTTTTTTTV
144	GCCGGTAATACGACTCACTATAGGGAGTTCTACAGTCCGACGATCNNNNNNCCAATCTTTTTTTTTTTTTTTTTTTTTTTTV
145	GCCGGTAATACGACTCACTATAGGGAGTTCTACAGTCCGACGATCNNNNNNAAGGTGTTTTTTTTTTTTTTTTTTTTTTTTV
146	GCCGGTAATACGACTCACTATAGGGAGTTCTACAGTCCGACGATCNNNNNNATCTCCTTTTTTTTTTTTTTTTTTTTTTTTV
147	GCCGGTAATACGACTCACTATAGGGAGTTCTACAGTCCGACGATCNNNNNNGTATCGTTTTTTTTTTTTTTTTTTTTTTTTV
148	GCCGGTAATACGACTCACTATAGGGAGTTCTACAGTCCGACGATCNNNNNNTGTTCCTTTTTTTTTTTTTTTTTTTTTTTTV
149	GCCGGTAATACGACTCACTATAGGGAGTTCTACAGTCCGACGATCNNNNNNGGTGTATTTTTTTTTTTTTTTTTTTTTTTTV
150	GCCGGTAATACGACTCACTATAGGGAGTTCTACAGTCCGACGATCNNNNNNTACTCCTTTTTTTTTTTTTTTTTTTTTTTTV
151	GCCGGTAATACGACTCACTATAGGGAGTTCTACAGTCCGACGATCNNNNNNATCAGCTTTTTTTTTTTTTTTTTTTTTTTTV
152	GCCGGTAATACGACTCACTATAGGGAGTTCTACAGTCCGACGATCNNNNNNAAGTGCTTTTTTTTTTTTTTTTTTTTTTTTV
153	GCCGGTAATACGACTCACTATAGGGAGTTCTACAGTCCGACGATCNNNNNNAGGCTATTTTTTTTTTTTTTTTTTTTTTTTV
154	GCCGGTAATACGACTCACTATAGGGAGTTCTACAGTCCGACGATCNNNNNNCCTATGTTTTTTTTTTTTTTTTTTTTTTTTV
155	GCCGGTAATACGACTCACTATAGGGAGTTCTACAGTCCGACGATCNNNNNNTATCGCTTTTTTTTTTTTTTTTTTTTTTTTV
156	GCCGGTAATACGACTCACTATAGGGAGTTCTACAGTCCGACGATCNNNNNNCGCTAATTTTTTTTTTTTTTTTTTTTTTTTV
157	GCCGGTAATACGACTCACTATAGGGAGTTCTACAGTCCGACGATCNNNNNNGTAACCTTTTTTTTTTTTTTTTTTTTTTTTV
158	GCCGGTAATACGACTCACTATAGGGAGTTCTACAGTCCGACGATCNNNNNNACATGGTTTTTTTTTTTTTTTTTTTTTTTTV
159	GCCGGTAATACGACTCACTATAGGGAGTTCTACAGTCCGACGATCNNNNNNCCGATATTTTTTTTTTTTTTTTTTTTTTTTV
160	GCCGGTAATACGACTCACTATAGGGAGTTCTACAGTCCGACGATCNNNNNNGGATACTTTTTTTTTTTTTTTTTTTTTTTTV

(continued)

Table 1
(continued)

161	GCCGGTAATACGACTCACTATAGGGAGTTCTACAGTCCGACGATCNNNNNNGTTAGGTTTTTTTTTTTTTTTTTTTTTTTTV
162	GCCGGTAATACGACTCACTATAGGGAGTTCTACAGTCCGACGATCNNNNNNTACGCATTTTTTTTTTTTTTTTTTTTTTTTV
163	GCCGGTAATACGACTCACTATAGGGAGTTCTACAGTCCGACGATCNNNNNNAGATGCTTTTTTTTTTTTTTTTTTTTTTTTV
164	GCCGGTAATACGACTCACTATAGGGAGTTCTACAGTCCGACGATCNNNNNNTTGCACTTTTTTTTTTTTTTTTTTTTTTTTV
165	GCCGGTAATACGACTCACTATAGGGAGTTCTACAGTCCGACGATCNNNNNNCAGGAATTTTTTTTTTTTTTTTTTTTTTTTV
166	GCCGGTAATACGACTCACTATAGGGAGTTCTACAGTCCGACGATCNNNNNNTCTAGGTTTTTTTTTTTTTTTTTTTTTTTTV
167	GCCGGTAATACGACTCACTATAGGGAGTTCTACAGTCCGACGATCNNNNNNGCTTCATTTTTTTTTTTTTTTTTTTTTTTTV
168	GCCGGTAATACGACTCACTATAGGGAGTTCTACAGTCCGACGATCNNNNNNTTGGTCTTTTTTTTTTTTTTTTTTTTTTTTV
169	GCCGGTAATACGACTCACTATAGGGAGTTCTACAGTCCGACGATCNNNNNNTACCGATTTTTTTTTTTTTTTTTTTTTTTTV
170	GCCGGTAATACGACTCACTATAGGGAGTTCTACAGTCCGACGATCNNNNNNCATTGGTTTTTTTTTTTTTTTTTTTTTTTTV
171	GCCGGTAATACGACTCACTATAGGGAGTTCTACAGTCCGACGATCNNNNNNCTCGAATTTTTTTTTTTTTTTTTTTTTTTTV
172	GCCGGTAATACGACTCACTATAGGGAGTTCTACAGTCCGACGATCNNNNNNGCTTACTTTTTTTTTTTTTTTTTTTTTTTTV
173	GCCGGTAATACGACTCACTATAGGGAGTTCTACAGTCCGACGATCNNNNNNATACGGTTTTTTTTTTTTTTTTTTTTTTTTV
174	GCCGGTAATACGACTCACTATAGGGAGTTCTACAGTCCGACGATCNNNNNNGTATGCTTTTTTTTTTTTTTTTTTTTTTTTV
175	GCCGGTAATACGACTCACTATAGGGAGTTCTACAGTCCGACGATCNNNNNNTGTAGCTTTTTTTTTTTTTTTTTTTTTTTTV
176	GCCGGTAATACGACTCACTATAGGGAGTTCTACAGTCCGACGATCNNNNNNCGTAAGTTTTTTTTTTTTTTTTTTTTTTTTV
177	GCCGGTAATACGACTCACTATAGGGAGTTCTACAGTCCGACGATCNNNNNNTTACGCTTTTTTTTTTTTTTTTTTTTTTTTV
178	GCCGGTAATACGACTCACTATAGGGAGTTCTACAGTCCGACGATCNNNNNNTACCACTTTTTTTTTTTTTTTTTTTTTTTTV
179	GCCGGTAATACGACTCACTATAGGGAGTTCTACAGTCCGACGATCNNNNNNCGCATATTTTTTTTTTTTTTTTTTTTTTTTV
180	GCCGGTAATACGACTCACTATAGGGAGTTCTACAGTCCGACGATCNNNNNNGCTAAGTTTTTTTTTTTTTTTTTTTTTTTTV

(continued)

Table 1
(continued)

181	GCCGGTAATACGACTCACTATAGGGAGTTCTACAGTCCGACGATCNNNNNNATCCAC TTTTTTTTTTTTTTTTTTTTTTTTV
182	GCCGGTAATACGACTCACTATAGGGAGTTCTACAGTCCGACGATCNNNNNNCCTTGA TTTTTTTTTTTTTTTTTTTTTTTTV
183	GCCGGTAATACGACTCACTATAGGGAGTTCTACAGTCCGACGATCNNNNNNAGCGAA TTTTTTTTTTTTTTTTTTTTTTTTV
184	GCCGGTAATACGACTCACTATAGGGAGTTCTACAGTCCGACGATCNNNNNNGGTTAG TTTTTTTTTTTTTTTTTTTTTTTTV
185	GCCGGTAATACGACTCACTATAGGGAGTTCTACAGTCCGACGATCNNNNNNGATTGC TTTTTTTTTTTTTTTTTTTTTTTTV
186	GCCGGTAATACGACTCACTATAGGGAGTTCTACAGTCCGACGATCNNNNNNCGTTCA TTTTTTTTTTTTTTTTTTTTTTTTV
187	GCCGGTAATACGACTCACTATAGGGAGTTCTACAGTCCGACGATCNNNNNNATCCGA TTTTTTTTTTTTTTTTTTTTTTTTV
188	GCCGGTAATACGACTCACTATAGGGAGTTCTACAGTCCGACGATCNNNNNNGCATTC TTTTTTTTTTTTTTTTTTTTTTTTV
189	GCCGGTAATACGACTCACTATAGGGAGTTCTACAGTCCGACGATCNNNNNNCCTGAA TTTTTTTTTTTTTTTTTTTTTTTTV
190	GCCGGTAATACGACTCACTATAGGGAGTTCTACAGTCCGACGATCNNNNNNGGAATC TTTTTTTTTTTTTTTTTTTTTTTTV
191	GCCGGTAATACGACTCACTATAGGGAGTTCTACAGTCCGACGATCNNNNNNTCAACG TTTTTTTTTTTTTTTTTTTTTTTTV
192	GCCGGTAATACGACTCACTATAGGGAGTTCTACAGTCCGACGATCNNNNNNAACACC TTTTTTTTTTTTTTTTTTTTTTTTV

first at the ratio of 1:4 to get a 10 μM intermediate 384-well plate (total volume: 4 μl, columns: 1–12, 192 wells). The 10 μM intermediate plate can then be used to prepare four 1 μM 384-well primer source plates each containing 10 μl of primer solution. Centrifuge the plates at 4 °C after each step. Store the 40 μM two 96-well plates at −80 °C and 1 μM plates at −20 °C. Note that only half of the 384-well plate, i.e., 192 wells will contain the RT primers and the rest half will be empty (*see* **Note 1**).

2. randomhexRT primer: 5′ GCCTTGGCACCCGAGAATTC-CANNNNNN.
3. 13 RNA PCR primers (sequences provided in Table 2)

   We keep randomhexRT and RNA PCR primer stocks (100 μM) at −80 °C. Working solutions (10 μM) are stored at −20 °C.

Table 2
Sequences of RP1 primer and 12 RPI1–12 primers (same as provided by Illumina, Inc.)

Primer name	Sequence
RNA PCR Primer (RP1)	AATGATACGGCGACCACCGAGATCTACACGTTCAGAGTTC TACAGTCCGA
RNA PCR Primer, Index 1 (RPI1)	CAAGCAGAAGACGGCATACGAGATCGTGATGTGACTGGAG TTCCTTGGCACCCGAGAATTCCA
RNA PCR Primer, Index 2 (RPI2)	CAAGCAGAAGACGGCATACGAGATACATCGGTGACTGGAG TTCCTTGGCACCCGAGAATTCCA
RNA PCR Primer, Index 3 (RPI3)	CAAGCAGAAGACGGCATACGAGATGCCTAAGTGACTGGAG TTCCTTGGCACCCGAGAATTCCA
RNA PCR Primer, Index 4 (RPI4)	CAAGCAGAAGACGGCATACGAGATTGGTCAGTGACTGGAG TTCCTTGGCACCCGAGAATTCCA
RNA PCR Primer, Index 5 (RPI5)	CAAGCAGAAGACGGCATACGAGATCACTGTGTGACTGGAG TTCCTTGGCACCCGAGAATTCCA
RNA PCR Primer, Index 6 (RPI6)	CAAGCAGAAGACGGCATACGAGATATTGGCGTGACTGGAG TTCCTTGGCACCCGAGAATTCCA
RNA PCR Primer, Index 7 (RPI7)	CAAGCAGAAGACGGCATACGAGATGATCTGGTGACTGGAG TTCCTTGGCACCCGAGAATTCCA
RNA PCR Primer, Index 8 (RPI8)	CAAGCAGAAGACGGCATACGAGATTCAAGTGTGACTGGAG TTCCTTGGCACCCGAGAATTCCA
RNA PCR Primer, Index 9 (RPI9)	CAAGCAGAAGACGGCATACGAGATCTGATCGTGACTGGAG TTCCTTGGCACCCGAGAATTCCA
RNA PCR Primer, Index 10 (RPI10)	CAAGCAGAAGACGGCATACGAGATAAGCTAGTGACTGGAG TTCCTTGGCACCCGAGAATTCCA
RNA PCR Primer, Index 11 (RPI11)	CAAGCAGAAGACGGCATACGAGATGTAGCCGTGACTGGAG TTCCTTGGCACCCGAGAATTCCA
RNA PCR Primer, Index 12 (RPI12)	CAAGCAGAAGACGGCATACGAGATTACAAGGTGACTGGAG TTCCTTGGCACCCGAGAATTCCA

2.3 Plate Preparation for Single Cell Sorting

1. DNA LoBind 384-well plates.
2. 1 μM 384-well primer plate containing 192 RT primers prepared in Subheading 2.2, **item 1**.
3. dNTPs: 10 mM.
4. 0.35% Triton-X-100 in RNase-free water: Prepare a 50 ml solution. Aliquot 1 ml in 1.5 ml tubes and store at −20 °C for future use.
5. ERCC RNA Spike-In Mix (Ambion): Dilute the mix at the ratio of 1:100,000 and store at −20 °C.
6. Vapor-Lock (Qiagen).

2.4 cDNA Synthesis and In Vitro Transcription

1. SuperScript II™ Reverse Transcriptase (Invitrogen).
2. RNaseOUT Recombinant Ribonuclease Inhibitor (Invitrogen).
3. Second Strand Buffer (Invitrogen).
4. *E. coli* DNA Polymerase I (Invitrogen).
5. *E. coli* DNA ligase (Invitrogen).
6. *E. coli* Ribonuclease H (Invitrogen).
7. MEGAscript T7 Transcription Kit (Ambion).
8. ExoSAP-IT For PCR Product CleanUp (Affymetrix).
9. NEBNext® Magnesium RNA Fragmentation Module (New England Biolabs).
10. AMPure XP beads (Beckman Coulter).
11. RNAClean XP beads (Beckman Coulter).
12. Agilent RNA 6000 Pico Kit.

2.5 Library Preparation and Sequencing

1. SuperScript II Reverse Transcriptase (Invitrogen).
2. RNaseOUT Recombinant Ribonuclease Inhibitor (Invitrogen).
3. AMPure XP beads (Beckman Coulter).
4. Phusion® High-Fidelity PCR Master Mix with HF Buffer (New England Biolabs).
5. randomhexRT primer: 10 μM.
6. RNA PCR primers: 10 μM.
7. Qubit dsDNA HS (High Sensitivity) Assay Kit.
8. Agilent High Sensitivity DNA kit.

2.6 Expression Quantification and Data Analysis

1. Quality control software tools, e.g., FastQC (http://www.bioinformatics.babraham.ac.uk/projects/fastqc/).
2. Sequence alignment tools, e.g., BWA software package for sequence alignment (http://bio-bwa.sourceforge.net/).
3. R (https://www.r-project.org/) and RStudio (optional, https://www.rstudio.com/).
4. RaceID2 and StemID algorithms (https://github.com/dgrun/StemID). The link contains two R files: RaceID2_StemID_class.R and RaceID2_StemID_sample.R as well as a PDF version of the reference manual explaining how to run the algorithms in a stepwise manner.

3 Methods

Always wear gloves and disinfect working area with 70% ethanol. Use RNase decontamination solution to keep the pipetting robot, working bench, pipettes, and other equipment RNase-free. Use LoBind RNase-free filter tips during all the steps. All centrifugation steps are performed at 4 °C.

3.1 Plate Preparation for Sorting

1. Thaw the reagents and prepare the following mix in a 2 ml LoBind tube (1 ml): 200 μl of 1:100,000 ERCC RNA Spike-In Mix, 100 μl of 10 mM dNTPs, and 700 μl of 0.35% Triton X-100 solution in water. Using a single-channel pipette, dispense 5 μl in columns 1–12 of a 384-well plate (192 wells). Seal the plate, briefly centrifuge at the maximum speed (we centrifuge for 1 min at 2000 × *g*) and keep it on ice.
2. Thaw the 1 μM 384-well primer source plate on ice and dispense 1 μl RT primer each from columns 1–12 into the columns 1–12 of the 384-well plate prepared in **step 1**. Mix thoroughly by pipetting up and down. Seal the plate, briefly centrifuge at the maximum speed and keep it on ice.
3. At the end of **step 2**, half of the 384-well plate (columns 1–12, 192 wells) will have 6 μl of primer-dNTP mix in 0.2% Triton X-100 solution in water. This primer mix plate will be used to prepare the final plates for single cell sorting using the pipetting robot, mosquito®HTS. The 6 μl solution in the primer mix plate is enough to prepare 8–9 plates ready to use for single cell sorting.
4. Take a new 384-well plate and dispense 22 μl of Vapor-Lock in columns 1–12 (192 wells). Seal the plate and briefly centrifuge at the maximum speed. Keep this plate aside on ice for later use. 22 μl Vapor-Lock in each well is enough to overlay eight plates containing 240 nl primer mix for single cell sorting.
5. Switch-on the pipetting robot (we are using mosquito®HTS from TTP Labtech), wipe the deck of the machine with disinfectant and RNase decontamination solution. Keep the 6 μl primer mix plate prepared in **step 2** in position 1 and four new 384-well plates in positions 2–5. Using the robot, dispense 240 nl from columns 1–12 of the primer-mix plate into the columns 1–12 and 13–24 of the four 384-well plates placed in positions 2–5. It is important to instruct the software to change the tips before aspirating different RT primers to avoid cell-barcode contamination. Using mosquito, you can prepare four sorting plates simultaneously. This step will yield four 384-well plates, each containing 240 nl of primer mix solution.
6. Replace the primer mix plate from position 1 of the robot with the 22 μl Vapor-Lock plate prepared in **step 4**. Seal the primer

mix plate and keep it on ice to prepare four more plates later. Using the robot, dispense 1.2 μl from the 22 μl Vapor-Lock plate kept at position 1 into all the 24 columns of four 384-well plates containing 240 nl of primer mix solution placed at positions 2–5. Change tips between the transfers to avoid cell-barcode contamination.

7. Once the transfer is finished, remove the Vapor-Lock plate from position 1 and keep it on ice for further use. Now remove four 384-well plates from positions 2–5, seal the plates, centrifuge at the maximum speed and directly proceed for sorting or store the plates at −20 °C for later use.
8. Repeat **steps 5–7** to make another four plates. Through this protocol you can prepare eight single cell sorting plates containing 240 nl of primer-mix solution and 1.2 μl Vapor-Lock. Plates can be stored at −20 °C for a few months for later use.

3.2 Single Cell Sorting

1. Prepare the single cell suspension of the cultured cells or tissue/organ of your interest (*see* **Note 2**).
2. Calibrate the flow cytometer to sort the single cells in a 384-well plate. It is important that the drop encapsulating the single cell falls as close as possible to the center of the well containing Vapor-Lock and primer mix solution rather than sticking to the inner wall of the well.
3. Run the flow cytometer in single cell sort mode and use trigger pulse width to exclude doublets.
4. After sorting, seal the plates, centrifuge at the maximum speed for 5 min, snap-freeze them in liquid nitrogen and store them at −80 °C until cDNA synthesis (*see* **Note 3**).

3.3 cDNA Synthesis and In Vitro Transcription

The following protocol describes the cDNA synthesis and IVT for a single plate. The protocol can be scaled up for more plates depending on the availability of thermal cyclers. We routinely process three plates at the same time.

1. Thaw the sorted plate on ice and centrifuge briefly at the maximum speed.
2. Incubate the plate at 95 °C for 3 min (Lid temperature: 105 °C) and afterward quickly put the plate on ice (*see* **Note 4**).
3. Prepare the first strand synthesis (FSS) mix in a 0.5 ml RNase-free tube as follows (136 μl): 68 μl of First Strand Buffer, 34 μl of 0.1 M DTT, 17 μl of SuperScript II™ Reverse Transcriptase, and 17 μl of RNaseOUT Recombinant Ribonuclease Inhibitor. Using a single-channel pipette, dispense 8 μl in column 1 (16 wells) of a new 384-well plate. Seal the plate, briefly centrifuge at the maximum speed and keep it on ice. This plate will serve as a source plate to dispense 160 nl FSS mix in

the sorted plate. Do not discard the plate after use. The remaining empty wells can be used as source wells to dispense second strand synthesis (SSS) mix.

4. Keep the 384-well plate containing the FSS mix in position 2 and the sorted plate in position 3 of mosquito. Dispense 160 nl from column 1 of the FSS mix plate into all the columns (1–24) of the sorted plate. Change the tips after each transfer to avoid cell-barcode contamination. Each well of the sorted plate will now be having 400 nl of total volume and 1.2 μl of Vapor-Lock on the top.
5. Seal the plate, shortly centrifuge at the maximum speed and incubate at 42 °C for 1 h (Lid temperature: 50 °C).
6. Heat inactivate the plate by incubating it at 70 °C for 10 min (Lid temperature: 85 °C). Keep the plate on ice afterward.
7. While the sorted plate is undergoing heat inactivation step (**step 6**), prepare the SSS mix in a 2 ml tube as follows (927 μl): 648 μl of RNase-free water, 212 μl of Second Strand Buffer, 21 μl of 10 mM dNTPs, 30 μl of *E. coli* DNA Polymerase I, 8 μl of *E. coli* DNA ligase, and 8 μl of *E. coli* Ribonuclease H.
8. Using a single-channel pipette, dispense 14 μl in each well of columns 2–5 (64 wells) of the FSS mix plate from **step 3**. Seal the plate, briefly centrifuge at the maximum speed and keep it on ice.
9. Keep the 384-well plate containing the SSS mix in position 2 and the sorted plate after FSS and heat inactivation in position 3 of mosquito. Dispense 2.2 μl from columns 2, 3, 4 and 5 of the SSS mix plate into columns 1–6, 7–12, 13–18, and 19–24 of the sorted plate respectively. Do not dispense SSS mix directly after heat inactivation. Let the plate cool down first for 5 min on ice. Change the tips before aspirating the SSS mix from a new column (four times) to avoid contamination (*see* **Note 5**).
10. Seal the plate, centrifuge at the maximum speed for 1 min and incubate at 16 °C for 2 h (Lid temperature off).
11. 30 min before the completion of SSS, take out AMPure XP beads (approximately 640 μl) from the refrigerator and let them prewarm at room temperature.
12. After the completion of SSS, remove the plate from the cycler and keep it on ice for pooling. Pool columns 1–6, 7–12, 13–18, and 19–24 each in a DNA LoBind 0.5 ml tube. After pooling six columns in each tube, four samples per 384-well plate each containing cDNA from 96 barcoded cells will be obtained (*see* **Note 6**). These four samples will undergo cDNA clean-up, IVT and library preparation at the same time.

13. Centrifuge the four tubes at the maximum speed for 1 min in a minicentrifuge to separate aqueous and Vapor-Lock phase. Take out aqueous phase from the bottom of each tube using a single channel pipette. The volume should be approximately 220 μl. Distribute this volume equally, i.e., 110 μl in each of two wells of a new 96-well plate.
14. Vortex AMPure XP Beads and add 80 μl to all the eight wells containing 110 μl of pooled sample. Note that there will be eight samples in total (two from each 0.5 ml tube). Mix beads and the sample thoroughly by pipetting up and down. Change tips while mixing samples coming from different tubes to avoid contamination.
15. Incubate the 96-well plate at room temperature for 10 min and afterward place it on the magnetic stand till all the beads are attached to the wall of the plate and the liquid is clear. This should take 2–3 min. Prepare 80% ethanol in DNase/RNase-free water during this incubation period.
16. At this stage, cDNA is attached to the beads. Remove and discard the liquid while keeping the plate on the magnetic stand.
17. Add 150 μl of freshly prepared 80% ethanol to each well, incubate at least for 30 s, remove and discard the ethanol without disturbing the beads. Do not remove the plate from the magnetic stand during the ethanol washing steps.
18. Repeat the washing step once more. Make sure to remove small droplets of ethanol at the bottom or the corner of the well using a single-channel pipette.
19. Air-dry beads for 10 min while keeping the plate on the magnetic stand.
20. Remove the plate from the magnetic stand and resuspend the beads of one well with 7 μl DNase/RNase-free water. Pipette the entire volume up and down several times to mix thoroughly. Transfer this solution to the samples in the second well corresponding to the same 0.5 ml tube from which it was taken and mix again thoroughly. Repeat the procedure for remaining six wells. At the end of this step, four samples per plate containing beads in 7 μl of water will be obtained.
21. Incubate the plate at room temperature for 2 min, place it back on the magnetic stand until the liquid appears clear. The liquid will contain the eluted cDNA. Transfer 6.4 μl of supernatant to a new 96-well plate while keeping the plate on the magnetic stand. Keep cDNA samples on ice.
22. While the beads are air drying, prepare the following IVT mix in a 0.5 ml RNase-free tube (42 μl): 7 μl each of ATP, GTP,

UTP, and CTP solution, 7 μl of 10× T7 reaction buffer and 7 μl of T7 enzyme mix.

23. Add 9.6 μl of this IVT mix in 6.4 μl of the eluted cDNA (four samples) obtained in **step 21**. Mix well by pipetting up and down. Centrifuge shortly at the maximum speed and incubate for 13 h at 37 °C (Lid temperature: 70 °C). Set the cycler to 4 °C after 13 h of IVT to prevent RNA degradation.
24. After IVT, move the amplified RNA (aRNA) on ice. Add 6 μl of ExoSAP-IT to each of the four samples and incubate them for 15 min at 37 °C (Lid temperature: 50 °C). During this incubation step, thaw the NEBNext® Magnesium RNA Fragmentation Module containing 10× fragmentation reaction buffer and 10× fragmentation stop solution. Also, take out the RNA-Clean XP beads (approximately 88 μl) from the refrigerator and let them prewarm at room temperature.
25. After ExoSAP treatment, bring the aRNA samples back on ice and add 2.45 μl of 10× fragmentation reaction buffer. Mix them well and briefly centrifuge at the maximum speed. Incubate at 94 °C (Lid temperature: 105 °C) for 3 min. After fragmentation, quickly transfer the samples on ice and add 2.45 μl of 10× fragmentation stop solution immediately. Mix them well and centrifuge briefly (*see* **Note 7**).
26. Vortex RNAClean XP Beads and add 22 μl to each of the four aRNA samples. Mix beads and aRNA samples thoroughly by pipetting up and down.
27. Incubate the 96-well plate containing the aRNA samples at room temperature for 8 min and afterward place it on the magnetic stand until all the beads are attached to the wall of the plate and the liquid is clear. Prepare 70% ethanol in DNase/RNase-free water during this incubation period.
28. At this stage, the aRNA is attached to the beads. Remove and discard the liquid while keeping the plate on the magnetic stand.
29. Add 150 μl freshly prepared 70% ethanol, incubate for at least 30 s, remove and discard the ethanol without disturbing the beads.
30. Repeat this washing step twice. After the three washing steps, make sure to remove small droplets of ethanol at the bottom or the corner of the well using a single-channel pipette.
31. Air dry beads for 8 min while keeping the plate on the magnetic stand.
32. Remove the plate from the magnetic stand and resuspend beads in 7 μl RNase-free water. Pipette the entire volume up and down several times to mix thoroughly.

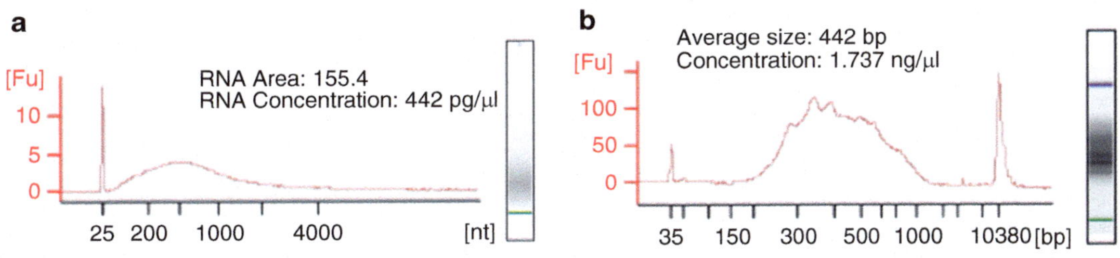

Fig. 2 Size distribution and concentration of amplified RNA (**a**) and the corresponding library (**b**) of 96 early thymic progenitor cells isolated from 6-week-old female mouse measured by Agilent 2100 Bioanalyzer. The size distribution of fragmented amplified RNA samples always peaks at 500 bp whereas the average library size is cell-type dependent. Usually, the average library size varies between 400 and 600 bp

33. Incubate the plate at room temperature for 2 min, place it back on the magnetic stand until the liquid appears clear and transfer 6 μl of the supernatant to a new 96-well plate while keeping the plate on the magnetic stand. Keep the aRNA on ice.
34. Use 1 μl of the aRNA to measure the concentration and check the size distribution using the Agilent 2100 Bioanalyzer and Agilent RNA 6000 Pico Kit (Assay class: Eukaryote Total RNA Pico, Fig. 2a) (*see* **Note 8**). Proceed for library preparation with the remaining 5 μl aRNA. Alternatively, samples can be stored at −80 °C.

3.4 Library Preparation and Sequencing

1. Thaw aRNA samples on ice (if frozen) and prepare the following primer-dNTP mix in a 0.5 ml RNase-free tube (7.5 μl): 5 μl of 10 μM randomhexRT primer and 2.5 μl of 10 mM dNTPs. Mix well by flicking and spin down briefly in a minicentrifuge. Add 1.5 μl to each of the four aRNA samples and incubate at 65 °C (Lid temperature: 80 °C) for 5 min. Transfer samples on ice.
2. During the 65 °C incubation step, prepare the following RT mix in a 0.5 ml RNase-free tube (20 μl): 10 μl of First Strand Buffer, 5 μl of 0.1 M DTT, 2.5 μl of SuperScript II™ Reverse Transcriptase, and 2.5 μl of RNaseOUT Ribonuclease Inhibitor.
3. Add 4 μl to each of the four aRNA samples. Mix well by pipetting up and down and centrifuge briefly at the maximum speed. Incubate the samples at 25 °C (Lid temperature: 37 °C) for 10 min and at 42 °C (Lid temperature: 50 °C) for 1 h.
4. Fifteen minutes before the completion of the RT reaction, prepare the following PCR mix in a 1.5 ml tube (171 μl): 49.5 μl of DNase/RNase-free water, 112.5 μl of Phusion® High-Fidelity PCR Master Mix with HF Buffer, and 9 μl of 10 μM RP1 primer.

5. After the completion of the RT reaction, transfer the samples on ice and add 38 μl of PCR mix in each sample. Afterward, add 2 μl of one of RPI1–12 primers to each sample separately, mix well by pipetting and centrifuge the plate briefly (*see* **Note 9** regarding the selection of RPI primers for multiplexing).
6. Use the following PCR cycle conditions to prepare the libraries: 98 °C (30 s), 11 cycles of: 98 °C (10 s), 60 °C (30 s), and 72 °C (30 s), final elongation at 72 °C for 10 min and infinite hold at 4 °C (*see* **Note 10** to determine the number of PCR cycles during library preparation).
7. After starting the PCR, take out the AMPure XP beads (approximately 240 μl) from the refrigerator and let them prewarm at room temperature.
8. Once the PCR is finished, remove the plate from the cycler, vortex AMPure XP Beads and add 40 μl to each library sample. Mix beads with the sample thoroughly by pipetting up and down. Change tips while mixing different libraries to avoid contamination. If library preparation was done in PCR tubes, add 40 μl AMPure XP Beads to each sample, mix well and transfer the sample and the beads in a 96-well plate for cleanup.
9. Incubate the 96-well plate at room temperature for 10 min and afterward place it on the magnetic stand until all the beads are attached to the wall of the plate and the liquid is clear. At this stage, DNA is attached to the beads. Remove and discard the liquid while keeping the plate on the magnetic stand. Prepare 80% ethanol in DNase/RNase-free water during this incubation period.
10. Add 150 μl freshly prepared 80% ethanol, incubate at least for 30 s, remove and discard the ethanol without disturbing the beads.
11. Repeat the washing step once more. After the washing steps, make sure to remove small droplets of ethanol at the bottom or the corner of the well using a single-channel pipette.
12. Air dry beads for 10 min while keeping the plate on the magnetic stand.
13. Remove the plate from the magnetic stand and resuspend beads in 26 μl DNase/RNase-free water. Pipette the entire volume up and down several times to mix thoroughly.
14. Incubate the plate at room temperature for 2 min, place it back on the magnetic stand until the liquid appears clear and transfer 25 μl of supernatant to the empty wells of the same 96-well plate.
15. Repeat the bead cleanup process again but this time with 20 μl of beads. Elute the DNA in 11 μl of DNase/RNase-free water.

16. Use 1–1 μl per library sample to measure the concentration by Qubit® Fluorometer using Qubit dsDNA HS (High Sensitivity) Assay Kit and size distribution by Agilent 2100 Bioanalyzer using High Sensitivity DNA kit (Assay class: High Sensitivity DNA Assay, Fig. 2b) respectively. The remaining 8 μl of the libraries can be stored at −20 °C.
17. Based on the concentration measured by Qubit® Fluorometer and the average library size measured by Agilent 2100 Bioanalyzer, calculate the molarity of each sample. Adjust the concentration to a particular molarity, pool the samples and submit them for pair-end sequencing. Consult your sequencing facility regarding the type of run and concentration needed by them to be loaded on the flow cell.

3.5 Expression Quantification and Data Analysis

1. After demultiplexing of the Illumina indices from the sequenced libraries, paired-end sequencing will generate two fastq files per library (96 cells) consisting of right and left mate of each read pair. Perform a quality check with standard QC tools such as FastQC. Filter or trim the reads, if necessary (*see* **Note 11**).
2. Align the right mate of the read pair to the reference transcriptome using standard mapping tools such as the BWA software package. Discard the reads mapping to multiple loci. To avoid loss of reads mapping to multiple transcripts, it is recommended to define gene loci comprising isoforms and group together loci of genes with high sequence identity such as paralogs. Reads are then mapped to these groups and only reads mapping to multiple groups are discarded.
3. Use left mate of the read pair to retrieve information about the cell barcode (first 6 base pairs) and UMIs (next 7–12 base pairs). The remaining read contains a ployT stretch and hence should not be mapped to the reference transcriptome.
4. For each cell barcode, count the number of UMIs per transcript, aggregate the identical UMIs across all transcripts derived from the same gene locus and count them only once to eliminate the amplification bias.
5. Use the following formula to convert the number of UMIs observed into the number of sequenced transcripts per cell (the formula accounts for the effect of random counting statistics, *see* ref. 18):

$$m_i = -K \ln\left(1 - \frac{k_{o,i}}{K}\right)$$

m_i denotes the number of sequenced transcripts, $k_{o,\,i}$ denotes the number of observed different UMIs for each gene i, and K denotes the total number of different UMIs.

6. After determining transcript counts for all the cells, determine the low-quality or stressed cells using various parameters such as low transcript count due to incomplete lysis or RNA degradation, relatively higher expression of various mitochondrial genes as well as long noncoding RNAs such as Kcnq1ot1. Remove such cells before data analysis (*see* **Note 12** for more details).
7. For data normalization and subsequent data analysis, use the RaceID2 algorithm implemented in R. To eliminate technical variability and batch effects, down-sample the number of transcripts to a particular chosen value (*see* **Note 13**).
8. To identify various subpopulations of cells in the dataset, infer the optimal number of clusters justifying the data set based on the saturation of the average within cluster dispersion. Using this number, perform k-medoids clustering. Check the robustness of the clusters using Jaccard's similarity score.
9. To identify rare cell types, implement the outlier identification method described in the algorithm. Check the quality of the background model fit by plotting variance as a function of the mean using the following command: `plotbackground(sc)`.
10. Visualize the identified subpopulation of cells and rare cell types by t-Distributed Stochastic Neighbor Embedding (t-SNE) algorithm. Functions to identify differentially expressed genes in each cluster are also provided with RaceID2.
11. To identify stem/progenitor cell populations among the identified subpopulations of cells and derive a lineage tree, use the StemID algorithm (*see* **Note 14**).

4 Notes

1. One can also design 384 RT primers keeping in mind that the 6 bp cell barcode present in the RT primers should differ from each other by at least two nucleotides in order to accommodate for the sequencing error which may lead to false-positive cell barcode identification after sequencing. Moreover, the GC content in the barcodes should be between 40% and 60%.
2. We routinely sort various cell types from different tissues such as lymphocytes and progenitors from the bone marrow and thymus, innate lymphoid cells, macrophages, lymph node stromal cells, thymic epithelial cells, cells from human liver and prostate, zebrafish pronephros, cells from early mouse embryos as well as from various cell lines. It is important to minimize the digestion time to get single cell suspensions from the tissues. The fraction of stressed or dead cells increases with longer isolation procedures, which leads to a loss of many cells during

downstream data analysis. Therefore, it is advisable to carefully optimize the isolation procedure and minimize the isolation and sorting time.

3. It is preferable to process the sorted plates as soon as possible. However, plates can be kept at −80 °C for up to 8–10 weeks without any evident signs of pronounced RNA degradation.
4. It is not necessary to incubate the plate at 95 °C. This step can be replaced by 65 °C incubation for 5 min as described in the original protocol. With certain smaller cell types such as double positive cells from the thymus, we have seen an enhanced recovery of aRNA with incubation at 95 °C for 3 min.
5. The volume range of mosquito®HTS is 25 nl–1.2 μl. Therefore, we dispense 2.2 μl SS mix in 2× 1.1 μl steps.
6. It is time consuming to pool 4 × 6 columns directly into four LoBind 0.5 ml tubes using a single-channel pipette. Therefore, using a multichannel pipette, we first pool columns 1–6, 7–12, 13–18, and 19–24 in columns 1, 7, 13, and 19, respectively. Afterward, using a single-channel pipette we pool the contents of columns 1, 7, 13, and 19 in four different 0.5 ml LoBind tubes.
7. It is important to be as fast as possible during the aRNA fragmentation step to avoid overfragmentation. Use of a multichannel pipette is recommended in cases with a large number of samples.
8. Depending on the cells of interest, one may get aRNA sample concentration ranging from <100 pg/μl for very small cells to >10 ng/μl for bigger cells, e.g., embryonic stem (ES) cells.
9. We use 12 RPI primers for multiplexing the libraries for sequencing. Our standard format is to sequence 12 libraries (12 × 96 = 1152 cells) per lane on an Illumina HiSeq2500 sequencer in High Output mode generating 200 million reads per lane (pair-end multiplexing run, 100 bp read length).
10. While working with cells having a large amount of RNA such as ES cells, a relatively higher concentration of aRNA (>10 ng/μl for 96 cells) can be obtained. Therefore, in this case only seven PCR cycles can be used to make the libraries. If the cells of interest have a very low amount of RNA and the aRNA concentration is in the range of 30–70 pg/μl as measured by the Bioanalyzer, the number of PCR cycles can be increased up to 14–15. We suggest trying increasing the PCR cycles for one or two libraries and quantify the concentration as well as size distribution with Qubit and Bioanalyzer, respectively, before processing other samples.
11. We advise to load diverse single cell sequencing libraries from different cell types on the flow cell. Exclusively loading the

libraries prepared from cells with lower number of transcripts on a flow cell may increase the fraction of low quality reads. We routinely load libraries from different cells types and species on all the eight lanes of HiSeq2500 flow cell and get good quality reads without any necessity for trimming.

12. We routinely observe relatively higher expression of several long noncoding RNAs such as Kcnq1ot1 in stressed cells. We notice that the expression increases with longer isolation procedures. We remove cells from the analysis if they express more than 2% of total transcripts as Kcnq1ot1 transcripts. Moreover, we also remove all the genes correlating with Kcnq1ot1 from the analysis (correlation coefficient >0.75 for mouse cells). These parameters need to be adjusted depending on the cell type and species under investigation.
13. Choosing a particular threshold to downsample the data can be tricky, especially if the dataset includes cells with a high as well as a low number of transcripts by the virtue of their biological origin. In such cases, care has to be taken in order not to choose a very high value as this will lead to the removal of cells with lower number of transcripts from the further downstream analysis which can be important for lineage tree inference.
14. Due to the limited space, we have described our analysis pipeline briefly. Readers are advised to download the reference manual describing the use of RaceID2 and StemID algorithms from the following link—https://github.com/dgrun/StemID. It is a step-by-step guide to identify subpopulations of cells and rare cell types by k-medoids clustering and outlier identification as well as to predict stem cell populations in the dataset and computationally derive lineage trees.

Acknowledgments

The authors would like to thank Thomas Boehm, Sebastian Hobitz, and Ulrike Bönisch for their help in developing the protocol.

References

1. Raj A, van den Bogaard P, Rifkin SA, van Oudenaarden A, Tyagi S (2008) Imaging individual mRNA molecules using multiple singly labeled probes. Nat Methods 5(10):877–879. https://doi.org/10.1038/nmeth.1253. nmeth.1253 [pii]
2. Citri A, Pang ZP, Sudhof TC, Wernig M, Malenka RC (2011) Comprehensive qPCR profiling of gene expression in single neuronal cells. Nat Protoc 7(1):118–127. https://doi.org/10.1038/nprot.2011.430. nprot.2011.430 [pii]
3. Luo L, Salunga RC, Guo H, Bittner A, Joy KC, Galindo JE, Xiao H, Rogers KE, Wan JS, Jackson MR, Erlander MG (1999) Gene expression profiles of laser-captured adjacent neuronal

subtypes. Nat Med 5(1):117–122. https://doi.org/10.1038/4806

4. Grun D, Lyubimova A, Kester L, Wiebrands K, Basak O, Sasaki N, Clevers H, van Oudenaarden A (2015) Single-cell messenger RNA sequencing reveals rare intestinal cell types. Nature 525(7568):251–255. https://doi.org/10.1038/nature14966. nature14966 [pii]
5. Tang F, Barbacioru C, Bao S, Lee C, Nordman E, Wang X, Lao K, Surani MA (2010) Tracing the derivation of embryonic stem cells from the inner cell mass by single-cell RNA-Seq analysis. Cell Stem Cell 6(5):468–478. https://doi.org/10.1016/j.stem.2010.03.015. S1934-5909(10)00114-1 [pii]
6. Treutlein B, Brownfield DG, Wu AR, Neff NF, Mantalas GL, Espinoza FH, Desai TJ, Krasnow MA, Quake SR (2014) Reconstructing lineage hierarchies of the distal lung epithelium using single-cell RNA-seq. Nature 509(7500):371–375. https://doi.org/10.1038/nature13173. nature13173 [pii]
7. Patel AP, Tirosh I, Trombetta JJ, Shalek AK, Gillespie SM, Wakimoto H, Cahill DP, Nahed BV, Curry WT, Martuza RL, Louis DN, Rozenblatt-Rosen O, Suva ML, Regev A, Bernstein BE (2014) Single-cell RNA-seq highlights intratumoral heterogeneity in primary glioblastoma. Science 344(6190):1396–1401. https://doi.org/10.1126/science.1254257. science.1254257 [pii]
8. Grun D, Muraro MJ, Boisset JC, Wiebrands K, Lyubimova A, Dharmadhikari G, van den Born M, van Es J, Jansen E, Clevers H, de Koning EJ, van Oudenaarden A (2016) De novo prediction of stem cell identity using single-cell transcriptome data. Cell Stem Cell 19(2):266–277. https://doi.org/10.1016/j.stem.2016.05.010. S1934-5909(16)30094-7 [pii]
9. Hashimshony T, Wagner F, Sher N, Yanai I (2012) CEL-Seq: single-cell RNA-Seq by multiplexed linear amplification. Cell Rep 2(3):666–673. https://doi.org/10.1016/j.celrep.2012.08.003. S2211-1247(12)00228-8 [pii]
10. Islam S, Kjallquist U, Moliner A, Zajac P, Fan JB, Lonnerberg P, Linnarsson S (2011) Characterization of the single-cell transcriptional landscape by highly multiplex RNA-seq. Genome Res 21(7):1160–1167. https://doi.org/10.1101/gr.110882.110.gr.110882.110 [pii]
11. Islam S, Zeisel A, Joost S, La Manno G, Zajac P, Kasper M, Lonnerberg P, Linnarsson S (2014) Quantitative single-cell RNA-seq with unique molecular identifiers. Nat Methods 11(2):163–166. https://doi.org/10.1038/nmeth.2772. nmeth.2772 [pii]
12. Klein AM, Mazutis L, Akartuna I, Tallapragada N, Veres A, Li V, Peshkin L, Weitz DA, Kirschner MW (2015) Droplet barcoding for single-cell transcriptomics applied to embryonic stem cells. Cell 161(5):1187–1201. https://doi.org/10.1016/j.cell.2015.04.044. S0092-8674(15)00500-0 [pii]
13. Macosko EZ, Basu A, Satija R, Nemesh J, Shekhar K, Goldman M, Tirosh I, Bialas AR, Kamitaki N, Martersteck EM, Trombetta JJ, Weitz DA, Sanes JR, Shalek AK, Regev A, McCarroll SA (2015) Highly parallel genome-wide expression profiling of individual cells using nanoliter droplets. Cell 161(5):1202–1214. https://doi.org/10.1016/j.cell.2015.05.002. S0092-8674(15)00549-8 [pii]
14. Picelli S, Bjorklund AK, Faridani OR, Sagasser S, Winberg G, Sandberg R (2013) Smart-seq2 for sensitive full-length transcriptome profiling in single cells. Nat Methods 10(11):1096–1098. https://doi.org/10.1038/nmeth.2639. nmeth.2639 [pii]
15. Ramskold D, Luo S, Wang YC, Li R, Deng Q, Faridani OR, Daniels GA, Khrebtukova I, Loring JF, Laurent LC, Schroth GP, Sandberg R (2012) Full-length mRNA-Seq from single-cell levels of RNA and individual circulating tumor cells. Nat Biotechnol 30(8):777–782. nbt.2282 [pii]. https://doi.org/10.1038/nbt.2282
16. Sasagawa Y, Nikaido I, Hayashi T, Danno H, Uno KD, Imai T, Ueda HR (2013) Quartz-Seq: a highly reproducible and sensitive single-cell RNA sequencing method, reveals non-genetic gene-expression heterogeneity. Genome Biol 14(4):R31. https://doi.org/10.1186/gb-2013-14-4-r31. gb-2013-14-4-r31 [pii]
17. Hashimshony T, Senderovich N, Avital G, Klochendler A, de Leeuw Y, Anavy L, Gennert D, Li S, Livak KJ, Rozenblatt-Rosen O, Dor Y, Regev A, Yanai I (2016) CEL-Seq2: sensitive highly-multiplexed single-cell RNA-Seq. Genome Biol 17:77. https://doi.org/10.1186/s13059-016-0938-8. 10.1186/s13059-016-0938-8 [pii]
18. Grun D, Kester L, van Oudenaarden A (2014) Validation of noise models for single-cell transcriptomics. Nat Methods 11(6):637–640. https://doi.org/10.1038/nmeth.2930. nmeth.2930 [pii]

Chapter 16

Functional Insulator Scanning of CpG Islands to Identify Regulatory Regions of Promoters Using CRISPR

Alice Grob, Masue Marbiah, and Mark Isalan

Abstract

The ability to mutate a promoter in situ is potentially a very useful approach for gaining insights into endogenous gene regulation mechanisms. The advent of CRISPR/Cas systems has provided simple, efficient, and targeted genetic manipulation in eukaryotes, which can be applied to studying genome structure and function.

The basic CRISPR toolkit comprises an endonuclease, Cas9, and a short DNA-targeting sequence, made up of a single guide RNA (sgRNA). The catalytic domains of Cas9 are rendered active upon dimerization of Cas9 with sgRNA, resulting in targeted double stranded DNA breaks. Among other applications, this method of DNA cleavage can be coupled to endogenous homology-directed repair (HDR) mechanisms for the generation of site-specific editing or knockin mutations, at both promoter regulatory and gene coding sequences.

A well-characterized regulatory feature of promoter regions is the high abundance of CpGs. These CpG islands tend to be unmethylated, ensuring a euchromatic environment that promotes gene transcription. Here, we demonstrate CRISPR-mediated editing of two CpG islands located within the promoter region of the MDR1 gene (Multi Drug Resistance 1). Cas9 is used to generate double stranded breaks across multiple target sites, which are then repaired while inserting the beta globin (β-globin) insulator, 5′HS5. Thus, we are screening through promoter regulatory sequences with a chromatin barrier element to identify functional regions via "insulator scanning." Transcriptional and functional assessment of MDR1 expression provides evidence of genome engineering. Overall, this method allows the scanning of CpG islands to identify their promoter functions.

Key words Genome engineering, CRISPR, CpG islands, DNA methylation, Insulator scanning, MDR1

1 Introduction

CRISPR/Cas systems (Clustered Regularly Interspaced Short Palindromic Repeats/CRISPR Associated genes) are RNA-guided genome editing tools. Originally identified as adaptive immune responses in bacteria and archaea, CRISPR/Cas systems have

Alice Grob and Masue Marbiah contributed equally to this work.

Tanya Vavouri and Miguel A. Peinado (eds.), *CpG Islands: Methods and Protocols*, Methods in Molecular Biology, vol. 1766, https://doi.org/10.1007/978-1-4939-7768-0_16,

since been adapted for genome engineering within a wide range of model organisms [1–3]. CRISPR-mediated genome editing most commonly utilizes a highly versatile and programmable endonuclease, Cas9, which gains specificity through dimerization with a single guide RNA (sgRNA) (*see* Fig. 1). Two nuclease domains within Cas9 are responsible for generating double-strand breaks within a targeted DNA sequence. Specifically, the HNH nuclease domain cleaves the complementary strand [4], while the Ruv-C like nuclease domain cleaves the non-complementary strand [5]. Cas9 cleavage sites are determined by a short conserved sequence known as the Protospacer Adjacent Motif (PAM), which has the consensus sequence "NGG." PAM sites allow CRISPR systems to differentiate between foreign DNA (containing PAM) and the host loci coding for the protospacer target region; Cas9 fails to cleave target sequences lacking the PAM site [6]. Therefore, while the PAM site is a necessary marker of any cleavable genomic target site, it is not included in the sgRNA sequence. sgRNA is itself a chimeric sequence comprising a 20–25 nt sequence (spacer) that forms base pairs with the target genomic sequence, a 42 nt hairpin scaffolding structure that facilitates Cas9 binding, and a 40 nt transcriptional terminator derived from *S. pyogenes.*

Cas9 cleavage efficiencies have been enhanced by modifications to the sgRNA scaffold. Among others, these include the disruption of four consecutive U's within the hairpin structure that modifies a putative RNA polymerase III termination sequence. This is purported to inhibit early termination of U6 polymerase III mediated transcription. In addition, the 42 nt hairpin can be extended by five base pairs to improve complex formation with Cas9 [7]. The resulting flipped and extended (FE) sgRNA improves the efficiency of Cas9 on-target cleavages.

Since its discovery in 2012, continual optimization of the genome editing toolkit has meant that CRISPR/Cas systems are now suitable for an extended range of applications. For example, CRISPR/Cas systems can be used to integrate ectopic synthetic DNA at a targeted site within the genome. Indeed, after CRISPR-targeted genome cleavage, cells provided with a synthetic repair template will integrate this DNA by homology-directed repair (HDR) into the cleavage site (*see* Fig. 1) [8]. CRISPR/Cas systems have also been adapted to modulate promoter activity and gene expression. Specifically, fusion of transcriptional activators or repressors to a nuclease-dead (ND) Cas9 is now a common method used to modulate the chromatin state of specific targeted promoter regions [9].

Some promoter regions are highly enriched in cytosine (C) and guanine (G) residues, which form clusters of CpG repeats, called CpG islands [10]. Indeed, 60–70% of all annotated promoter regions contain CpG islands [10]. These islands are mostly unmethylated, and thus ensure a euchromatin status of promoter

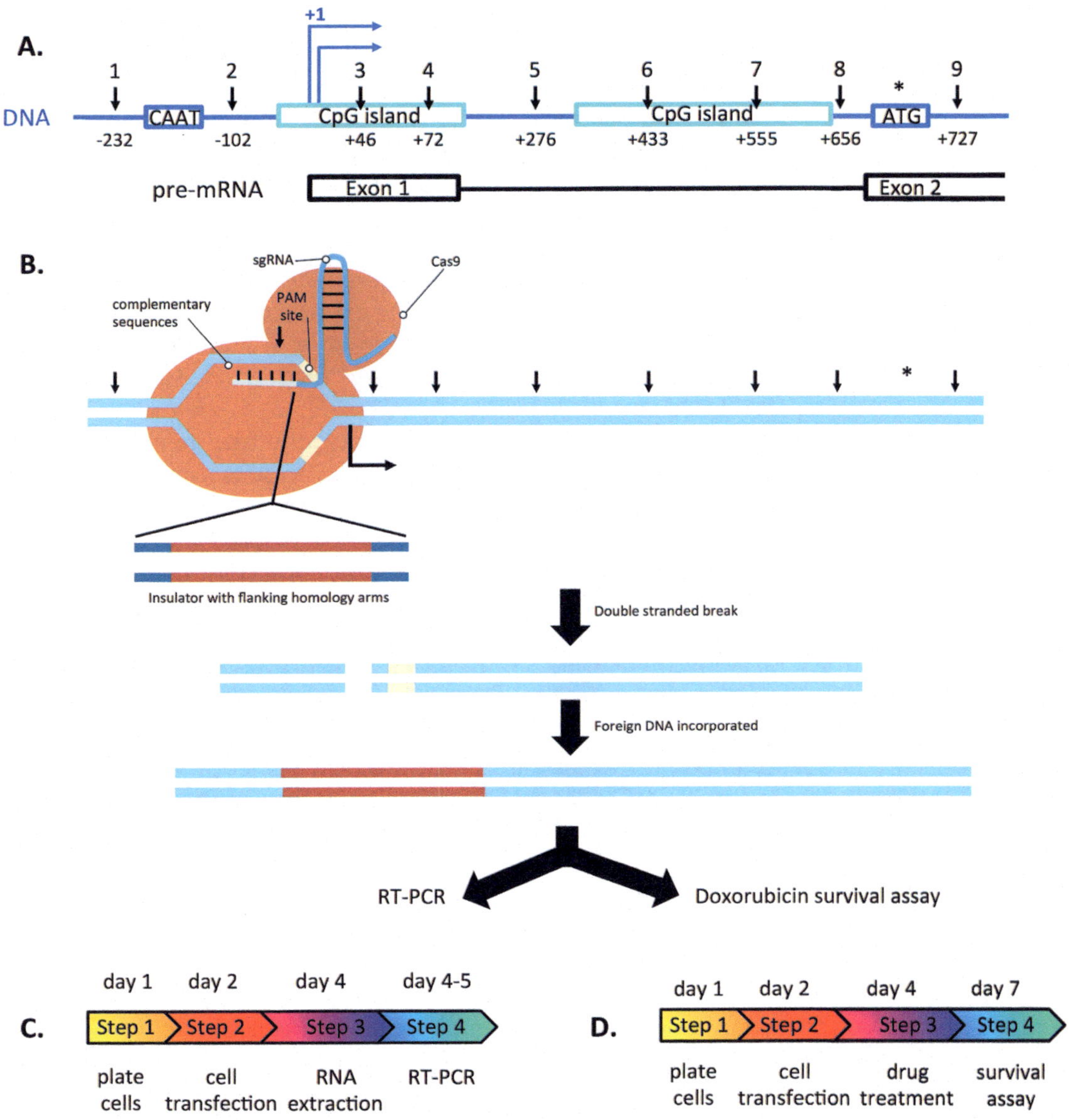

Fig. 1 Schematic representation of functional insulator scanning performed on CpG islands within the MDR1 promoter region. (**a**) Graphical representation of the MDR1 promoter region with designed CRISPR-targeted sites 1–9 indicated by arrows. The CAAT box (−116:−113 nt), transcription start site (elbow arrow, +1 and +4 nt) and ATG translation start site (star: *, +704:+707 nt) are highlighted. The CpG islands span across −60:+141 nt and +310:+649 nt of the MDR1 promoter region. The pre-mRNA resulting from MDR1 transcription is represented below, with Exon 1 coded by +1:+134 nt and Exon 2 coded from +698 nt. (**b**) Upon formation of the DNA/sgRNA/Cas9 complex, Cas9 generates a double stranded break 3 nt upstream of the PAM site. Cells are provided with a repair template composed of the β-globin 5′HS5 insulator, flanked by homology arms corresponding to the sequences that surround the PAM site. Thus, following HDR, β-globin 5′HS5 insulator sequences are integrated in the genome while the PAM site is removed. The effects of insulator scanning can be assessed either at a transcriptional level by RT-PCR or at a functional level by a doxorubicin survival assay. (**c**) A schematic showing the major steps and timeline for transcriptional validation of CRISPR-mediated genome editing by RT-PCR without doxorubicin drug treatment. (**d**) A schematic showing the major steps and timeline for a functional doxorubicin survival assay following CRISPR-mediated genome editing

regions, leading to efficient gene transcription. In the context of CRISPR/Cas-targeted cleavages, the abundance of guanine residues in CpG islands correlates with a high frequency of PAM sites. This provides an ideal setting to target multiple, consecutive sites within these promoter regions (*see* Fig. 1). In theory, this should allow us to determine which CpG island is contributing to transcription. Indeed, by coupling CRISPR-cleavages with the insertion of a chromatin insulator element, it should be possible to interfere with the CpG-dependant regulation of promoter chromatin state and thus the level of gene expression. The transcriptional changes that result from such "insulator scanning" of promoter regions can be easily assessed by reverse transcription PCR (RT-PCR).

In the example that follows, the promoter region of Multi Drug Resistance protein 1 (MDR1) is selected as a model promoter containing CpG islands. MDR1 is a member of the ATP-binding cassette transporter proteins that are responsible for energy-dependent xenobiotic efflux (including molecules such as doxorubicin). Thus, MDR1 is implicated in multidrug resistance as it decreases intracellular accumulation of toxic drugs such as doxorubicin [11]. Here, we aim to perform insulator scanning across two CpG islands within the MDR1 promoter region [12] by inserting the β-globin 5′HS5 insulator element. We posit that our insulator scanning protocol will affect the promoter function of CpG islands, decreasing MDR1 expression, and rendering cells more susceptible to doxorubicin toxicity. Cell viability is assessed in a microplate reader by measuring the reducing rate of PrestoBlue, a resazurin blue reagent, which is reduced in living cells into resarufin, a red fluorophore.

In conclusion, we describe a protocol for using the CRISPR/Cas9 system to generate double stranded breaks across the MDR1 promoter region, in order to integrate insulator elements. Transcriptional and functional assays (*see* Fig. 2) are then used to determine the effect of insulator scanning on promoter activities, allowing the user to map out CpG island functionality.

2 Materials

2.1 CRISPR Genome Engineering System and Insulator Repair Templates

1. Human codon-optimized Cas9 coding plasmid as a bacterial "stab" in agar (Addgene, plasmid 41815).
2. FE-sgRNA-cloning plasmid (a kind gift from Prof. B. McStay) [13].
3. pSF-CAG-Ub-Puro plasmid containing the β-globin 5′HS5 insulator sequences (Oxford genetics, OG600).

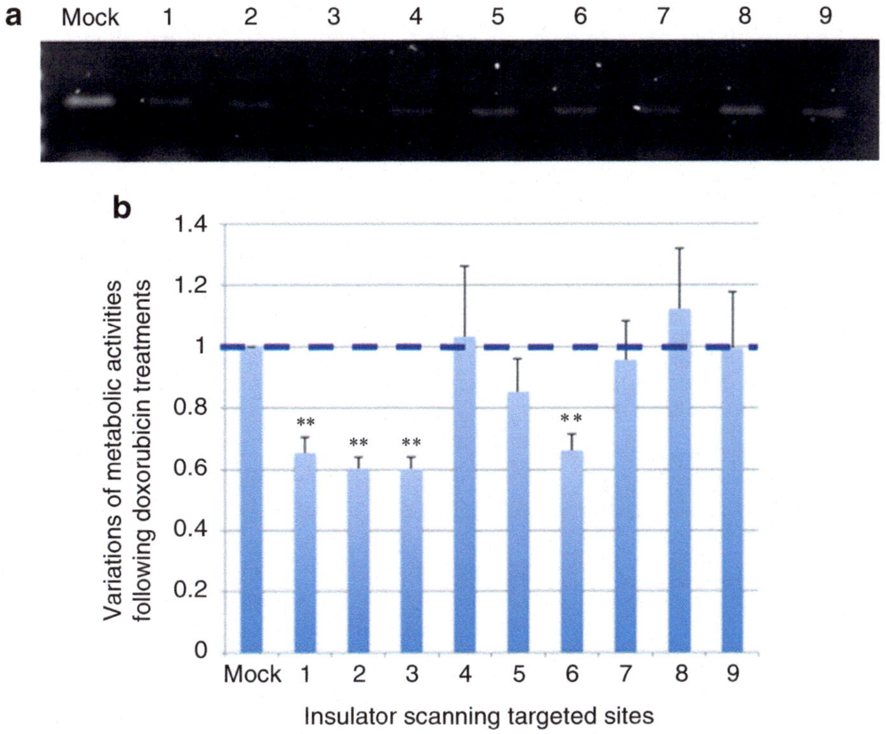

Fig. 2 Assessment of CRISPR-Cas genome editing on MDR1 transcript and function. (**a**) RT-PCR to estimate *mdr1* expression levels. HT-1080 cells were transfected with Cas9 plasmid, FE-sgRNA plasmid and insulator repair template DNA so as to integrate the β-globin 5′HS5 insulator at selected loci across the MDR1 promoter region. Two days later, total RNAs were extracted and processed by RT-PCR. A representative example of gel electrophoresis with the amplified cDNAs is shown here. Image enhanced using nonlinear transformation to aid visualization. (**b**) Variations in metabolic activities following insulator scanning and doxorubicin treatment. HT-1080 cells were transfected with Cas9 plasmid, FE-sgRNA plasmid and insulator repair template fragment DNA in order to integrate the β-globin 5′HS5 insulator sequences across 9 loci of the MDR1 promoter region. Two days later, cells were treated with either 0 or 100 nM of doxorubicin for 72 h. Plotted here are the variations of metabolic activities that have been normalized to the variations observed for the mock treated cells. The mock treated cells correspond to cells transfected with Cas9 plasmid DNA only. $n = 21$ samples per treatment (biological replicates) and the error bars are 1 s.d. The smaller the bar, the more the CpG locus activity is disrupted by insulator integration. Significant disruptions, as assessed by *t*-test values inferior to 0.01, are indicated by two stars (**). Disrupting some positions (e.g., 1, 2, 3, and 6) consistently increases cell sensitivity to doxorubicin (up to 40%), while targeting other positions (e.g., 4, 5, and 7–9) appears to have no deleterious effect on cell survival with doxorubicin (N.S). Thus, position 1 and 2, surrounding the CAAT box, as well as positions 3 and 6, in the 5′ regions of CpG islands, are likely more important in regulating *mdr1* expression levels

4. Primers encoding FE-sgRNA spacer sequences (*see* Table 1) and primers to amplify targeted insulator repair templates (*see* Table 2).
5. Phusion® hot start flex DNA polymerase kit, Q5® hot start high-fidelity 2× master mix and Gibson Assembly® master mix (NEB).

Table 1
Forward and Reverse primer pairs to clone FE-sgRNA for each targeted Cas9-cleavage site (1–9)

	cleavage site	spacer-PAM sequences	Forward & Reverse FE-sgRNA primers
1	-232	CGCGCATCAGCTGAATCATT-GGG	**F** TTTCTTGGCTTTATATATCTTGTGGAAAGGACGAAACACC*GCGCGCATCAGCTGAATCATT* **R** CTGTTTCCAGCATAGCTCTTAAAC *AATGATTCAGCTGATGCGCGC*
2	-102	AGCATTCAGTCAATCCGGGC-CGG	**F** TTTCTTGGCTTTATATATCTTGTGGAAAGGACGAAACACC*GAGCATTCAGTCAATCCGGGC* **R** CTGTTTCCAGCATAGCTCTTAAAC *GCCCGGATTGACTGAATGCTC*
3	46	GACCTAAAGGAAACGAACAG-CGG	**F** TTTCTTGGCTTTATATATCTTGTGGAAAGGACGAAACACC*GGACCTAAAGGAAACGAACAG* **R** CTGTTTCCAGCATAGCTCTTAAAC *CTGTTCGTTTCCTTTAGGTCC*
4	72	AGAAGATACTCCGACTTTAG-TGG	**F** TTTCTTGGCTTTATATATCTTGTGGAAAGGACGAAACACC*GCTGTTCGTTTCCTTTAGGTC* **R** CTGTTTCCAGCATAGCTCTTAAAC *GACCTAAAGGAAACGAACAGC*
5	276	AGTCTAGATCTAACCCCACT-TGG	**F** TTTCTTGGCTTTATATATCTTGTGGAAAGGACGAAACACC*AGTCTAGATCTAACCCCACT* **R** CTGTTTCCAGCATAGCTCTTAAAC *AGTGGGGTTAGATCTAGACTC*
6	433	GAAGCATCGTCCGCGGCGAC-TGG	**F** TTTCTTGGCTTTATATATCTTGTGGAAAGGACGAAACACC*GGAAGCATCGTCCGCGGCGAC* **R** CTGTTTCCAGCATAGCTCTTAAAC *GTCGCCGCGGACGATGCTTCC*
7	555	CCAGCTGCTCTGGCCGCGAT-GGG	**F** TTTCTTGGCTTTATATATCTTGTGGAAAGGACGAAACACC*GCCAGCTGCTCTGGCCGCGAT* **R** CTGTTTCCAGCATAGCTCTTAAAC *ATCGCGGCCAGAGCAGCTGGC*
8	656	TAGCCAAATGCATGAGCCTC-AGG	**F** TTTCTTGGCTTTATATATCTTGTGGAAAGGACGAAACACC*GTAGCCAAATGCATGAGCCTC* **R** CTGTTTCCAGCATAGCTCTTAAAC *GAGGCTCATGCATTTGGCTAC*
9	727	GGTTTCTCTTCAGGTCGGAA-TGG	**F** TTTCTTGGCTTTATATATCTTGTGGAAAGGACGAAACACC*GGGTTTCTCTTCAGGTCGGAA* **R** CTGTTTCCAGCATAGCTCTTAAAC *TTCCGACCTGAAGAGAAACCC*

Positions of cleavage sites as well as spacer and PAM sequences are also indicated. Targeting sites within CpG islands are highlighted in grey

6. MAX Efficiency® DH10β™ competent cells (Invitrogen™ by life technologies™, 18297-010).
7. LB and LB agar.
8. Ampicillin and kanamycin antibiotics.
9. NucleoSpin® Plasmid miniprep and NucleoBond® Xtra midiprep kits (Macherey-Nagel).
10. BamHI, EcoRI, XbaI and AflII restriction enzymes and dedicated buffers (NEB).
11. NucleoSpin® Gel and PCR Clean-up kit (Macherey-Nagel).
12. Pure sterile water (e.g., MilliQ).
13. TE (10:1), pH 7.15: 10 mM Tris–HCl pH 7.5, 0.1 mM EDTA pH 8.

2.2 Cell Culture

Store frozen aliquots of mammalian cells in a liquid nitrogen tank; DMEM stock and culture medium at 4 °C; Trypsin–EDTA solution at −20 °C (or 4 °C while in use); doxorubicin hydrochloride solutions in the dark at −20 °C.

1. HT-1080, human fibrosarcoma-derived cell line (LGC, ATCC-CCL-121).

Table 2
Forward and Reverse primer pairs to PCR-amplify the insulator repair templates

		Forward & Reverse targeted insulator repair primers
1	F	atgtgaactttgaaagacgtgtctacataagttgaaatgtCATCTTGGACCATTAGCTCC
	R	agaaagggcaagtagagaaacgcgcatcagctgaatcattGAGAGGTAGCTGAAGCTGC
2	F	tcgcagtttctcgaggaatcagcattcaGTCAATccgggcCATCTTGGACCATTAGCTCC
	R	TTCCTGCCCagccaatcagcctcaccacagatgactgctcGAGAGGTAGCTGAAGCTGC
3	F	cAttcgagtagcggctcttccaagctcaaagaagcagaggCATCTTGGACCATTAGCTCC
	R	tactccgactttagtggaaagacctaaaggaaacgaacagGAGAGGTAGCTGAAGCTGC
4	F	caaagaagcagaggccgctgttcgtttcctttaggtctttCATCTTGGACCATTAGCTCC
	R	ccaagacgtgaaattttggaagaagatactccgactttagGAGAGGTAGCTGAAGCTGC
5	F	ggcgtggatagtgtgaagtcctctggcaagtccatggggaCATCTTGGACCATTAGCTCC
	R	cgctgctccaggagctcctgagtctagatctaaccccactGAGAGGTAGCTGAAGCTGC
6	F	ccgcgggcggtgggtgggaggaagcatcgtccgcggcgacCATCTTGGACCATTAGCTCC
	R	aaccgggagggagaatcgcactggcggcgggcaaagtccaGAGAGGTAGCTGAAGCTGC
7	F	agatgctggagaccccgcgcacaggaaagcccCTGCAGtgCATCTTGGACCATTAGCTCC
	R	gagcgcccgccgttgatgccccagctgctctggccgcgatGAGAGGTAGCTGAAGCTGC
8	F	cttcgacgggggactagaggttagtctcacctccagcgcgCATCTTGGACCATTAGCTCC
	R	gaagagaaaccgcagctcattagccaaatgcatgagcctcGAGAGGTAGCTGAAGCTGC
9	F	gcatttggctaatgagctgcggtttctcttcaggtcggaACATCTTGGACCATTAGCTCC
	R	ATCTTGAAGGGGACCGCAATGGAGGAGCAAAGAAGAAGAAGAGAGGTAGCTGAAGCTGC

β-globin 5'HS5 insulator
CATCTTGGACCATTAGCTCCACAGGTATCTTCTTCCCTCTAGTGGTCATAACAGCAGCTTCAGCTACCTCTC

β-globin 5′HS5 insulator sequences are also indicated below the table. Targeting sites within CpG islands are highlighted in grey

2. Culture medium: 500 ml DMEM medium (Gibco® by life technologies™, 41965-039), 10% (v/v) heat-inactivated FBS (Labtech.com, FCS-SA).
3. Dulbecco's Phosphate Buffered Saline (PBS) solution (Sigma, D8537-500ML).
4. 0.25% (v/v) Trypsin–EDTA solution (Gibco® by Life Technologies™, 25200-056).
5. Scepter™ 2.0 cell counter (Merck Millipore, PHCC20040) and Sensor Scepter™ 2.0 40 μm particle size range 3–18 μm (Merck Millipore, PHCC40050).
6. Doxorubicin hydrochloride (Sigma, D1515-10MG) resuspended in sterile DMSO (Sigma, D2438).

2.3 Transfection by Calcium Phosphate Precipitation

Prepare all solutions using pure sterile water (e.g., MilliQ). Adjust the pH of all solutions carefully. Filter-sterilize TE (10:1) solution; autoclave calcium and phosphate buffers. TE (10:1) solution is stored at 4 °C; calcium and phosphate buffers are stored at −20 °C.

1. TE (10:1), pH 7.15: 10 mM Tris–HCl pH 7.5, 1 mM EDTA pH 8.
2. Calcium buffer, pH 7.2: 2.5 M $CaCl_2$ (MW 110.98), 10 mM HEPES (MW 238.3).
3. Phosphate buffer, pH 7.05: 275 mM NaCl (MW 58.44), 10 mM KCl (MW 74.55), 1.4 mM Na_2HPO_4 (MW 141.96), 11 mM dextrose (MW 180.16), 35 mM HEPES (MW 283.3).

2.4 RNA Extraction and RT-PCR

1. Ribonucleoside Vanadyl Complexes (Sigma, R3380).
2. NucleoSpin® RNA extraction kit (MN, 740955.50).
3. ProtoScript® II First Strand cDNA Synthesis Kit (NEB, E6560).
4. Primers to PCR-amplify cDNA of interest.
5. Phusion® hot start flex DNA polymerase kit (NEB).
6. Tris-Borate-EDTA (TBE) 10× buffer and Agarose (Sigma) for DNA gel electrophoresis.

2.5 Tecan Measurement of Metabolic Activities

1. Infinite® M200Pro microplate reader and Gas Control Module (Tecan).
2. PrestoBlue® Reagent (Invitrogen™ by life technologies™, A13262).

3 Methods

3.1 Human Codon-Optimized Cas9 Plasmid

1. Obtain plasmid from Addgene (No. 41815).
2. Use a sterile loop to scrape the bacterial "stab" provided and streak it onto an LB agar plate containing 100 μg/ml ampicillin. Incubate the plate overnight (O/N) at 37 °C.
3. Pick a single colony from the plate to inoculate a 10 ml preculture in LB medium containing 100 μg/ml ampicillin. Grow the bacterial clone for 6 h at 37 °C with a 250 rpm orbital shaking (*see* **Note 1**).
4. Inoculate 300 ml of LB containing 100 μg/ml ampicillin with the preculture. Grow the culture O/N at 37 °C with a 250 rpm orbital shaking.
5. Use a NucleoBond® Xtra midiprep kit to purify between 1 and 3 μg of Cas9 plasmid DNA from the bacterial culture. The resulting Cas9 plasmid DNA is ready to be used in transfection by calcium phosphate precipitation.

6. Verify Cas9 plasmid integrity by BamHI/EcoRI restriction digestion, which should generate a 7.7 kb and a 1.8 kb fragment.

3.2 FE-sgRNA Plasmid

1. Kindly request and obtain the FE-sgRNA-cloning plasmid from Prof. Brian McStay (NUIG, Ireland) (*see* **Note 2**). FE-sgRNA-cloning plasmid DNA can be purified using the NucleoBond® Xtra midiprep kit from bacteria grown in 300 ml LB containing 50 μg/ml kanamycin. Its integrity can be verified by a BamHI/XbaI restriction digest generating 3.4, 0.35, 0.1, and 0.07 kb fragments.
2. Using an online CRISPR sgRNA design software (http://crispr.mit.edu/), identify all 22 bp genomic sites of 5′-N19-NGG-3′, which are suitable target sites for FE-sgRNA/DNA hybridization and Cas9 nuclease activity. Favor targeting sites with high score and low off-target binding sites (*see* **Note 3**).
3. Incorporate the first 19 nt and their reverse complement into the FE-sgRNA forward (TTTCTTGGCTTTATATATCTTG TGGAAAGGACGAAACACC-GN19) and reverse (CTGTTT CCAGCATAGCTCTTAAAC-N19C) primers respectively (*see* Table 1 and **Note 4**). Order lyophilized primers from a gene synthesis company such as Sigma. Resuspend FE-sgRNA forward and revers primers at 1 μg/μl in TE (10:0.1).
4. In a thermocycler, anneal and fill-in the forward and reverse primers together (0.02 μg/μl of each primer, 0.2 mM dNTPs, 0.02 U/μl Phusion® hot start flex DNA polymerase and 1× Phusion buffer), with the following single cycle: 30 s at 98 °C, 30 s at 50 °C, 10 min at 72 °C, and hold at 4 °C.
5. Digest the FE-sgRNA-cloning plasmid with AflII restriction enzyme and purify it on a column using the NucleoSpin® Gel and PCR Clean-up kit.
6. Insert the double-stranded DNA fragment of FE-sgRNA spacer sequences (prepared in **step 4**) into the AflII-digested FE-sgRNA-cloning plasmid (prepared in **step 5**) by Gibson Assembly (GA) using a Gibson Assembly® master mix. Typical GA reactions should contain a 1:15 ng ratio of FE-sgRNA spacer sequences to AflII-digested FE-sgRNA-cloning plasmid. Following a 15 min incubation at 50 °C, reactions are diluted threefold with pure sterile water (*see* **Note 5**).
7. Transform 1/15 of the diluted GA reactions into DH10β before plating the bacteria onto LB agar plate(s) containing 50 μg/ml kanamycin. Incubate plates O/N at 37 °C.
8. Pick 4–10 single colonies from the plates to inoculate 4 ml cultures in LB containing 50 μg/ml kanamycin. Grow bacterial clones O/N at 37 °C with a 250 rpm orbital shaking.

9. Use NucleoSpin® Plasmid miniprep kit to purify plasmid DNA from 2 ml of the miniprep cultures. Store the remaining bacterial culture at 4 °C.
10. Verify the FE-sgRNA plasmids by BamHI/XbaI digestions, which should generate 3.4, 0.39, 0.1 and 0.07 kb fragments (*see* **Note 6**).
11. Further verify that the correct spacer sequences have been inserted into the FE-sgRNA cloning plasmid by sending an aliquot of the resulting FE-sgRNA plasmid DNA for sequencing (e.g., to a DNA sequencing service such as GATC Biotech).
12. Inoculate 300 ml of LB containing 50 μg/ml kanamycin with 1 ml of the selected miniprep culture that contain the correct sequence-verified FE-sgRNA plasmid. Grow the resulting culture O/N at 37 °C with a 250 rpm orbital shaking.
13. Use a NucleoBond® Xtra midiprep kit to purify between 1 and 3 μg of FE-sgRNA plasmid DNA from bacterial culture. The resulting FE-sgRNA plasmid DNA is ready to be used in transfection by calcium phosphate precipitation.

3.3 Targeted Insulator Repair Templates

1. Design primers to amplify the β-globin 5′HS5 insulator sequences with homology arms to direct their genomic insertion by HDR to the CRISPR-targeted cleavage site (*see* Table 2 and **Note 7**). To this end, forward primers should include the 40 nt sequences upstream of the CRISPR cleavage site together with 20 nt corresponding to the 5′ end of the insulator sequences. Reverse primers should include the 40 nt sequences downstream of the CRISPR cleavage site together with 20 nt corresponding to the 3′ end of the insulator sequences. Order both forward and reverse primers from a gene synthesis company, such as Sigma.
2. PCR-amplify insulator repair templates using Q5® hot start high-fidelity 2× master mix. Typical reactions contain 1 ng/μl of pSF-CAG-Ub-Puro plasmid DNA template, 0.5 μM of Forward primer, 0.5 μM of Reverse primer, and 1× Q5® hot start high-fidelity master mix. Reactions are incubated in a thermocycler following this program: 3 min 98 °C, [30 s 98 °C, 30 s 65 °C, 1 min 72 °C] × 25 cycles, 5 min 75 °C, and hold at 4 °C (*see* **Note 8**).
3. Purify the PCR fragments using NucleoSpin® Gel and PCR Clean-up kit. The resulting insulator targeted repair template DNA fragments are ready to be used in transfection by calcium phosphate precipitation.

3.4 Cell Culture

Work under a tissue culture hood in sterile conditions.

1. Defrost HT-1080 cells by resuspending the frozen aliquot in 5 ml culture medium. Spin down the defrosted cells at 350 × *g* for 5 min and resuspend the cell pellet in 2 ml culture medium. Add this 2 ml-cell-suspension to a T75 flask containing 10 ml culture medium.
2. Leave cells to adhere and grow in a 37 °C, 5% CO_2 incubator for typically 2–3 days until 80–100% cell confluency is reached.
3. Wash cells with 5 ml PBS solution.
4. Detach cells from the T75 flask surface by a 3–5 min 37 °C incubation with 5 ml of 0.25% (v/v) Trypsin-EDTA solution.
5. Once cells are fully detached, inhibit Trypsin activity by addition of 5 ml culture medium. Pipet up and down several times to obtain a homogenized single cell suspension.
6. Seed culture stock with 1/10 of the cells in a T75 flask containing 12 ml culture medium. Put the stock culture back into the 37 °C, 5% CO_2 incubator for 2–3 days until cells are confluent and need "splitting" again.
7. Count the remaining cells using the Scepter™ 2.0 cell counter and sensors.
8. Plate 0.2 to 2 × 10^5 cells/well in 12-well plates (*see* **Note 9**). Put the plates into the 37 °C, 5% CO_2 incubator for 24 h before proceeding with the transfections by calcium phosphate precipitations.

3.5 Transfection by Calcium Phosphate Precipitation

Transfections are carried out 24 h after splitting the cells, when cells are in exponential phase at 25–50% confluency (*see* **Note 10**). Quantities indicated here are suitable to transfect 1 well of a 12-well plate seeded with HT-1080 cells. Always include a mock transfection control where cells are only transfected with Cas9 plasmid DNA.

1. Thaw calcium and phosphate buffers at 37 °C to defrost.
2. Prepare TE (10:1) solutions containing a 1:1:3 ratio of Cas9 plasmid and insulator repair template DNA to FE-sgRNA plasmid DNA. Typically, prepare 31.5 μl TE (10:1) containing 115 ng Cas9 plasmid DNA, 115 ng Insulator PCR fragment and 340 ng FE-sgRNA plasmid DNA (*see* **Note 2**). Vortex.
3. Add 3.5 μl of calcium buffer. Pipet up and down several times until solutions are well homogenized.
4. Prepare 1.5 ml eppendorf tubes containing 35 μl of phosphate buffer.

5. Gently add the TE/DNA/calcium solution mix dropwise from the top of the tube onto the phosphate buffer, with regular flicking of the tubes.
6. Incubate samples for 10 min in the hood and occasionally flick the tubes to mix.
7. Add the TE/DNA/calcium/phosphate solution mix dropwise to the plated cells, covering as much surface as possible (*see* **Note 11**).
8. Incubate the transfection plate for 5–7 h in a 37 °C, 5% CO_2 incubator.
9. Remove the media and perform 2–3 washes with 1 ml PBS solution, before adding fresh culture media to the cells.
10. Incubate transfected cells back in the 37 °C, 5% CO_2 incubator for 2 days.

3.6 RNA Extraction and RT-PCR

1. Design RT-PCR primer pairs to specifically amplify MDR1 cDNA (*see* **Note 12**). Order them from a gene synthesis company, like Sigma.
2. Transfect Cas9 plasmid, insulator targeted repair templates and FE-sgRNA plasmid DNA using the calcium phosphate precipitation protocol described above in Subheading 3.5.
3. Harvest cells by trypsinization with 1 ml of 0.25% (v/v) Trypsin-EDTA solution as described in **steps 4** and **5** of the cell culture Subheading 3.4.
4. Wash cells with prechilled PBS solution containing 20 mM Ribonucleoside Vanadyle Complexes to prevent excessive RNA degradation.
5. Extract total RNA from transfected cells using the NucleoSpin® RNA extraction kit.
6. Use 1 μg of total RNAs to produce cDNAs using an oligodT primer, with the ProtoScript® II First Strand cDNA Synthesis Kit. Include control reactions for DNA contamination by replacing the reverse transcriptase with sterile water.
7. PCR-amplify, with Phusion® hot start flex DNA Polymerase, 1 μl cDNA per 50 μl PCR reactions (*see* **Note 13**).
8. Run resulting samples using 0.8% (w/v) Agarose/TBE gel electrophoresis to compare MDR1 mRNA levels following the insertion of β-globin 5′HS1 insulator sequences, at different positions across the MDR1 promoter region (*see* Fig. 2a).

3.7 Drug Treatment

1. Transfect Cas9 plasmid, insulator repair template and FE-sgRNA plasmid DNA using the Calcium Phosphate precipitation described above in Subheading 3.5.

2. Two days post-transfection, change the culture medium for medium containing either 1% (v/v) DMSO (i.e., 0 nM doxorubicin) or 100 mM doxorubicin (*see* **Note 14**).
3. Incubate cells with drug in a 37 °C, 5% CO_2 incubator for 3 days.

3.8 Tecan Measurement of Metabolic Activities

1. Following the drug treatment described above in Subheading 3.7, add PrestoBlue® Reagent to the cells in a 1:10 ratio to the culture medium.
2. Incubate cells at 37 °C, in a 5% CO_2 incubator, for 30 min.
3. Measure the resorufin levels produced by viable cells using the Tecan microplate reader. For typical measurements, the 12-well plates are placed in the reader set at 37 °C with 5% CO_2. Perform measurement for 2 h with the following cycle every 10 min: 20 s orbital shaking of 1.5 mm amplitude, 5 s incubation time, measurement of resorufin fluorescence at 590 nm following an excitation at 560 nm with 25 flashes, a gain set at 60 and 4 × 4 square multiple reads per well from the top. Set a control well as reference for the z-position of measurements (*see* **Note 15**).
4. Export resorufin fluorescence measurements to an Excel file to get graphs of resorufin levels plotted against time. Add trendlines to the curves to reveal their equations. Report all the slope measurement obtained from the graphs into a table. These measurements correspond to the metabolic activities of the viable cells and are proportional to the number of viable cells in each well under each condition. Values for multiple biological repeats of each condition should be obtained and the average of these values should be calculated and normalized to the values obtained for cells treated with 0 nM doxorubicin. Plot these average values against the dose of drug treatment to compare cell survival under the different conditions. To visualize variations across the scanned promoter, obtain the slope of the metabolic activities plotted against drug doses and normalize it to the mock transfection. Plot these normalized values of metabolic activity variations against CRISPR-targeted site (*see* Fig. 2b).

4 Notes

1. A glycerol stock of the preculture containing 80% bacterial culture and 20% glycerol should be made and stored at −80 °C to provide an archive. The glycerol stock can then be used directly to start a fresh bacterial preculture and to prepare more Cas9 plasmid DNA if needed.

2. sgRNA-cloning plasmid lacking the FE modifications can be obtained from Addgene (plasmid 41824). This plasmid can then be mutated into the FE-sgRNA-cloning plasmid by site directed mutagenesis according to the mutations described in [7]. The FE-sgRNA interaction with Cas9 is stabilized, thus improving the efficiency of Cas9 targeted nuclease activity. If sgRNAs lacking FE modifications are to be used instead of FE-sgRNA, the sgRNA:Cas9 ratio used to transfect mammalian cells needs to be calibrated for optimal Cas9 activity.
3. Identify multiple Cas9 cleavage sites at regular intervals across the region of interest containing CpG islands. This will enable the functional assessment of this region in promoter activity by insulator scanning. Cleavage sites can be on either DNA strand. It is essential that the FE-sgRNAs resulting from this design minimize the off target activity of the CRISPR/Cas system. On line sgRNA design tools, such as http://crispr.mit.edu/, indicate potential off target cleavage sites with the position of mismatches between off and on targets. Choose FE-sgRNA spacer regions with a quality score over 70, off targets (if any) in noncoding regions of the genome and mismatches to off targets within the 9 bp adjacent to the PAM sites. Indeed, mismatches adjacent to the PAM site tend to prevent Cas9 cleavages, ensuring specificity. PCR-amplified genomic DNA of the region of interest can be sequenced to ensure the presence of FE-sgRNA spacer hybridization sites within the genome of the specific cell line used.
4. To clone sgRNA spacer sequences into the sgRNA-cloning plasmid (Addgene, plasmid 41824), use the FE-forward primer to incorporate the first 19 nt, while using the sgRNA-specific reverse primer (GACTAGCCTTATTTTAACTTGCTATTTC TAGCTCTAAAAC-N19C) to incorporate the 19 nt reverse complement.
5. Diluting GA reactions with sterile water prior to their transformation into DH10β™ competent cells increases transformation efficiency by decreasing toxicity.
6. FE-sgRNA and sgRNA plasmids contain, respectively, 40 and 60 bp more than their parental cloning plasmids. Thus, when comparing BamHI/XbaI restriction digests of the original cloning plasmids to their resulting GA plasmids, a shift in fragment size should be detected. BamHI/XbaI digestion of FE-sgRNA-cloning plasmid generates: 3.4 kb + 0.35 kb + 0.1 kb + 0.07 kb fragments; while FE-sgRNA plasmid digestion generates: 3.4 kb + 0.39 kb + 0.1 kb + 0.07 kb fragments. Similarly, BamHI/XbaI digestion of sgRNA-cloning plasmid generates: 3.4 kb + 0.32 kb + 0.1 kb + 0.07 kb fragments; while sgRNA

plasmid digestion generates: 3.4 kb + 0.38 kb + 0.1 kb + 0.07 kb fragments.

7. Suitably designed repair templates should ensure that the spacer hybridization and PAM sites are destroyed following genome integration to avoid further CRISPR-targeted cleavages. Homology arms of as little as 40 nt have been successfully used to easily generate repair templates for CRISPR-targeted genome integration [14], although arms of 1 kbp are more typical.
8. Optimal annealing temperatures for the PCR program should first be established by testing the PCR-amplification efficiency with a temperature gradient at the annealing step. Here, we found that 65 °C was the optimal annealing temperature for our primer pairs.
9. Primary tests are required to establish the optimal number of cells to be seeded in order to reach 25–50% confluence after 24 h. Variable factors to take into account are cell size, culture surfaces, and experiment time following transfections. We recommend seeding 2×10^4 cells/well in 12-well plates for optimal transfection of HT-1080 cells and their subsequent culture over 3–5 days without exceeding 100% cell confluence. Using plates with larger surfaces, such as 12-well plates compared to 48-well plates, facilitates homogenous adherence of cells across the well. We also recommend gently shaking the plates with a cross motion when putting them into the incubator, in order to avoid concentration of cells to the well periphery under a centrifuge force.
10. Transfection by calcium phosphate precipitation is a relatively cheap, nontoxic protocol to efficiently transfect established adherent cell lines. However, transfection efficiency may differ according to cell lines. While the HT-1080 cell line is easy to transfect, cell lines that have more primary cell-like phenotype, such as the hTERT-RPE1 cell line, are notably more difficult to transfect. The method described here is optimal to transfect HT-1080 cells plated in 12-well dishes. Parameters like transfected DNA ratios and quantity of reagent used should be optimized for other cell lines and other culture dishes. These parameters can be optimized by transfecting a plasmid encoding for a fluorescent protein and monitoring the percentage of transfected fluorescent cells under a fluorescent microscope. The chosen fluorescent protein can either be an easy-to-express protein, like GFP, or one that requires more complex folding, like a fluorescently tagged Cas9 protein. Conditions established to transfect a fluorescently tagged Cas9 protein will be more closely related to the experiments described in this insulator scanning protocol. Furthermore, the pHs of various

solutions required for the calcium phosphate precipitation are essential for transfection efficiency. Other transfection protocols, like electroporation or lipid-based transfection, should be tested if the chosen cell line is unsuccessfully transfected by calcium phosphate precipitation.

11. Under a light microscope, make sure that a fine "black-dotted" precipitate is visible in the culture medium. It is a good indication of successful transfection. Following the 5–7 h incubation in a 37 °C, 5% CO_2 incubator, this fine black-dotted precipitate should be at the bottom of the well and mostly at the cell periphery. Washes with a PBS solution are required to remove precipitates that did not enter the cells, thus removing a large potential source of cellular toxicity.

12. Multiple RT-PCR primer pairs should be designed to provide different combinations and to maximize the chance of successful RT-PCR. These primers can either hybridize to introns present in the pre-mRNA only, or to exonic regions present in the pre-mRNA, as well as mature mRNA. RT-PCR-amplification of regions close to the transcription start in the 5′ UTR gene region will reflect more closely variations in gene transcription level. Here, we selected the following primer pair: RT-F, GAG CAG AAG TTT GTT GGC TGA and RT-R, AGG CAC ACC AAG ACT AAG GG.

13. Samples are normalized by using the same concentration of RNA in each reverse transcription reaction. To maintain accuracy, equal volumes of cDNA are used for PCR since quantification by NanoDrop also measures "free dNTPs," RNA, and cDNA. The cDNA concentration range provided is an estimation based on average NanoDrop measurements and theoretical calculations which assume 1:1 conversion of RNA:cDNA. Concentrations of cDNA varied from 5 to 50 ng/μl for the test samples, up to 800 ng/μl for the mock sample.

14. Primary tests are required to establish the optimal dose of doxorubicin treatment in order to use the minimal amount of drug while observing toxicity. It is necessary to establish this optimal dose for each cell line. For HT-1080 cells, we tested 1 nM, 10 nM, 100 nM, and 1 μM doxorubicin hydrochloride treatment for 24 and 48 h. We decided to use 100 nM doxorubicin treatment for 48 h. As doxorubicin is resuspended in DMSO, it is essential to include the same amount of DMSO in the control wells treated with 0 nM doxorubicin. Here, 1% (v/v) DMSO was used as a control treatment for 100 nM doxorubicin hydrochloride treatment.

15. Primary tests are required to establish the optimal resorufin fluorescence measurement program for the Tecan microplate reader. The optimal gain for resorufin fluorescence measurement should be determined. We recommend using a gain of

60 on the Tecan Infinite® M200Pro microplate reader. The seeding range of cell number to maintain a good correlation between produced resorufin levels and cell numbers also needs to be established. For a 12-well plate, we recommend seeding up to 2×10^5 HT-1080 cells/well. The duration of the measurement program can be optimized. Indeed, resorufin fluorescence levels reach a plateau over time and the measurement program should only last until such a plateau is reached. In fact, only the slopes (over time) of the initial linear part of resorufin fluorescence levels are relevant to reflect the metabolic activities of viable cells.

Acknowledgments

The authors were funded by Wellcome Trust UK New Investigator Award No. WT102944 (MI, AG) and a BBSRC-Innovate UK Industrial Biotechnology Catalyst Grant BB/M028933/1 (MM). Alice Grob and Masue Marbiah contributed equally to this work.

References

1. Mali P, Esvelt KM, Church GM (2013) Cas9 as a versatile tool for engineering biology. Nat Methods 10:957–963
2. Wang T, Wei JJ, Sabatini DM et al (2013) Genetic screens in human cells using the CRISPR-Cas9 system. Science 343 (6166):80–84
3. Shan Q, Wang Y, Li J et al (2013) Targeted genome modification of crop plants using a CRISPR-Cas system. Nat Biotechnol 31 (8):686–688
4. Jinek M, Chylinski K, Fonfara I et al (2012) A programmable dual-RNA-guided DNA endonuclease in adaptive bacterial immunity. Science 337(6096):816–821
5. Gasiunas G, Barrangou R, Horvath P et al (2012) Cas9-crRNA ribonucleoprotein complex mediates specific DNA cleavage for adaptive immunity in bacteria. Proc Natl Acad Sci U S A 109(39):E2579–E2586
6. Sternberg SH, Redding S, Jinek M et al (2014) DNA interrogation by the CRISPR RNA-guided endonuclease Cas9. Nature 507 (7490):62–67
7. Chen B, Gilbert LA, Cimini BA et al (2013) Dynamic imaging of genomic loci in living human cells by an optimized CRISPR/Cas system. Cell 155(7):1479–1491
8. Shrivastav M, De Haro LP, Nickoloff JA (2008) Regulation of DNA double-strand break repair pathway choice. Cell Res 18 (1):134–147
9. McDonald JI, Celik H, Rois LE et al (2016) Reprogrammable CRISPR/Cas9-based system for inducing site-specific DNA methylation. Biol Open 5(6):866–874
10. Illingworth RS, Bird AP (2009) CpG islands – 'a rough guide'. FEBS Lett 583 (11):1713–1720
11. Park JG, Lee SK, Hong IG et al (1994) MDR1 gene expression: its effect on drug resistance to doxorubicin in human hepatocellular carcinoma cell lines. J Natl Cancer Inst 86 (9):700–705
12. Baker E, El-Osta A (2009) Epigenetics regulation of multidrug resistance 1 gene expression: profiling CpG methylation status using Bisulphite sequencing. In: Zhou J (ed) Multi-drug resistance in cancer. Humana Press, Totowa, NJ
13. Van Sluis M, McStay B (2015) A localized nucleolar DNA damage response facilitates recruitment of the homology-directed repair machinery independent of cell cycle stage. Genes Dev 29(1):1151–1163
14. Stewart-Ornstein J, Lahav G (2016) Dynamics of CDKN1A in single cells defined by an endogenous fluorescent tagging toolkit. Cell Rep 14(7):1800–1811

Chapter 17

An Application-Directed, Versatile DNA FISH Platform for Research and Diagnostics

Eleni Gelali, Joaquin Custodio, Gabriele Girelli, Erik Wernersson, Nicola Crosetto, and Magda Bienko

Abstract

DNA fluorescence in situ hybridization (DNA FISH) has emerged as a powerful microscopy technique that allows a unique view into the composition and arrangement of the genetic material in its natural context—be it the cell nucleus in interphase, or chromosomes in metaphase spreads. The core principle of DNA FISH is the ability of fluorescently labeled DNA probes (either double- or single-stranded DNA fragments) to bind to their complementary sequences in situ in cells or tissues, revealing the location of their target as fluorescence signals detectable with a fluorescence microscope. Numerous variants and improvements of the original DNA FISH method as well as a vast repertoire of applications have been described since its inception more than 4 decades ago. In recent years, the development of many new fluorescent dyes together with drastic advancements in methods for probe generation (Boyle et al., Chromosome Res 19:901–909, 2011; Beliveau et al., Proc Natl Acad Sci U S A 109:21301–21306, 2012; Bienko et al., Nat Methods 10:122–124, 2012), as well as improvements in the resolution of microscopy technologies, have boosted the number of DNA FISH applications, particularly in the field of genome architecture (Markaki et al., Bioessays 34:412–426, 2012; Beliveau et al., Nat Commun 6:7147, 2015). However, despite these remarkable steps forward, choosing which type of DNA FISH sample preparation protocol, probe design, hybridization procedure, and detection method is best suited for a given application remains still challenging for many research labs, preventing a more widespread use of this powerful technology. Here, we present a comprehensive platform to help researchers choose which DNA FISH protocol is most suitable for their particular application. In addition, we describe computational pipelines that can be implemented for efficient DNA FISH probe design and for signal quantification. Our goal is to make DNA FISH a versatile and streamlined technique that can be easily implemented by both research and diagnostic labs.

Key words DNA fluorescence in situ hybridization, High-definition DNA FISH, Superresolution microscopy, Genome architecture, Cytogenetics

1 Introduction

Depending on the application, different strategies for DNA FISH can be identified. In diagnostics, in addition to specificity and sensitivity, the ease of production and cost of probes together with the size and brightness of the generated signal are the most

Tanya Vavouri and Miguel A. Peinado (eds.), *CpG Islands: Methods and Protocols*, Methods in Molecular Biology, vol. 1766, https://doi.org/10.1007/978-1-4939-7768-0_17, © Springer Science+Business Media, LLC, part of Springer Nature 2018

important factors that need to be taken into account while choosing which DNA FISH protocol to use. Traditionally, genomic DNA fragments cloned in bacterial artificial chromosomes (BAC) have been the main source of DNA FISH probes for diagnostics, and most commercial DNA FISH probes used in routine cytogenetic diagnostics are BAC-based. However, with the number of new biomarkers being discovered growing at incredibly fast pace following the advent of next-generation sequencing technologies, many diagnostic labs are now facing the need to quickly and cost-effectively produce DNA FISH probes and develop protocols for validating newly discovered biomarkers, and BAC-based probes are not an efficient solution. We have identified PCR as the most agile and cost-effective method that any diagnostic laboratory can easily implement to rapidly generate effective DNA FISH probes for targets for which commercial probes are not available [1].

On the other hand, in many research applications, particularly in the study of genome architecture, the most important aspects that need to be considered while choosing a DNA FISH protocol are, in addition to specificity and sensitivity, the throughput and the spatial resolution with which a given set of DNA loci can be identified in cells with as little perturbation of the nuclear morphology as possible (ideally, in living cells). For this purpose, the development in recent years of in silico-designed probes with well-defined composition and optimal thermodynamic properties—together with the availability of new microscopy technologies and image processing algorithms enabling high-resolution (ideally, sub-diffraction) localization of fluorescence signals—represent a major avenue of improvement of DNA FISH technologies. Several methods for producing high quantities of single-stranded DNA oligonucleotides that can be used as DNA FISH probes are now available [2–7], and oligonucleotide-based, single-stranded DNA (ssDNA) FISH probes currently represent the most promising tool for both sensitive and specific detection of DNA loci at (sub)diffraction resolution. Unfortunately, the relatively small number of available fluorescent dyes with nonoverlapping spectra limits the multiplexing capacity of DNA FISH assays. However, the use of few, spectrally distinct dyes coupled with sequential cycles of probe hybridization, sample imaging and probe stripping is a powerful strategy to achieve high multiplexing [8–11].

In addition to choosing a protocol for sample preparation, probe production, and probe hybridization, a crucial aspect in setting up a DNA FISH assay is deciding on which microscopy platform is best suited for a given application. In general, fluorescence microscopes equipped with basic charged coupled device (CCD) cameras and standard filter sets are used in routine cytogenetic diagnostics to visualize bright and relatively large DNA FISH signals (spots >500 nm in diameter). In research applications, the use of high-performance CCD or scientific complementary metal-oxide-semiconductor (sCMOS) cameras and of microscopes equipped with motorized stages for three-dimensional (3D) sample scanning is indispensable, for example, to observe fine details of the genomic 3D architecture in single cells. The choice of the microscope will also depend on how many individual DNA loci one wishes to detect at once. If multiple loci need to be imaged simultaneously (in the order of hundreds per cell) or if the loci are physically very close to each other in 3D, the use of superresolution microscopes (i.e., microscopes that can image the sample at a resolution higher than the optical diffraction limit) is recommended. However, superresolution microscope designs are more limited in the number of suitable dyes in comparison to more standard fluorescence microscope settings, such as wide-field or confocal microscopes (for an overview of superresolution microscopy platforms, *see* [12]).

Finally, approaches have also been developed to perform DNA FISH in living cells [13–15]. These methods are based on the observation that DNA bases can pair with each other not only via the classical Watson–Crick base pairing, but also via the so-called Hoogsteen base pairing [16]. By carefully designing ssDNA oligonucleotides that bind to selected genomic sequences via Hoogsteen base pairing, and by labeling them with fluorescent dyes, one can visualize selected DNA loci directly in living cells. This method, called COMBO-FISH [13, 14] is currently restricted to short stretches of homopurines and homopyrimidines, as these are most amenable to Hoogsteen base pairing. Unfortunately, the relatively low frequency of these sequences in the human genome precludes the use of COMBO-FISH to detect most of genomic regions. Noteworthy, the use of oligonucleotide probes containing peptide nucleic acid (PNA) [15] or locked nucleic acid (LNA) [17] bases has helped and might further help in the future to improve this as well as other DNA FISH methods. One such approach, which is based on the combination of PNA probes with a padlock probe [18] and rolling circle amplification [19], allows for detection of very short DNA stretches (a few dozen bases) and discrimination of sequences that differ by as little as two nucleotides (nt) [20].

In conclusion, the choice of DNA FISH probe type (as exemplified in Fig. 1) and protocol for probe production, sample preparation, and hybridization, as well as the choice of microscope setup and image processing pipeline strictly depends on the actual application. Here, we describe a comprehensive set of step-by-step protocols encompassing: (1) the production and use of PCR probes, following the High-Definition DNA FISH (HD-FISH) method, which we previously described [1]; and (2) the use of ssDNA oligonucleotide probes carrying universal barcode sequences that are suitable for multiplexing the number of loci visualized simultaneously (uniFISH). We have successfully developed and applied the uniFISH approach for the detection of

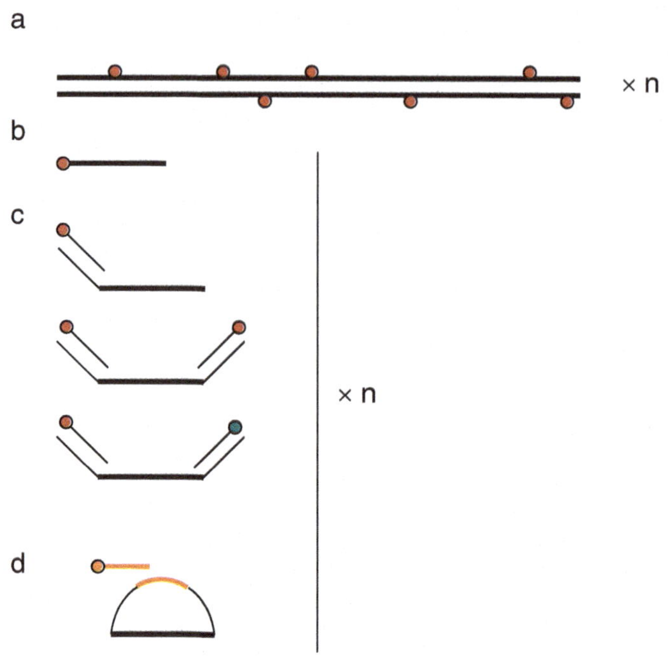

Fig. 1 Scheme of different design strategies for DNA FISH probes. (**a**) HD-FISH probes are composed of a pool of $n = 48–96$ amplicons, ~200 bp in length. Every amplicon is labeled post-PCR using platinum-based chemistry, which allows for the conjugation of a fluorophore of interest to guanine residues present in each amplicon. The optimal degree of labeling is 1–3% (i.e., 1–3 fluorophores incorporated every 100 nt). (**b**) A pool of fluorescently labeled oligonucleotides of defined length (typically, 40–100 nt) can also constitute a DNA FISH probe. Generally, $n = 48–96$ oligonucleotides are sufficient to visualize the target DNA sequence. (**c**) In uniFISH, a probe consists of a pool of usually $n = 48–96$ primary oligonucleotides of defined length, each containing a target-specific sequence (thick line) and a universal sequence barcode (thin line) to which a secondary fluorescently labeled oligonucleotide can then hybridize. (**d**) Padlock probes can also serve as DNA FISH probes. A padlock probe consists of a linear oligonucleotide, which upon specific annealing to the target sequence undergoes in situ ligation, leading to its circularization. This circularized probe is then amplified in situ using rolling circle amplification, which produces thousands of copies of the detection sequence (yellow) to which a complementary fluorescently labeled oligonucleotide can bind. Typically one padlock probe is used to detect a given target DNA locus. Filled-in circles: fluorescent dyes. For considerations on the choice of n, *see* **Note 3**)

both DNA and RNA (unpublished data), and other groups recently reported the independent development of a similar strategy [4, 6]. We present an overview of both the HD-FISH and uniFISH methods in Fig. 2. In addition to experimental protocols, we also provide considerations on optimal probe design (*see* **Note 1**) as well as image processing and analysis (*see* **Note 2**). Lastly, to facilitate the

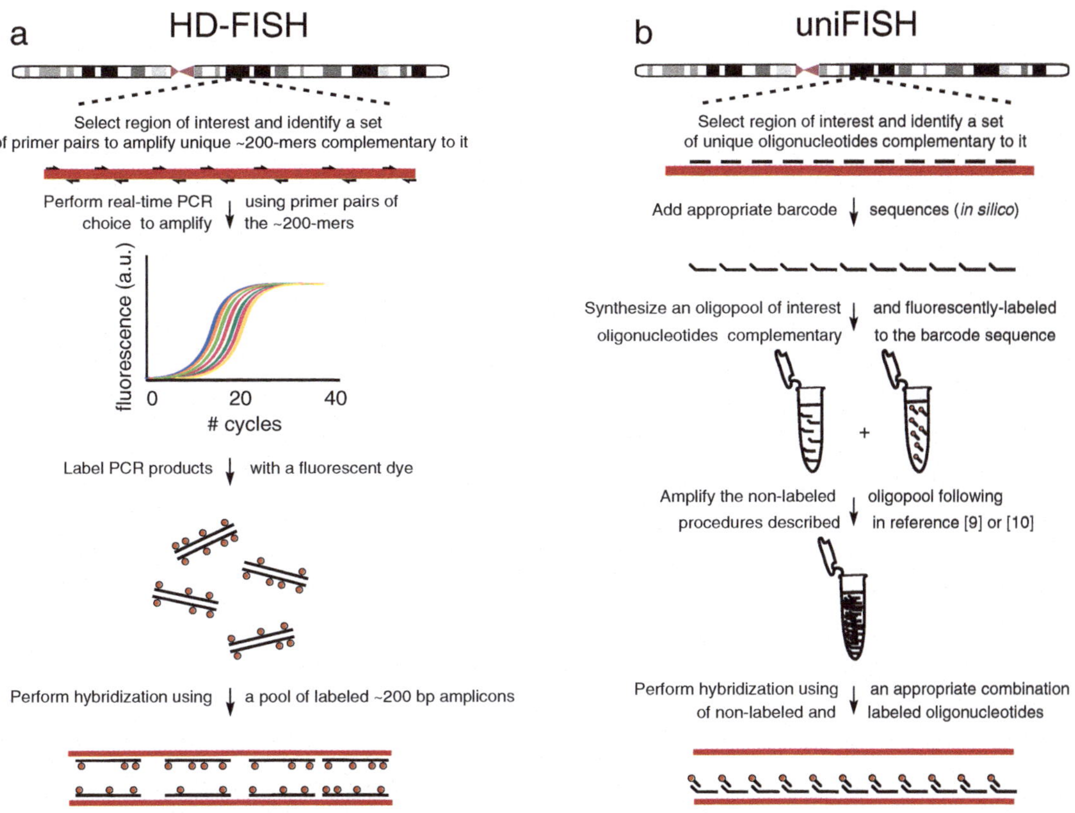

Fig. 2 Scheme of HD-FISH and uniFISH. (**a**) In HD-FISH, a set of primer pairs (typically 48–96 per probe) is synthesized and used to selectively amplify ~200 bp long unique dsDNA fragments. Amplification is performed using real-time PCR in order to monitor amplification kinetics of each of the amplicons, and only amplicons showing similar amplification kinetics are pooled together to compose the probe. The purified PCR product is fluorescently labeled using platinum-based chemistry, which allows for conjugation of fluorescent dyes to guanine residues present in the amplicons. During the hybridization, both the probe and the targeted DNA need to be denatured to be able to anneal to each other. (**b**) In uniFISH, a set of primary oligonucleotides (usually 48–96) of defined length (typically, 30–80 nt) is designed to specifically target a complementary DNA region. Afterward, a universal sequence barcode (usually, 20 nt long) is added in silico to the target-specific sequences, and a pool of such oligonucleotides is synthesized using oligo-arrays. In parallel, fluorescently labeled secondary oligonucleotides complementary to the universal barcode are synthesized. Different barcodes can be used to multiplex the number of fluorescent dyes that are docked to the target-specific oligos. The hybridization occurs in two steps: (1) a pool of primary, nonlabeled oligonucleotides is added to the sample; (2) after washing of unbound primary oligonucleotides, secondary fluorescently labeled oligonucleotides are hybridized to their complementary barcode sequence

choice between different DNA FISH methods, in Fig. 3 we provide a flowchart summarizing different options and recommended applications.

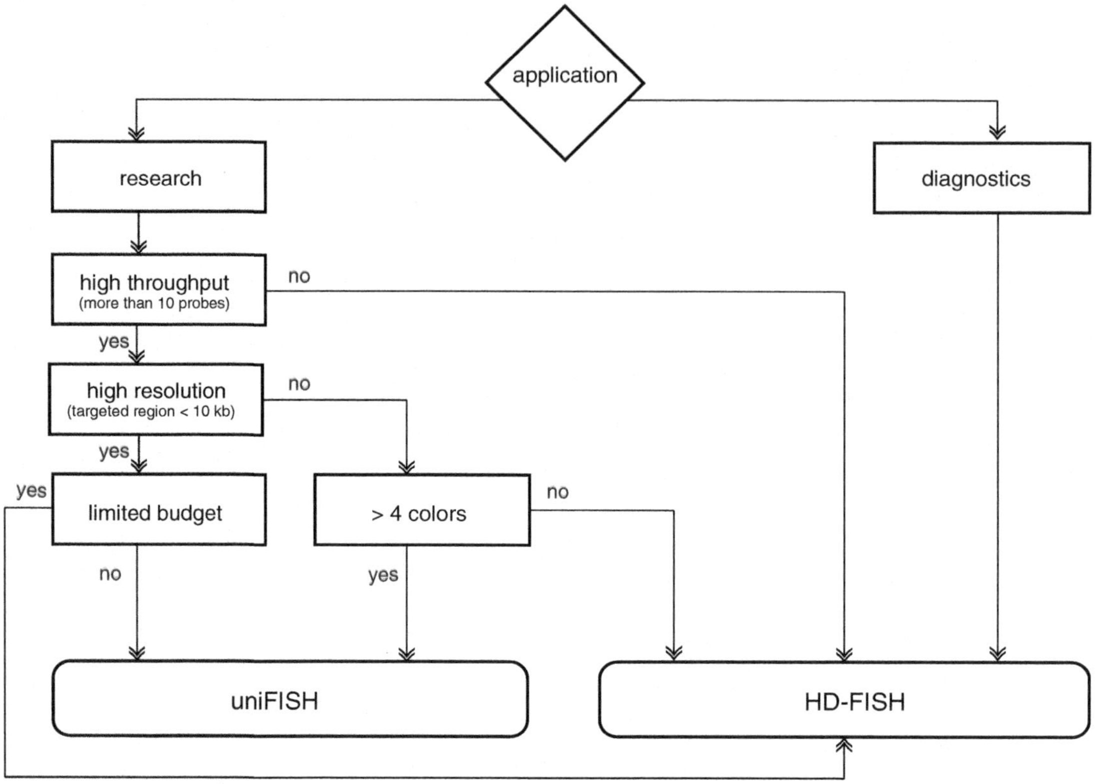

Fig. 3 Flowchart guiding the choice of DNA FISH strategy depending on the specific application

2 Materials

2.1 Reagents

1. Absolute ethanol (EtOH) (VWR, cat. no. 20821.310).
2. Catalase from *Aspergillus niger* (Sigma-Aldrich, cat. no. 3515-10MG).
3. 50× Denhardt's solution (Thermo Fisher Scientific, cat. no. 750018).
4. Dextran sulfate (Sigma-Aldrich, cat. no. D8906-10G).
5. Dulbecco's phosphate buffered saline (with $MgCl_2$ and $CaCl_2$) (Sigma-Aldrich, cat. no. D8662-500ML).
6. EDTA (Thermo Fisher Scientific, cat. no. AM9261).
7. Formamide (Deionized) (Thermo Fisher Scientific, cat. no. AM9344).
8. Glacial acetic acid (Sigma-Aldrich, cat. no. 320099-500ML).
9. Glucose (Thermo Fisher Scientific, cat. no. 15023021).
10. Glucose oxidase from *Aspergillus niger* (Sigma-Aldrich, cat. no. G2133-10KU).
11. Glycerol (Acros Organics, cat. no. 327255000).

12. Glycine (Sigma-Aldrich, cat. no. 50046-250G).
13. Glycogen (Sigma-Aldrich, cat. no. 10901393001).
14. Hoechst 33342 (Thermo Fisher Scientific, cat no. H1399).
15. Human Cot-1 DNA (Thermo Fisher Scientific, cat. no. 15279011).
16. Human genomic DNA, 100 μg (Promega, cat. no. G3041).
17. (±)-6-Hydroxy-2,5,7,8-tetramethylchromane-2-carboxylic acid (Trolox) (Sigma-Aldrich, cat. no. 238813).
18. KaryoMAX® Colcemid™ Solution in PBS (Thermo Fisher Scientific, cat. no. 15212-012).
19. KCl (Sigma-Aldrich, cat. no. P9541-500G).
20. Methanol (Sigma-Aldrich, cat. no. 34860-1L-R).
21. Nuclease-free TE buffer (Thermo Fisher Scientific, cat. no. AM9849).
22. Nuclease-free water (Thermo Fisher Scientific, cat. no. 4387936).
23. 16% methanol-free paraformaldehyde (PFA) (EMS, cat. no. 15710).
24. Phosphate-buffered saline (PBS) tablets (Thermo Fisher Scientific, cat. no. 003002).
25. Poly-L-lysine (PLL) solution (Sigma-Aldrich, cat. no. P8920-100ML).
26. Ribonucleic acid, transfer from *Escherichia coli* (Sigma-Aldrich, cat. no. R1753).
27. RNAse A (Qiagen, cat. no. 19101).
28. 3 M sodium acetate, pH 5.5 (Thermo Fisher Scientific, cat. no. AM9740).
29. 20× SSC (Thermo Fisher Scientific, cat. no. AM9763).
30. SYBR SELECT master mix for CFX (Life Technologies, cat. no. 4472942).
31. Triton® X-100 (Promega, cat.no. H5142).
32. Tween® 20 (Promega, cat. no. H5152).
33. UltraPure™ 1 M Tris–HCI Buffer, pH 7.5 (Thermo Fisher Scientific, cat. no. 15567-027).
34. UltraPure™ BSA (50 mg/mL) (Thermo Fisher Scientific, cat. no. AM2616).
35. UltraPure™ Salmon Sperm DNA Solution (Thermo Fisher Scientific, cat. no. 15632011).
36. Ulysis™ Alexa Fluor® Nucleic Acid Labeling Kit (Thermo Fisher Scientific, cat. no. U-21650 for Alexa Fluor® 488/U-21654 for Alexa Fluor® 594/U-21660 for Alexa Fluor® 647) or FISHBright Labeling Kit for PlatinumBright 547 (Leica Biosystems, cat. no. GLK002).

2.2 Consumables

1. CultureWell MultiSlip™ Coverglass Inserts, Silicone Supported (Grace Bio-Labs, cat. no. SKU: 104412).
2. Eppendorf® RNA/DNA LoBind microcentrifuge tubes 1.5 mL (Sigma-Aldrich, cat. no. Z666548).
3. Fixogum Rubber Cement (Leica Biosystems, cat. no. LK071A).
4. Whatman® qualitative filter paper for technical use (Sigma-Aldrich, cat. no. WHA10538877).
5. KREApure columns (Leica Biosystems, cat. no. KP050).
6. Parafilm M® (Bemis, cat. no. PM992).
7. Sapphire Filter tips, low retention (Greiner Bio-One, cat. no. 771265, 773265, 738265, 750265).

2.3 Equipment

1. Nalgene® Polyethylene Dewar Flask (Electron Microscopy Sciences, cat. no. 62038-01).
2. Incubator (Binder incubator, Model KB 53).
3. Incubating orbital shaker (Troemner, cat. no. 980184-C).
4. Multichannel pipettes.
5. Real-time PCR machine.
6. Tabletop centrifuge (Eppendorf® Microcentrifuge 5424).
7. ThermoBrite (Leica Biosystems, cat. no. 3800-004852-002).

3 Methods

3.1 Cell Culturing

Adherent cells are typically grown on multiple cover glasses immobilized on a silicone support (CultureWell MultiSlip™ system) (*see* **Note 4**). For suspension cells, the cover glasses of the CultureWell MultiSlip™ system are coated with poly-L-lysine and the cells are spotted onto them before fixation.

3.2 Fixation Procedures (See Note 5)

3.2.1 Classical Cell Sample Preparation

1. Wash the cells with 1× PBS for 2 min at room temperature (RT).
2. Aspirate the PBS solution from the dish.
3. Gently add a freshly prepared solution of methanol–acetic acid 3:1 (v/v) at RT.
4. Incubate the cover glasses in methanol–acetic acid for 15 min at RT.
5. Aspirate the methanol–acetic acid solution.
6. Briefly rinse the cells in 1× PBS/0.05% Triton X-100 at RT.
7. Wash the cover glasses with 1× PBS/0.05% Triton X-100, three times, 10 min each at RT, while shaking.

8. Aspirate the PBS solution.
9. Add 100 μg/mL RNAse A/1× PBS at RT.
10. Incubate for 60 min at 37 °C.
11. Briefly wash the cells in 1× PBS at RT.
12. Aspirate the PBS solution.
13. Add ice-cold 70% ethanol (EtOH).
14. Incubate for 2 min at RT.
15. Exchange to 85% EtOH at RT.
16. Incubate for 2 min at RT.
17. Exchange to 100% EtOH at RT.
18. Incubate for 2 min at RT.
19. Air-dry the cover glasses for at least 3 h (preferably overnight) at RT.
20. Store the cover glasses at RT until ready for hybridization (*see* **Note 6**).

*3.2.2 Preparation of Metaphase Spreads (See **Note 7**)*

1. Grow cells in a cell culture dish or flask.
2. Add 0.02–0.04 μg/mL of colcemid to the medium.
3. Incubate for 16–18 h at 37 °C.
4. Rinse the cells a few times with 75 mM KCl prewarmed to 37 °C (about 10 mL are needed for a 10 cm dish).
5. Incubate the cells with fresh 75 mM KCl for 40 min at 37 °C.
6. After the incubation, pipette the solution up and down 10–15 times against the bottom of the dish to detach the metaphase cells.
7. Transfer the solution into a 15 mL tube.
8. Slowly add 1 mL of a freshly prepared solution of methanol–acetic acid 3:1 (v/v) at RT.
9. Gently invert the tube 5–6 times.
10. Centrifuge at 1000 × *g* for 10 min at RT.
11. Aspirate the supernatant down to 2 mL.
12. Disaggregate the pellet by tapping the bottom of the tube 5–6 times.

    Attention: swelled metaphase cells are fragile and should be treated gently.
13. Slowly add 8 mL of the methanol–acetic acid solution at RT.
14. Gently mix by pipetting up and down 3–4 times.
15. Centrifuge at 1000 × *g* for 10 min at RT.
16. Repeat **steps 11–15** four times in total.

17. Remove the supernatant.
18. Add 1 mL of the methanol–acetic acid solution at RT.
19. Gently mix by pipetting up and down 3–4 times.
20. Transfer the cells into a 1.5 mL tube.
21. Store the sample at 4 °C up to a month.
22. Warm a microscope slide by placing it for 2 min on a surface directly above a water bath set to 50 °C.
23. From a distance of 20–30 cm, drop 20 μL of fixed metaphase cells onto the prewarmed slide.
24. Quickly drop 20 μL of the methanol–acetic acid solution on the spot where the metaphases were dropped.
25. Incubate for 1 min.
26. Place the slide in a slidebox.
27. Dry the slides for at least 2–3 days at RT before performing DNA FISH.

3.2.3 Preparation of Morphologically Preserved Cell Samples for 3D-FISH (Adapted from [21])

1. Wash the cells with 1× PBS with Ca^{2+} and Mg^{2+} and prewarmed to 37 °C.
2. Incubate for 2 min at RT.
3. Aspirate the PBS solution from the dish.
4. Add 0.4× PBS with Ca^{2+} and Mg^{2+} at RT.
5. Incubate for 1 min at RT.
6. Aspirate the PBS solution.
7. Add 4% formaldehyde/0.4× PBS at RT.
8. Incubate for 10 min at RT.
9. Aspirate the fixative solution from the dish.
10. Quench the residual formaldehyde by adding 125 mM glycine/1× PBS.
11. Incubate for 5 min at RT.
12. Wash the cells in 1× PBS/Triton X-100 0.05% three times, 5 min each at RT, while shaking.
13. Aspirate the PBS solution and then add 1× PBS/Triton X-100 0.5% at RT.
14. Incubate for 20 min at RT.
15. Briefly wash the cells in 1× PBS at RT.
16. Add 100 μg/mL RNAse A/1× PBS at RT.
17. Incubate for 60 min at 37 °C.
18. Briefly wash the cells in 1× PBS at RT.
19. Briefly rinse the cells in 20% glycerol/1× PBS at RT.
20. Incubate the samples in 20% glycerol/1× PBS overnight at RT.

21. Fill a cryo-resistant Dewar flask with liquid N_2.
22. Remove the glycerol solution from the dish (leave a thin layer on top of the cells).
23. Place the dish on the surface of liquid N_2 for 20–30 s (until the cracking stops).
24. Place the dish at RT and let the cover glasses thaw gradually, making sure that they do not dry.
25. Once all the cover glasses are thawed, fill the dish with fresh 20% glycerol/1× PBS at RT.
26. Incubate for 3 min at RT.
27. Repeat **steps 22–26** four times.
28. Wash the samples in 1× PBS/0.05% Triton X-100 three times, 5 min each at RT, while shaking.
29. Briefly rinse the samples in 0.1 M HCl at RT.
30. Incubate the samples for 5 min in 0.1 M HCl at RT.

    Attention: the samples should be kept in HCl for a maximum of 5 min including the quick rinsing step.
31. Briefly rinse the samples in 1× PBS/0.05% Triton X-100 at RT.
32. Wash the samples in 1× PBS/0.05% Triton X-100 two times, 5 min each at RT, while shaking.
33. Wash the samples 5 min in 2× SSC.
34. Briefly rinse the samples with 50% formamide/2× SSC/50 mM sodium phosphate (*see* **Note 8**).
35. Aspirate the solution.
36. Exchange to fresh 50% formamide/2× SSC/50 mM sodium phosphate.
37. Incubate overnight at RT.
38. Transfer the samples to 4 °C.
39. Store the samples for 1 week at 4 °C.
40. Exchange the formamide solution to 2× SSC and proceed to hybridization or store in 2× SSC for up to 3 months at 4 °C.

3.3 Preparation of Buffers

3.3.1 HD-FISH Prehybridization Buffer (1× HDPHB)

1. In a 15 mL tube, mix the following reagents in the indicated order:

(a) 20× SSC	850 μL
(b) 1 M sodium phosphate buffer	500 μL
(c) 50× Denhardt's solution	1 mL
(d) 0.5 M EDTA	20 μL
(e) 10 mg/mL ssDNA	350 μL
(f) 100% formamide (*see* **Note 8**)	6.8 mL

2. Add nuclease-free water up to 10 mL final volume.
3. Add 50 μL of 37% HCl at RT.
4. Rotate the tube for at least 10 min at RT.
5. Centrifuge the tube briefly and measure the solution pH.

   Attention: the pH should range between 7 and 7.5. If not, adjust it by gradually adding HCl or NaOH at RT.
6. Dispense the buffer in 500 μL aliquots in 1.5 mL tubes.
7. Store the aliquots at −20 °C up to 2 months.

3.3.2 HD-FISH Hybridization Buffer (1× HDHB)

1. In a 1.5 mL tube, weight 1 g of dextran sulfate.
2. In a 15 mL tube, add 1 mL of nuclease-free water and 850 μL of 20× SSC buffer.
3. Gradually add the dextran sulfate powder to the 15 mL tube.

   Attention: add only a small amount of the powder at a time and vortex; continue by adding small amounts of dextran sulfate and vortexing, until all the powder has been added.
4. Rotate the tube at RT until all the powder is dissolved.
5. Briefly centrifuge the tube to remove the liquid from the lid.
6. Add the following reagents in the indicated order:

(a) 1 M sodium phosphate buffer	500 μL
(b) 50× Denhardt's solution	1 mL
(c) 0.5 M EDTA	20 μL
(d) 10 mg/mL ssDNA	350 μL
(e) 100% formamide (*see* **Note 8**)	6.8 mL

7. Add nuclease-free water up to 10 mL final volume.
8. Add 50 μL of 37% HCl at RT.
9. Rotate the tube for at least 10 min at RT.
10. Centrifuge the tube briefly and measure the solution pH.

    Attention: the pH should range between 7 and 7.5. If not, adjust it by gradually adding HCl or NaOH at RT.
11. Dispense the buffer in 500 μL aliquots in 1.5 mL tubes.
12. Store the aliquots at −20 °C up to 2 months.

3.3.3 uniFISH Prehybridization Buffer (1× UPHB)

1. In a 15 mL tube, mix the following reagents in the indicated order:

(a) 20× SSC	1 mL
(b) 1 M sodium phosphate buffer	500 μL
(c) 50× Denhardt's solution	1 mL
(d) 0.5 M EDTA	20 μL
(e) 10 mg/mL ssDNA	350 μL
(f) 100% formamide (*see* **Note 8**)	5 mL

2. Add nuclease-free water up to 10 mL final volume.
3. Add 50 μL of HCl 37% at RT.
4. Rotate the tube for at least 10 min at RT.
5. Centrifuge the tube briefly and measure the solution pH.

   Attention: the pH should range between 7 and 7.5. If not, adjust it by gradually adding HCl or NaOH.
6. Dispense the buffer in 500 μL aliquots in 1.5 mL tubes.
7. Store the aliquots at −20 °C up to 2 months.

3.3.4 uniFISH 1st Hybridization Buffer (1.1× UHB-1)

1. In a 1.5 mL tube, weight 1.1 g of dextran sulfate.
2. In a 15 mL tube, add 1 mL of nuclease-free water and 2.22 mL of 20× SSC buffer.
3. Gradually add the dextran sulfate powder to the 15 mL tube.

   Attention: add only a bit of powder at a time and vortex; continue by adding small amounts of dextran sulfate and vortexing, until all the powder has been added.
4. Rotate the tube at RT until all the powder is dissolved.
5. Briefly centrifuge the tube to remove the liquid from the lid.
6. Add the following reagents in the indicated order:

(a) 1 M sodium phosphate buffer	555 μL
(b) 50× Denhardt's solution	1.11 mL
(c) 0.5 M EDTA	22.2 μL
(d) 10 mg/mL ssDNA	388 μL
(e) 100% formamide (*see* **Note 8**)	5.55 mL

7. Add nuclease-free water up to 10 mL final volume.
8. Add 50 μL of 37% HCl at RT.
9. Rotate the tube for at least 10 min at RT.
10. Centrifuge the tube briefly and measure the solution pH.

Attention: the pH should range between 7 and 7.5. If not, adjust it by gradually adding HCl or NaOH at RT.

11. Dispense the buffer in 500 μL aliquots in 1.5 mL tubes.
12. Store the aliquots at −20 °C up to 2 months.

3.3.5 uniFISH 2nd Hybridization Buffer (1× UHB-2)

1. In a 1.5 mL tube, weight 1 g of dextran sulfate.
2. In a 15 mL tube, add 4 mL of nuclease-free water and 1 mL of 20× SSC buffer.
3. Gradually add the dextran sulfate powder to the 15 mL tube.

   Attention: add only a bit of powder at a time and vortex; continue by adding small amounts of dextran sulfate and vortexing, until all the powder has been added.
4. Rotate the tube at RT until all the powder is dissolved.
5. Briefly centrifuge the tube to remove the liquid from the lid.
6. Add the following reagents in the indicated order:

(a) BSA	40 μL
(b) tRNA	500 μL
(c) 100% formamide (*see* **Note 8**)	2.5 mL

7. Rotate the tube for at least 10 min at RT.
8. Centrifuge the tube briefly.
9. Add nuclease-free water up to 10 mL final volume.
10. Dispense the buffer in 500 μL aliquots in 1.5 mL tubes.
11. Store the aliquots at −20 °C up to 2 months.

3.3.6 Universal Wash Buffer (UWB)

1. In a 50 mL tube, mix 12.5 mL of 100% formamide with 5 mL of 20× SSC buffer.
2. Add nuclease-free water up to 50 mL final volume.
3. Store at 4 °C up to a year.

3.3.7 Trolox Working Solution

1. In a 1.5 mL tube, weight 50 mg of the Trolox powder.
2. Add 1 mL of 100% EtOH.
3. Store at −20 °C.

3.3.8 Glucose Oxidase Working Solution

1. Dissolve the powder in 50 mM sodium acetate buffer, pH 5.5, to a final concentration of 3.7 mg/mL.
2. Dispense in 50 μL aliquots in 1.5 mL tubes.
3. Store the aliquots at −20 °C up to a year.

3.3.9 Imaging Buffer Without Enzymes or Trolox (IMB −ET)

Attention: the IMB buffer should be prepared fresh before each imaging session.

1. In a 15 mL tube prepare the following buffer (glucose buffer):

(a) 20× SSC	1 mL
(b) 40% glucose	100 μL
(c) 1 M Tris–HCl, pH 7.5	100 μL
(d) Nuclease-free water	8.8 mL

3.3.10 Imaging Buffer with Enzymes and Trolox (IMB +ET)

1. In a 1.5 mL tube, add the following reagents in the indicated order:

(a) IMB –ET buffer	93 μL
(b) 200 mM Trolox	5 μL
(c) 3.7 mg/mL glucose oxidase	1 μL
(d) Catalase	1 μL

Attention: after adding Trolox, vortex the solution vigorously for 5–10 s before adding the enzymes.

3.4 HD-FISH Procedure

*3.4.1 Probe Preparation (See **Note 3**)*

1. In a 96-well PCR plate, using a multichannel pipette dispense 90 μL of nuclease-free water in each well.
2. Using a multichannel pipette, dispense 5 μL of each forward and 5 μL of each corresponding reverse primer at 100 μM per well (for example, 5 μL of forward primer in well A1 should be mixed with 5 μL of the corresponding reverse primer in well A1).

   Attention: the concentration of each primer is now at 5 μM.
3. In each well of a new 96-well PCR plate, dispense 25 μL of the following master mix:

(a) Nuclease-free water	10 μL
(b) Human genomic DNA 50 ng/μL	0.5 μL
(c) PCR master mix 2×	12.5 μL

4. To each well add 2 μL of the forward and reverse primers mix at 5 μM.
5. Seal the 96-well plate and wrap it in aluminum foil until it is ready to be placed in the PCR instrument.
6. Using a real-time PCR instrument, perform the following cycles:
   (a) 95 °C, 5 min.
   (b) 95 °C, 10 s.

(c) 55 °C, 10 s.

(d) 72 °C, 10 s.

(e) Repeat steps two to four 25–30 times.

(f) 4 °C, hold.

7. Select the wells that exhibit a similar amplification kinetics and pool their contents into a 2 mL tube.

   Attention: at most 400 μL of PCR product should be pooled into a 2 mL tube.

8. To each sample, add 11.1 μL of 3 M sodium acetate, pH 5.5, for every 100 μL of pooled PCR product (10% v/v).

9. To each sample, add 2.5 volumes of ice-cold 100% EtOH and immediately vortex thoroughly.

10. Incubate the samples overnight at −20 °C.

11. The following day, equally split the contents of each sample into two 1.5 mL tubes, then centrifuge at 25,000–30,000 × *g* for 30 min at 4 °C.

    Attention: at the end of centrifugation, a white pellet should be visible at the bottom of the tube.

12. Gently aspirate the supernatant and resuspend the DNA pellet in 500 μL of ice-cold 70% EtOH.

13. Vigorously vortex the samples for at least 5 s.

14. Centrifuge the samples again at 25,000–30,000 × *g* for 10 min at 4 °C.

15. Repeat the wash in 70% EtOH and centrifugation.

16. Gently aspirate the supernatant and air-dry the DNA pellet for 20–60 min at RT.

    Attention: avoid overdrying the pellet by monitoring it periodically.

17. Dissolve each DNA pellet in 20 μL of nuclease-free water.

18. Merge all the samples corresponding to the same probe into one 1.5 mL tube.

19. Incubate the samples for 1–2 h at RT to allow DNA to dissolve.

20. Measure the DNA concentration, for example using NanoDrop.

    Checkpoint: confirm that the size of the probe is correct by analyzing an aliquot of the purified PCR product by agarose gel electrophoresis.

21. Store the purified PCR products at −20 °C.

    Attention: when not in use, the plates containing the initial primers as well as the plate containing the primers mix can be stored at −20 °C.

3.4.2 Probe Labeling

1. In a 200 μL PCR tube, add a volume of purified PCR product corresponding to 1 μg.
2. Add 2 μL of the 10× labeling buffer from the KREAtech kit.
3. Add nuclease-free water up to 17 μL.
4. Denature the DNA by incubating the sample for 5 min at 95 °C.
5. Quickly spin the sample and transfer it on ice.
6. Add 3 μL of dye (*see* **Note 9**).
7. Quickly mix by pipetting up and down 3–4 times.
8. Incubate for 30 min at 85 °C.
9. Remove the unbound dye using a KREApure column according to the manufacturer's instructions.
10. Calculate the degree of labeling (DoL) (*see* **Note 10**).
11. Stored the labeled probe at −20° up to a year (*see* **Note 11**).

3.4.3 Probe Precipitation

1. In a 1.5 mL tube, add the following reagents in the indicated order:

(a) Nuclease-free water	Up to 100 μL
(b) 20 mg/mL glycogen	2.5 μL
(c) 10 mg/mL Cot-1	7.5 μL
(d) Labeled probe	20–200 ng (*see* **Note 12**)

1. Add 34 μL of 3 M sodium acetate, pH 5.5, and 900 μL of ice-cold 100% EtOH.
2. Briefly vortex the tube for 5 s.
3. Incubate for 2 h at −20 °C.
4. Centrifuge 30 min at 30,000 × *g* at 4 °C.
5. Carefully remove the supernatant.
6. Wash the DNA pellet with 600 μL of ice-cold 70% EtOH.
7. Centrifuge for 15 min at 30,000 × *g* at 4 °C.
8. Wash the DNA pellet with 600 μL of ice-cold 70% EtOH.
9. Centrifuge for 15 min at 30,000 × *g* at 4 °C.
10. Air-dry the DNA pellet for 20–60 min at RT.

    Attention: avoid overdrying the pellet by monitoring it periodically.
11. Resuspend the pellet in 1× HDHB buffer.
12. Store the probe solution at 4 °C (protected from light) for up to 2 months (*see* **Note 11**).

3.4.4 Hybridization in Classically Prepared Cell Samples

1. Remove a dry cover glass from the silicone support and place it in a dish containing 70% formamide/2× SSC (preheated to 75 °C) placed on top of a water bath set to 75 °C (*see* **Note 8**).
2. Incubate for 2–5 min at 75 °C (*see* **Note 13**).
3. Transfer the cover glass to a dish containing ice-cold 70% EtOH.
4. Incubate for 2 min at RT.
5. Exchange to 85% EtOH at RT.
6. Incubate for 2 min at RT.
7. Exchange to 100% EtOH at RT.
8. Incubate for 2 min at RT.
9. Air-dry the cover glasses for at least 20–30 min at RT.
10. Denature the appropriate amount of the probe solution for 5 min at 75 °C.
11. Once denatured, quickly dispense a droplet of HD-FISH probe solution (for 12 × 12 mm cover glasses, dispense 10 μL; for 18 × 18 mm cover glasses, dispense 22 μL; for 22 × 22 mm cover glasses, dispense 25 μL) on a microscope slide and place the cover glass on it (*see* **Note 12**).
12. Remove the excess of liquid by gently passing a filter paper around the edge of the cover glass.
13. Seal the cover glass with fixogum rubber cement.

    Attention: from this point, the sample should be protected from light whenever possible.
14. Place the slide in a humidity chamber.
15. Incubate overnight at 37 °C in darkness.
16. Immerse the slide in a 10 cm dish containing 2× SSC.
17. Remove the fixogum with tweezers.
18. Gently remove the cover glass and place it in a 6-well plate. (The cells should now face up.)
19. Wash the cover glass in 2× SSC/0.2% Tween, three times, 10 min each at 37 °C, while shaking.
20. Wash the cover glass in 0.2× SSC/0.2% Tween (prewarmed to 60 °C), two times, 7 min each at 60 °C in the water bath.
21. Briefly rinse the cover glass in 4× SSC/0.2% Tween at RT.
22. Rinse the cover glass two times with 2× SSC at RT.
23. Incubate with 1 ng/μL Hoechst/2× SSC for 5 min at RT.
24. Rinse the cover glass with 2× SSC.
25. Rinse the cover glass with IMB –ET buffer.
26. Mount with IMB +ET buffer.
27. Image the sample.

3.4.5 Hybridization in Metaphase Spreads

1. Incubate a slide with metaphase spreads on in 100 μg/mL RNAse A/1× PBS for 60 min at 37 °C.
2. Briefly wash the slide in 1× PBS at RT.
3. Aspirate the PBS solution.
4. Incubate in 2% pepsin/0.01 M HCl for 1 h at RT.
5. Wash the slide in 1× PBS for 5 min at RT, while shaking.
6. Wash the slide in 50 mM $MgCl_2$/1× PBS for 5 min at RT.
7. Fix with 1% formaldehyde/1× PBS (+Mg^{2+}) solution for 5 min at RT.
8. Wash with 1× PBS for 5 min at RT.
9. Exchange the PBS to 70% EtOH at RT.
10. Incubate for 2 min at RT.
11. Exchange to 85% EtOH at RT.
12. Incubate for 2 min at RT.
13. Exchange to 100% EtOH at RT.
14. Incubate for 2 min at RT.
15. Air-dry the slide for at least 20–30 min at RT.
16. Immerse the slide in a dish containing 70% formamide/2× SSC (preheated to 80 °C) placed on top of a water bath set to 80 °C (*see* **Note 8**).
17. Incubate for 5 min at 80 °C.
18. Transfer the slide to a dish containing ice-cold 70% EtOH.
19. Incubate for 2 min at RT.
20. Exchange to 85% EtOH at RT.
21. Incubate for 2 min at RT.
22. Exchange to 100% EtOH at RT.
23. Incubate for 2 min at RT.
24. Air-dry the cover glasses for at least 20–30 min at RT.
25. Denature the appropriate amount of the probe solution for 5 min at 75 °C.
26. Once denatured, quickly dispense 25 μL of HD-FISH probe solution on the slide onto the area containing the metaphase spreads and place the 22 x 22 mm cover glass on it (*see* **Note 12**).
27. Remove the excess of liquid by gently passing a filter paper around the edge of the cover glass.
28. Seal the cover glass with fixogum rubber cement.

    Attention: from this point, the sample should be protected from light whenever possible.
29. Place the slide in a humidity chamber.

30. Incubate for 48 h at 37 °C in darkness.
31. Immerse the slide in a 10 cm dish containing 2× SSC.
32. Remove the fixogum with tweezers.
33. Gently remove the cover glass.
34. Wash the slide in 2× SSC/0.2% Tween, three times, 10 min each at 37 °C, while shaking.
35. Wash the slide in 0.2× SSC/0.2% Tween (prewarmed to 60 °C), two times, 7 min each at 60 °C in the water bath.
36. Briefly rinse the slide in 4× SSC/0.2% Tween at RT.
37. Rinse the slide two times with 2× SSC at RT.
38. Incubate with 1 ng/μL Hoechst/2× SSC for 5 min at RT.
39. Rinse the slide with 2× SSC.
40. Rinse the slide with IMB –ET buffer.
41. Mount with IMB +ET buffer.
42. Image the sample.

3.4.6 Hybridization in Morphologically Preserved Cell Samples (3D-FISH)

1. Cut a square of Parafilm M® and place it inside a 10 cm dish.
2. All along the inner edge of the 10 cm dish, bend a piece of absorbing tissue soaked in distilled water (to build a humidity chamber).
3. On top of the Parafilm M® piece, dispense 100 μL of 1× HDPHB buffer.
4. With a Whatman filter paper, gently dry the edges of the cover glass (*see* **Note 14**).
5. Place the cover glass onto the droplet of 1× HDPHB buffer.
6. Seal the humidity chamber with Parafilm M®.
7. Incubate for 1 h at 37 °C.
8. On a microscope slide, dispense a droplet of HD-FISH probe solution (for 12 × 12 mm cover glasses, dispense 10 μL; for 18 × 18 mm cover glasses, dispense 22 μL; for 22 × 22 mm cover glasses, dispense 25 μL) (*see* **Note 12**).
9. Gently lift the cover glass with cells from the piece of Parafilm M®.
10. With a Whatman filter paper, gently dry the edges of the cover glass (*see* **Note 14**).
11. Place the cover glass onto the droplet of HD-FISH probe.
12. Remove the excess of liquid by gently passing a Whatman filter paper around the edge of the cover glass.
13. Seal the cover glass with fixogum rubber cement.

Attention: from this point, the sample should be protected from light whenever possible.

14. Allow the fixogum to solidify in darkness at RT.
15. Denature the slide on a heating block (ThermoBrite) for 2–5 min at 75 °C (*see* **Note 13**).
16. Remove the slide from the heating block and place it in a humidity chamber.
17. Seal the humidity chamber with Parafilm M®.
18. Incubate overnight at 37 °C in darkness.
19. Immerse the slide in a 10 cm dish containing 2× SSC.
20. Remove the fixogum with tweezers.
21. Gently remove the cover glass and place it in a 6-well plate. (The cells should now face up.)
22. Wash the cover glass three times, 10 min each, in 2× SSC/0.2% Tween at 37 °C, while shaking.
23. Wash the cover glass two times, 7 min each, in 0.2× SSC/0.2% Tween (prewarmed to 60 °C) at 60 °C in the water bath.
24. Briefly wash the cover glass in 4× SSC/0.2% Tween at RT.
25. Rinse the cover glass two times with 2× SSC at RT.
26. Incubate in 1 ng/μL Hoechst/2× SSC for 5 min at RT.
27. Rinse the cover glass with 2× SSC.
28. Rinse the cover glass with IMB –ET buffer.
29. Mount with IMB +ET buffer.
30. Image the sample.

3.5 uniFISH Procedure

3.5.1 Hybridization in 3D Preserved Cell Samples (3D-uniFISH)

1. Cut a square of Parafilm M® and place it inside a 10 cm dish.
2. All along the inner edge of the 10 cm dish, bend a piece of absorbing tissue soaked in distilled water (to build a humidity chamber).
3. On top of the Parafilm M® piece, dispense 100 μL of 1× UPHB buffer.
4. With a Whatman filter paper, gently dry the edges of the cover glass (*see* **Note 14**).
5. Place the cover glass onto the droplet of 1× UPHB buffer.
6. Seal the humidity chamber with Parafilm M®.
7. Incubate for 1 h at 37 °C.
8. In the meantime, prepare the uniFISH probe solution by adding 1 μL of the 1:500 working solution of the probe to 9 μL of the 1.1× UHB-1 in an Eppendorf® RNA/DNA LoBind 1.5 mL tube (*see* **Note 12**).
9. Vortex vigorously.

10. Centrifuge at 30,000 × *g* for 10 min at RT.
11. On a microscope slide, dispense a droplet of the uniFISH probe solution (for 12 × 12 mm cover glasses, dispense 10 μL; for 18 × 18 mm cover glasses, dispense 22 μL; for 22 × 22 mm cover glasses, dispense 25 μL) (*see* **Note 12**).
12. Gently lift the cover glass with cells from the piece of Parafilm M®.
13. With a Whatman filter paper, gently dry the edges of the cover glass (*see* **Note 14**).
14. Place the cover glass onto the droplet of the probe.
15. Remove the excess of liquid by gently passing a Whatman filter paper around the edge of the cover glass.
16. Seal the cover glass with fixogum rubber cement.
17. Allow the fixogum to solidify.
18. Denature the slide on a heating block (ThermoBrite) for 2–5 min at 75 °C (*see* **Note 13**).
19. Remove the slide from the heating block and place it in a humidity chamber.
20. Seal the humidity chamber with Parafilm M®.
21. Incubate overnight at 37 °C.
22. Immerse the slide in a 10 cm dish containing 2× SSC.
23. Remove the fixogum with tweezers.
24. Gently remove the cover glass and place it in a 6-well plate. (The cells should now face up.)
25. Wash the cover glass in 2× SSC/0.2% Tween, three times, 10 min each at 37 °C, while shaking.
26. Wash the cover glass in 0.2× SSC/0.2% Tween (prewarmed to 60 °C), two times, 7 min each at 60 °C in the water bath.
27. Briefly wash the cover glass in 4× SSC/0.2% Tween at RT.
28. Rinse the cover glass two times with 2× SSC at RT.
29. Rinse the cover glass with UWB buffer.
30. In the meantime thaw the 1:50 solution of the fluorescently labeled secondary oligonucleotide, vortex it and briefly spin down making sure not to expose it to light (*see* **Note 12**).
31. Prepare a hybridization solution by mixing 1 μL of the 1:50 dilution of the fluorescently labeled oligonucleotide with 99 μL of 1× UHB-2 buffer.
32. Centrifuge at 30,000 × *g* for 10 min at RT.
33. Prepare a humidity chamber containing a piece of the Parafilm M®.
34. On top of the Parafilm M® dispense 100 μL of hybridization solution.

35. With a Whatman filter paper, gently dry the edges of the cover glass (*see* **Note 14**).
36. Place the cover glass onto the droplet of hybridization solution.
37. Seal the humidity chamber with Parafilm M®.
38. Incubate for 2 h at 30 °C in darkness.
39. Wash the cover glass for 30 min with UWB buffer at 30 °C in darkness.
40. Wash the cover glass for 30 min with 1 ng/mL Hoechst/UWB buffer at 30 °C in darkness.
41. Rinse the cover glass with 2× SSC.
42. Rinse the cover glass with IMB –ET buffer.
43. Mount with IMB +ET buffer.
44. Image the sample.

4 Notes

1. *Design of uniFISH probes.*

   A uniFISH probe is typically composed of 48–96 oligonucleotides with a length that can vary from 50 up to 100 nt (30–80 nt for the target-specific sequence and 20 nt for the barcode sequence) depending on the application (Fig. 1). Since the oligonucleotides that constitute a uniFISH probe have to hybridize simultaneously to the same target, they must be selected to have as similar thermodynamic properties as possible. Regardless of their length, the most important feature of the oligonucleotides is their specificity for the locus of interest: an oligonucleotide should only hybridize to its target locus in order to avoid the formation of an unspecific signal. This specificity can be verified using either BLAT (https://genome.ucsc.edu) or nucleotide BLAST (https://blast.ncbi.nlm.nih.gov). An oligonucleotide can be considered specific for a genomic region when there are no other regions with a similarity higher than a given threshold (e.g., we typically accept 60% as maximum tolerated similarity). Once a set of specific oligonucleotides targeting a locus of interest has been identified, a subset with homogeneous characteristics is selected based on the GC content and melting temperature. At the same time, the oligonucleotides are further selected to avoid overlapping target sequences and to homogeneously cover the targeted locus. Oligonucleotides that contain homopolymer stretches, self-dimerize or form stable secondary structures are also discarded. It is also recommended to check that the selected oligonucleotides do not dimerize with each other. When the sequences that are complementary to the target region are ready, a 20 nt barcode sequence is added to them. This barcode sequence serves as target for secondary fluorescently labeled oligonucleotides (Fig. 1), which are hybridized in the second hybridization step. Barcode sequences should be orthogonal to the genome of the cells under study, and they can be for example obtained from [24]. It is crucial that the secondary oligonucleotides do not self-dimerize, contain stable secondary structures, or dimerize with each other. Depending on the sequence of barcodes that are appended to the target-specific oligonucleotides, multiplexing by combinatorial fluorescent labeling is also possible. In single-color uniFISH probes, every labeled secondary oligonucleotide recognizes the same sequence on all the target-bound oligonucleotides. In contrast, multicolored uniFISH probes consist of primary oligonucleotides that are tagged with different combinations of barcodes, each of which is then bound by a secondary oligonucleotide labeled with a specific fluorophore. In the latter case, the different barcodes are designed with similar thermodynamic properties to allow for the simultaneous hybridization of the different labeled secondary oligonucleotides. At the same time, the distribution of the different fluorophores (different barcodes) over the probed region is critical in achieving a good overlap of their signals. A naive approach to this issue would be to distribute the barcodes in an alternating pattern. Still, differences in the fluorophore intensities and in the architecture of the locus of interest might influence the outcome, thus requiring more ad hoc designs and empirical optimization.

2. *Principles of image analysis.*

   Both HD-FISH and uniFISH data can be processed through a dedicated image analysis pipeline that consists of several modules: (a) correction of chromatic aberrations; (b) nuclei segmentation; (c) cell cycle phase identification depending on the integrated Hoechst signal intensity in the nucleus; (d) dot detection; (e) dot characterization; (f) dot classification; (g) visualization/verification of detected dots; (h) application of specific analysis. In general, DNA FISH signals do not have to be extremely bright to be analyzed. On the other hand, it is essential that the imaging sensor is not saturated at the dot locations, since this will degrade subsequent spatial localization attempts and other signal measurements. Chromatic aberrations or frequency-dependent optical behavior are unavoidable even in the most advanced microscope setups. These aberrations can be identified as local displacements between the different image channels, and they can be modeled and canceled with high precision using a polynomial function of low order in the lateral plane, and a constant offset in the axial direction [25]. To do so, calibration beads with emission in the same frequency range as the probes should routinely be imaged before the FISH samples. If the probes are small, the dots can be sub diffraction limited. If that is the case, isolated dots can be localized with high precision, well below the diffraction limit knowing the noise characteristics of the sensor and the point spread function [26]. Dots with some overlap can also be localized but at a higher computational cost [27]. Whenever the dots are small, wide-field microscopy is preferred over confocal. In this case, the localization precision, which depends on the amount of registered photons, will be better and there will be less photobleaching. Larger probes will typically be imaged as large, bright regions or blobs. In this case, the point spread function (PSF) cannot be fitted to the probes and the analysis has to proceed with identifying these larger regions. In this case, it can be beneficial to deconvolve the images and use a high-pass filter to suppress background variations due to uneven illumination and autofluorescence. While it is possible to distinguish signals from noise in images by statistical tests, it is not always possible to make a strict distinction between probes that are hybridized correctly and nonspecifically hybridized probes, autofluorescence, or dots that bleed through from other channels. These issues can be intricate and require ad hoc adjustments to the analysis pipeline, depending on the particular application.

3. *Primer design for HD-FISH probes.*

   HD-FISH probes consist of pools of PCR amplicons. We have generated a database of primer pairs (www.hdfish.eu), which can be used to amplify unique 200–220 bp genomic DNA fragments. We have chosen the length of the PCR-amplicons to be ~200 bp, however shorter (down to 100 bp) or longer (up to 500 bp) amplicons may also work. This database was designed with uniqueness in mind, meaning that the primer pairs should uniquely amplify the fragments of choice, and that the amplified 200-mers are complementary to a unique region of either the human or the mouse genome. Primer pairs yielding amplicons with 70% or higher similarity outside the target locus were discarded. The number of amplicons per probe (n) is arbitrary and depends on the size of the target to be visualized. In cases where resolution is not critical, for example in diagnostic assays where sensitivity instead is crucial, we recommend using $n = 96$ or more amplicons, given that the sensitivity and specificity of the detection scales with n. On the other hand, the minimal number of amplicons needed to sensitively detect a certain locus is not standard and needs to be determined empirically. We have managed to detect the ERBB2 locus in human cells with as few as 15 × 200-mer amplicons targeting a locus spanning ~3 kb. However, this number is likely locus-specific and influenced by the degree of protein occupancy of the locus as well as by the level of cellular autofluorescence. For this purpose, information about DNA accessibility of a locus of interest in a given cell type might be useful in optimizing *n*, although we do not currently have a way to predict *n* based on DNA accessibility.

4. *Cell culture.*

   The protocols described here have been adapted for the use of the CultureWell MultiSlip™ system (Grace Bio-Labs). This system is suitable for high-throughput applications that require simultaneous processing of many samples. However, many different formats can be used. For example, cells can be cultured and fixed on top of standard #1.5 cover glasses (PLL-coated in the case of suspension cells) placed in 6- or 12-well plates.

5. *Choice of fixation protocol.*

   The cell fixation protocols described here are compatible with both the HD-FISH and uniFISH procedures. The classical fixation protocol uses a methanol–acetic acid mixture of 3:1 (v/v) to fix the sample by coagulating proteins present in the cell (action of methanol) and by precipitation of nucleic acids (action of acetic acid). This solution acts by replacing water in the sample, disrupting both hydrophilic and hydrophobic interactions. Thanks to these actions, the mix of methanol and acetic acid exposes DNA to the FISH probe very efficiently. This protocol yields bright signals and for this reason is frequently used in diagnostic applications. However, since methanol–acetic acid disrupts the 3D morphology of the cell, this fixation procedure is not suitable when studying genome architecture. Instead, whenever the nuclear morphology needs to be preserved, the 3D-FISH fixation protocol should be used instead. For considerations on how well the 3D-FISH procedure preserves the nuclear morphology *see* [22, 23].

6. *Storage of samples.*

   Cells fixed in methanol–acetic acid can be air-dried and stored at RT in a dry format. Alternatively, they can be stored in 70% EtOH at 4 °C for up to 1 year. Cells fixed according to the 3D-FISH fixation protocol should be stored in 50% formamide/2× SSC/50 mM phosphate buffer for one week at 4 °C, after which the solution should be exchanged to 2× SSC. The samples can then be stored in 2× SSC at 4 °C for up to 3 months.

7. *Synchronization of cells in metaphase.*

   Metaphase spreads are prepared from mitotic cells that have been arrested in metaphase. Here, we use colcemid for the cell cycle arrest of adherent cells, but the same protocol can be easily adapted to suspension cells. Colcemid arrests cells in metaphase by depolymerizing the microtubules and inhibiting the spindle fiber formation. Exposure of cells to concentrations of colcemid higher than the one recommended or excessive incubation time in colcemid may result in the formation of supercondensed chromosomes. The duration of the incubation in hypotonic solution (75 mM KCl) might need to be adjusted to the cell type. In our experience, 30–40 min incubation result in optimal cell swelling without causing cell bursting.

8. *Handling of formamide-containing buffers.*

   Formamide is a teratogen that is easily absorbed through the skin, thus it should be handled with extreme care. Before using formamide, make sure that you have read the material safety data sheet and that the appropriate personal protective equipment (gloves, laboratory coat, etc.) is in place. Formamide-containing solutions should be handled under a chemical fume hood. Moreover, when taken out of the fridge/freezer, all buffers containing formamide should be allowed to reach room temperature before being opened. Formamide is also light-sensitive: always store the buffers containing formamide in darkness.

9. *Choice of fluorescent dye and amount of dye for labeling HD-FISH probes.*

   In our hands, PlatinumBright 547 from Leica Biosystems works better than the equivalent Alexa Fluor® 546 available from Thermo Fisher Scientific, while for all other channels we rely on the Alexa Fluor® dyes. For the Alexa Fluor® dyes, we typically use 3 μL of freshly prepared Alexa Fluor® 647 or Alexa Fluor® 488, or 2 μL of Alexa Fluor® 594 in a final reaction volume of 20 μL. We typically use 2–3 μL of PlatinumBright 547 in a final 20 μL volume. Once prepared, the Alexa Fluor® 647 is stable for 2–3 months, while the Alexa Fluor® 594 and the PlatinumBright 547 dyes are stable for a longer period (up to 1 year). As time passes, however, more dye is usually needed to achieve a good labeling efficiency.

10. *Degree of labeling of HD-FISH probes.*

    The degree of labeling (DoL) of HD-FISH probes is calculated as the ratio between 34 times the dye concentration (pmol/μL) divided by the nucleic acid concentration (ng/μL) as measured by NanoDrop. For HD-FISH, we recommend a DoL between 1% and 3%, since probes with lower DoL values do not produce strong enough signals, while probes with higher DoL might not hybridize efficiently.

11. *Storage and handling of probes.*

    Both labeled HD-FISH probes and the fluorescently labeled uniFISH oligonucleotides are sensitive to photobleaching and thus should be kept in darkness whenever possible. Labeled HD-FISH probes in labeling buffer may be stored at −20 °C for several years. However, upon precipitation of the desired probe amount and its resuspension in the HDHB buffer, the hybridization mix should be stored at 4 °C (protected from light) for up to 2 months and should never be frozen. For uniFISH probes, lyophilized oligonucleotides should be dissolved in nuclease-free water or TE buffer. The fluorescently labeled oligonucleotides should be aliquoted in small volumes immediately upon receipt, since repeated cycles of freeze–thaw may result in decreased signal intensity.

12. *Amount of probe for hybridization.*

    For HD-FISH, the concentration of the probe in the hybridization mix can vary between 5 and 50 nM. The optimal probe amount should be determined experimentally taking into consideration how many amplicons compose a given probe (the more the amplicons, the higher the concentration should be). In the case of uniFISH, the final concentration of each oligonucleotide in the hybridization mix is ~0.4 nM. To achieve this, for a probe consisting of 48 oligonucleotides each at 100 μM stock concentration, we pool 1 μL of each oligonucleotide to a final volume of 48 μL. From this solution, we prepare a 1:500 working solution in nuclease-free water. In order to prepare a ready-to-use probe solution, we mix 1 μL of the 1:500 dilution with 9 μL of 1.1× UHB-1 buffer. Similarly, for a probe consisting of 96 oligonucleotides, after pooling 1 μL of each oligonucleotide to a final volume of 96 μL, we prepare a 1:250 working solution. In this way, after mixing 1 μL of this dilution with 9 μL of 1.1× UHB-1 buffer, the final concentration of each oligonucleotide is also ~0.4 nM. When pooling multiple probes each consisting of 48 oligonucleotides or more, it is likely that above a certain threshold of the number of probes the final concentration of each oligonucleotide needs to be adjusted and be lower than 0.4 nM. However, we have not yet determined the threshold and how the concentration of each oligo should be scaled down. The stock solution of the fluorescently labeled 20 nt oligonucleotide is at 100 μM, and we dilute it in nuclease-free water to a 1:50 working solution. We use the fluorescently labeled oligonucleotide at 20 nM final concentration by diluting 1:100 the 1:50 working solution in 1× UHB-2 buffer. We store the working solutions of both the nonlabeled and the labeled oligonucleotides at −20 °C.

13. *Optimization of denaturation and hybridization conditions.*

    The denaturation time can vary between different samples and thus it needs to be optimized for every batch of cover glasses. Depending on the sample, 2–5 min of denaturation at 75° are enough for the probe to hybridize (in the sense that a proper signal is observed, while the nuclear morphology remains unaffected). When studying genome architecture, special consideration should be taken on optimizing the denaturation time, since preservation of the 3D structure of the nucleus is critical. Denaturation times longer than 5 minutes can lead to disruption of the chromatin structure, drastically diminishing the quality of the sample.

14. *General considerations on sample handling.*

    Avoid drying the sample. During the entire HD-FISH and uniFISH procedure (except for the alcohol dehydration steps), special care should be taken in order to avoid drying of the samples, since cells that dried at any point during the procedure show increased autofluorescence. To this end:

    (a) The cells should be monitored during the freeze–thaw step of the 3D-FISH fixation protocol and never let dry. Leaving a thin layer of glycerol solution before placing the plate in the Dewar flask containing liquid nitrogen is necessary to prevent drying of the sample during the thawing step.

    (b) All the solutions must be quickly exchanged during the washes. Make sure that cells never dry between two consecutive washes.

    (c) When placing a cover glass on top of the hybridization mix, make sure that as much as possible of the prehybridization buffer is removed without allowing the sample to dry.

       Avoid photobleaching. Following the hybridization of the HD-FISH probe or of the secondary uniFISH oligonucleotide, perform all the steps which do not require light for manipulation (incubations, washes, etc.) in darkness, since fluorophores are readily photobleached by exposure to light. Moreover, when handling the sample, try to reduce exposure to light as much as possible (cover plates with aluminum foil, place microscope slides with the mounted cover glass inside an opaque box while waiting for the fixogum to dry, etc.).

References

1. Bienko M, Crosetto N, Teytelman L, Klemm S, Itzkovitz S, van Oudenaarden A (2013) A versatile genome-scale PCR-based pipeline for high-definition DNA FISH. Nat Methods 10:122–124
2. Boyle S, Rodesch MJ, Halvensleben HA, Jeddeloh JA, Bickmore WA (2011) Fluorescence in situ hybridization with high-complexity repeat-free oligonucleotide probes generated by massively parallel synthesis. Chromosom Res 19:901–909
3. Beliveau BJ, Joyce EF, Apostolopoulos N, Yilmaz F, Fonseka CY, McCole RB, Chang Y, Li JB, Senaratne TN, Williams BR, Rouillard J-M, Wu C-T (2012) Versatile design and synthesis platform for visualizing genomes with Oligopaint FISH probes. Proc Natl Acad Sci U S A 109:21301–21306
4. Beliveau BJ, Boettiger AN, Avendaño MS, Jungmann R, McCole RB, Joyce EF, Kim-Kiselak C, Bantignies F, Fonseka CY, Erceg J, Hannan MA, Hoang HG, Colognori D, Lee JT, Shih WM, Yin P, Zhuang X, Wu C-T (2015) Single-molecule super-resolution imaging of chromosomes and in situ haplotype visualization using Oligopaint FISH probes. Nat Commun 6:7147
5. Dahl F, Banér J, Gullberg M, Mendel-Hartvig-M, Landegren U, Nilsson M (2004) Circle-to-circle amplification for precise and sensitive DNA analysis. Proc Natl Acad Sci U S A 101:4548–4553
6. Schmidt TL, Beliveau BJ, Uca YO, Theilmann M, Da Cruz F, Wu C-T, Shih WM (2015) Scalable amplification of strand subsets from chip-synthesized oligonucleotide libraries. Nat Commun 6:8634
7. Moffitt JR, Zhuang X (2016) RNA imaging with multiplexed error-robust fluorescence in situ hybridization (MERFISH). Methods Enzymol 572:1–49
8. Lubeck E, Cai L (2012) Single-cell systems biology by super-resolution imaging and combinatorial labeling. Nat Methods 9:743–748
9. Lubeck E, Coskun AF, Zhiyentayev T, Ahmad M, Cai L (2014) Single-cell in situ RNA profiling by sequential hybridization. Nat Methods 11:360–361
10. Chen KH, Boettiger AN, Moffitt JR, Wang S, Zhuang X (2015) RNA imaging. Spatially resolved, highly multiplexed RNA profiling in single cells. Science 348:aaa6090
11. Wang S, Su J-H, Beliveau BJ, Bintu B, Moffitt JR, Wu C-T, Zhuang X (2016) Spatial organization of chromatin domains and compartments in single chromosomes. Science 353:598–602
12. Sydor AM, Czymmek KJ, Puchner EM, Mennella V (2015) Super-resolution microscopy: from single molecules to supramolecular assemblies. Trends Cell Biol 25:730–748
13. Hausmann M, Winkler R, Hildenbrand G, Finsterle J, Weisel A, Rapp A, Schmitt E, Janz S, Cremer C (2003) COMBO-FISH: specific labeling of nondenatured chromatin targets by computer-selected DNA oligonucleotide probe combinations. BioTechniques 35(564–70):572–577
14. Schmitt E, Schwarz-Finsterle J, Stein S, Boxler C, Müller P, Mokhir A, Krämer R, Cremer C, Hausmann M (2010) Combinatorial oligo FISH: directed labeling of specific genome domains in differentially fixed cell material and live cells. In: Bridger JM, Volpi EV (eds) Fluorescence in situ Hybridization (FISH). Humana, Louisville, KY, pp 185–202
15. Müller P, Rößler J, Schwarz-Finsterle J, Schmitt E, Hausmann M (2016) PNA--COMBO-FISH: from combinatorial probe design in silico to vitality compatible, specific labelling of gene targets in cell nuclei. Exp Cell Res 345:51–59
16. Honig B, Rohs R (2011) Biophysics: flipping Watson and Crick. Nature 470:472–473
17. Silahtaroglu A, Pfundheller H, Koshkin A, Tommerup N, Kauppinen S (2004) LNA-modified oligonucleotides are highly efficient as FISH probes. Cytogenet Genome Res 107:32–37
18. Nilsson M, Malmgren H, Samiotaki M, Kwiatkowski M, Chowdhary BP, Landegren U (1994) Padlock probes: circularizing oligonucleotides for localized DNA detection. Science 265:2085–2088
19. Banér J, Nilsson M, Mendel-Hartvig M, Landegren U (1998) Signal amplification of padlock probes by rolling circle replication. Nucleic Acids Res 26:5073–5078
20. Yaroslavsky AI, Smolina IV (2013) Fluorescence imaging of single-copy DNA sequences within the human genome using PNA-directed padlock probe assembly. Chem Biol 20:445–453
21. Solovei I, Cremer M (2010) 3D-FISH on cultured cells combined with immunostaining. Methods Mol Biol 659:117–126
22. Markaki Y, Smeets D, Fiedler S, Schmid VJ, Schermelleh L, Cremer T, Cremer M (2012) The potential of 3D-FISH and super-resolution structured illumination microscopy

for studies of 3D nuclear architecture: 3D structured illumination microscopy of defined chromosomal structures visualized by 3D (immuno)-FISH opens new perspectives for studies of nuclear architecture. BioEssays 34:412–426

23. Solovei I, Cavallo A, Schermelleh L, Jaunin F, Scasselati C, Cmarko D, Cremer C, Fakan S, Cremer T (2002) Spatial preservation of nuclear chromatin architecture during three-dimensional fluorescence in situ hybridization (3D-FISH). Exp Cell Res 276:10–23
24. Xu Q, Schlabach MR, Hannon GJ, Elledge SJ (2009) Design of 240,000 orthogonal 25mer DNA barcode probes. Proc Natl Acad Sci U S A 106:2289–2294
25. Kozubek M, Matula P (2000) An efficient algorithm for measurement and correction of chromatic aberrations in fluorescence microscopy. J Microsc 200:206–217
26. Abraham AV, Ram S, Chao J, Ward ES, Ober RJ (2009) Quantitative study of single molecule location estimation techniques. Opt Express 17:23352–23373
27. Stetson PB (1987) DAOPHOT – a computer program for crowded-field stellar photometry. PASP 99:191

Index

Tanya Vavouri and Miguel A. Peinado (eds.), *CpG Islands: Methods and Protocols*, Methods in Molecular Biology, vol. 1766, https://doi.org/10.1007/978-1-4939-7768-0,

D

E

F

G

H

S

T

U

V

W

X

Z

Printed by Printforce, the Netherlands